T0276197

Integrative Approaches for Health

Biomedical Research, Ayurveda and Yoga

Dedication

Creative writing gives immense pleasure to authors but creates a testing time for many dear ones.

Authors dedicate this book to their spouses for their life long support and inspiration for our work…

The late Dr Mukta Mutalik
Bhagyada Patwardhan
Neelima Tillu

Integrative Approaches for Health

Biomedical Research, Ayurveda and Yoga

Bhushan Patwardhan, PhD, FAMS
Interdisciplinary School of Health Sciences
Savitribai Phule Pune University
Ganeshkind, Pune, India

Gururaj Mutalik, MD, FAMS
Jijnyasa Foundation for Education and Research
Sarasota, FL, USA

Girish Tillu, MD
Interdisciplinary School of Health Sciences
Savitribai Phule Pune University
Ganeshkind, Pune, India

AMSTERDAM • BOSTON • HEIDELBERG • LONDON
NEW YORK • OXFORD • PARIS • SAN DIEGO
SAN FRANCISCO • SINGAPORE • SYDNEY • TOKYO
Academic Press is an imprint of Elsevier

Academic Press is an imprint of Elsevier
125 London Wall, London EC2Y 5AS, UK
525 B Street, Suite 1800, San Diego, CA 92101-4495, USA
225 Wyman Street, Waltham, MA 02451, USA
The Boulevard, Langford Lane, Kidlington, Oxford OX5 1GB, UK

ISBN: 978-0-12-801282-6

British Library Cataloguing in Publication Data
A catalogue record for this book is available from the British Library

Library of Congress Catalog Number
A catalog record for this book is available from the Library of Congress

For information on all Academic Press publications
visit our website at http://store.elsevier.com

Working together
to grow libraries in
developing countries

www.elsevier.com • www.bookaid.org

Publisher: Mica Haley
Acquisition Editor: Stacy Masucci
Editorial Project Manager: Shannon Stanton
Production Project Manager: Julia Haynes
Designer: Matt Limbert

Typeset by TNQ Books and Journals
www.tnq.co.in

Printed and bound in the United States of America

Contents

About the Authors

The authors represent the three generations of scientific progress. The first generation of Dr Gururaj Mutalik has seen dreadful epidemics like plague and small pox when there were no antibiotics. The second generation of Dr Bhushan Patwardhan has seen rapid progress of biomedical sciences and emergence of omics technologies. The new generation of Dr Girish Tillu has grown with information technology and artificial intelligence, when the world is realizing the value of systems view, ethics, and peace. The authors also bring a unique blend of domain expertise in each of the three components of this book *Biomedical Research, Ayurveda and Yoga*.

BIOMEDICAL SCIENCES

Bhushan Patwardhan, PhD, received his degrees in biochemistry. He is a professor and the Director of the Interdisciplinary School of Health Sciences, Savitribai Phule Pune University, India, where he is engaged in research in biomedical sciences, pharmaceutical biology, and evidence-based Ayurveda. He is a Fellow of the National Academy of Medical Sciences (India) and the recipient of many prestigious orations. He is founder and Editor in Chief of the *Journal of Ayurveda and Integrative Medicine*. He has published over 100 scientific papers, over 4000 citations and holds eight international patents. His research on drug discovery and genomics has received high acclaim.

MODERN MEDICINE

Gururaj Mutalik, MD, received formal education in modern medicine and is a distinguished physician with insights in Ayurveda, who has witnessed transformations in clinical medicine for about six decades. He was a dean and professor of medicine at Byramjee Jeejeebhoy Medical College, Pune, India. He directed health services as well as medical education and research departments for the State of Maharashtra. He worked as a postdoctoral fellow in human genetics at Johns Hopkins University School of Medicine. He served as the Director of the World Health Organization at the United Nations in the New York City. He was Chief Executive Officer of International Physicians for the Prevention of Nuclear War, a Boston-based, international organization of physicians, which received Nobel Peace Prize in 1985. Presently, he is engaged in research in Indian knowledge systems.

AYURVEDA AND YOGA

Girish Tillu, MD, is trained in Ayurvedic medicine, and pursuing research at the Interdisciplinary School of Health Sciences, Savitribai Phule University of Pune, India. He worked as a scientist at the Center for Development of Advanced Computing, Pune and was instrumental in the development of the innovative software suite, *AyuSoft*, which is a decision-support system based on logic, and knowledge base of Ayurveda. He is involved in the study of Ayurveda and Yoga through interdisciplinary approaches comprising epidemiology, clinical pharmacology, and informatics. He is recipient of *Vaidya* Scientist Fellowship.

Foreword

As this book is being published, I am celebrating the seventy-sixth year of my close encounter with biomedicine, and Ayurveda and Yoga—as a school boy, student, researcher, practicing physician, chairman of a department of internal medicine, dean of a medical school, and public health administrator. I worked at national and international levels, on three different continents. I have been a patient, and medical doctor at a primary health center in a developing country, and at one of the leading hospitals in the United States. Several valuable experiences and the lessons I learned during this period are the foundation for this book and the major inspiration for my imagining the shape of future health care.

I was born in 1929 into a traditional Indian family, in a small village with no electricity, no piped water, no sanitation, and where every house was constructed with thick mud walls. There were no modern doctors in the village. The only one who provided medical services to the community was my father. My father Vaidyaraj Shrinivasrao taught me Yoga and Sanskrit, enabling me to understand basic tenets of Ayurveda. The intimate exposure I had in childhood to Ayurveda and Yoga left me with a lasting perspective: while the age-old systems lack remedies for acute infections and conditions requiring surgery, they have remedies for common illnesses—remedies which are safe, widely accessible, and effective. Ayurveda and Yoga in addition contain in their core, proven principles of prevention of disease, promotion of health, restoration of vitality—a recipe for fitness and longevity.

Two childhood episodes are firmly rooted in my memory. The first was when I was 9 years old, and my entire village was evacuated because of a raging epidemic of bubonic plague, which in that pre-antibiotic era killed 70% of those who contracted the disease. The WHO estimates that this disease has claimed some 200 million lives in the course of its history! The other experience which left its impression upon me was the cases of smallpox that I would see in my father's medical practice. Most of those who contracted the disease either succumbed or bore lifelong scars on their faces and elsewhere. This disease, too, is estimated to have killed some 500 million people in the twentieth century alone!

It was not until 1978 that smallpox was vanquished in a "military style," global campaign, using a simple piece of technology like the bifid needle. The vaccine was effective, but the epidemiological strategy of identifying and immunizing all the contacts of each patient was crucial. We have not been so fortunate in dealing with HIV/AIDS, and even less so with the current challenge of Ebola; it is hoped that public health education, an understanding of epidemiology, and breakthrough discoveries in the genomics of the viruses will soon make a difference in these challenges as well.

I was admitted to the medical college in Pune in 1947—the year of India's independence. In 1949, I contracted typhoid fever. There were no antibiotics available. As a treatment, I received a once-a-day, 200 mL, intravenous injection of glucose solution, which produced exceedingly painful widespread thrombosis of the veins. This was how modern medicine was being practiced in a developing country like India. Finally, my body's immunity won, spontaneous remission took place, the threat to life was over, and destiny gave me an opportunity to contribute to the health and welfare of society.

I received my MBBS (Bachelor of Medicine and Bachelor of Surgery—1952) and MD (Doctor of Medicine—1955). In 1956 I married Mukta my classmate I was courting for 5 years—who provided me lifelong love and inspiration to give my best in whatever I undertook. I worked as a medical officer, and experienced the realities of life in rural India. After my work as a medical officer, I became an assistant professor of medicine and 3 years later a professor at Sassoon Hospital, a historic medical institution in Pune. This was a time when modern medicine was developing new arsenals of antibiotics, and drugs to deal with cardiovascular and neurological diseases; and making great advances in surgery and other specialties.

INFLUENCE OF JOHNS HOPKINS

In 1965, I studied medical genetics under Professor Victor McKusick at the Moore Clinic at Johns Hopkins School of Medicine, Baltimore—a world-renowned center of excellence. I was exposed to basic techniques of human genetics: cytogenetics, biochemical genetics, and population genetics. I returned to India and established a human genetics center at my own institute. This work was modeled on the exemplary studies that Professor McKusick did on the Amish community at Lancaster County, PA. The center offered genetic counseling services and cytogenetic studies, in addition to epidemiological studies on hemoglobinopathies among tribal population in endemic areas. I was instrumental in establishing the Human Genetic Society of India, and served as its first secretary. I organized the first international conference at my institute at Pune. Professor Victor McKusick was the guest of honor, as a visiting scientist. I mention this episode to illustrate how modern medicine, and its specialties, are spread globally by individuals who have the opportunity to receive training at global centers of excellence, and how the influence of such exchanges endures.

I learned much from the grand rounds at Johns Hopkins, especially the value of clinical acumen, which we have discussed in this book. In Baltimore, the School of Medicine, and the School of Public Health at Johns Hopkins are separated by Wolf Street. Professor E.A. Murphy, an Irish physician with characteristic Irish humor, once said "… more than a street stretches between the School of Health and the School of Public Health!" This medicine/public health disconnect continues to remain the one of the greatest challenge for us to overcome in the modern world. This divide is similar to that which exists between modern medicine and traditional medicine like Ayurveda and Yoga. It is my belief that the significant move toward these disciplines' convergence is the beginning of a paradigm shift that augurs well for the betterment of global health. We have discussed these aspects in the book.

BACK TO INDIA

In 1971, I became dean of the medical school at the Pune University. Among other projects, I established a research unit to study using traditional medicine in the treatment of chronic diseases, such as diabetes and rheumatoid arthritis. Working with a team of medical doctors and senior Ayurvedic physicians was a great challenge. The medical team insisted on statistical sampling, double-blind trials, and evaluation based on statistical testing. The Ayurvedic team, despite their perseverance, could not comprehend or utilize the methods of the medical team; they worked from empirical evidence, instinct, and the patients' reporting of subjective feelings of better or worse. This divide in methodologies has yet to be overcome. We have dwelt upon this subject in the book.

Later, I worked as the Director of Health Services and director of medical education of the State of Maharashtra. Here, I witnessed the dichotomy and distance between the medical schools and the health services establishments, including district and subdistrict hospitals and numerous primary health care centers at the village level. It was then, and is now, my belief that this divide is artificial and counterproductive. In an initiative to bring together medical education and public health services, we trained village housewives as primary health workers to work as barefoot doctors, emulating the Chinese experiment. Three years later, such experiments were numerous in India, and similar work was being done in Bangladesh, Thailand, and other South Asian countries. During this period, I was fortunate to be able to participate in the historic International Conference on Primary Health Care, where the Declaration of Alma-Ata was adopted in 1978. I subsequently joined the WHO as a staff member.

AT THE WORLD HEALTH ORGANIZATION

I served the WHO for two decades, working at its regional office New Delhi (1975–1980), and then at its headquarters in Geneva, Switzerland (1980–1985). I was Director of the Office of the WHO at the United Nations, in New York City (1985–1991). During my time with the WHO, I was afforded insight into the gross disparity in health care that exists across the world. Sovereign nations inevitably treated advice from these bodies as mere policy advice—accepting the global mandates at their assemblies, but delaying implementation. In only one instance—during the smallpox eradication campaign—were the governments persuaded to treat the mandates of the WHO as a supranational dictate. The result of this dictate was the eradication of a terrible, historic disease, which killed and maimed millions in many parts of the world. In my position as a state director of health, I did play a small role in this historic saga.

During my tenure at the Southeast Asia office, I contributed to advancing the role of traditional medicine in the global health care system. From the regional office, we supported a research project at an Ayurvedic center in Kerala, which was a clinical trial of Ayurvedic and Yoga therapy for rheumatoid arthritis. This led the global office of the WHO at Geneva to establish a department of traditional medicine to promote traditional systems of medicine in health care.

In 1991, I retired from the WHO and worked as a consultant to a Carnegie project designed to provide a platform for dialogue between United States health care leaders and those in other countries. During the 1990s, I served 5 years as executive director and later as CEO, to the International Physicians for the Prevention of Nuclear War (IPPNW), an organization which had received the Nobel Peace Prize (1985) for its advocacy against spread of nuclear weapons; this organization had positioned its advocacy as a public health issue. Indeed, public health is not only concerned with health and disease, but has ramifications in development, disarmament, public safety, disaster relief, human rights, and the environment.

Modern medicine has powerful healing powers, bordering almost on the miraculous, but even the best specialists can make a misdiagnosis if they are focused only on their own superspecialty. I have learned this from my own experience when I was misdiagnosed in the United States. Had I undergone the wrong procedure, in hindsight, I can see the possibility of total paralysis, of revisiting the surgery, of being confined to a wheelchair, or even facing untimely death due to the progression of tumor. It cost my insurance $65,000 to get me back on my feet—from the time I was admitted to hospital to when I was discharged. I wondered what would happen to an uninsured person (and also a lay person) who landed in a similar situation. Had I not insisted on another MRI, and had I followed the doctor's advice, what would the result have been? These questions reflect on the need to rectify the major shortcomings of the

Unites States' health care system, which—while still being the most sophisticated in the world—has multiple shortcomings from the patient's perspective.

Medicine has always been regarded as a noble profession, both in the West and the East. It is spiritually rooted in the urge to heal and reduce suffering. Today's practice of medicine has departed from the teachings of Hippocrates, and Osler. Ayurveda and Yoga evolved from the precious teachings of the great sages, shaped by the perennial philosophy of the *Bhagavad Gita* and the spiritual genius of Patanjali, who gave Yoga to the world. These disciplines, too, have fallen prey to the commercialization of the West, to the rigidity of empiricism, and to a reliance on blind faith and dogma.

Leading medical centers in the West have adopted Yoga and meditation to complement their arsenals of healing. Integrative medicine indeed is on the way. The demand from the public and the pressure from consumers will ensure the continued momentum toward integration.

In the last 6 years I have had much to do with projects involving health care of the elderly, and advocating for the use of Ayurveda and Yoga. I work with a nonprofit organization, the Janaseva Foundation, which provides services to over 40 villages near Pune.

Today, at the age of 85, I am physically and mentally fit. I work even more hours daily than I did in my preretirement career, thanks to the practice of Yoga and a disciplined life that my father bestowed upon me through his example and instruction.

Over the years, I have seen curative medicine marching with newer and more and more powerful drugs, which in their wake kill as much as they cure and sometimes reduce the quality of life. The basic secret of health and wellness in many ways still eludes us. Ayurveda and Yoga have these secrets embedded in their very core, but they need multidisciplinary scientific research to unearth their systemic details for further progress. As we have emphasized in our book, a paradigm shift is in the making, but can only be hastened if the mainstream medical and biomedical scientists work in concert with experts in Ayurveda and yogic science. This cooperation will transform the very fabric of medical science in the near future, for the benefit of humanity.

Writing this book has been a labor of love. I hold Dr Bhushan Patwardhan in very high esteem. He is one of the foremost biomedical scientists in India today. No one has done more for the advancement of Ayurveda and Yoga in the modern age. In his career, he has endeavored to bring the spirit of science to his advocacy for the ancient systems to evolve their own evidence base, consistent with its epistemology. I have also been fortunate to be associated with Girish Tillu, an extraordinary individual who knows Ayurveda in depth, and practices it as well. His humility and deep desire to contribute to research, and his expertise in ancient Sanskrit sourcebooks of Ayurveda and Yoga have been great assets to us.

Even after the arduous effort of writing this book, we are already discussing the need for another book. This book would explore the life sciences through a perspective which melds science with the principles of ancient philosophy, and what the ancients have to tell us about the nature of life and the human system. In our flights of fancy, we imagined calling the book *From Genome to OM*; whether this idea will remain a dream or become reality is another story.

Dr Gururaj Mutalik
Sarasota, Florida, United States

Preface

The year 2014 marked my thirtieth year in academia. I received my doctor of philosophy degree in biochemistry, and subsequently spent 20 more years teaching at university. I studied protein structures, enzymology, bioenergetics, pharmacology, genetics, and molecular biology. While biotechnology was emerging as a new discipline, unlike my colleagues, I decided to work on traditional medicine, which was then out of fashion. This gave me the unique opportunity to study multiple disciplines in the biomedical sciences. I worked in the pharmaceutical industry, and was involved in natural product drug discovery and development. I also participated in a few interesting clinical studies.

I interacted with medical doctors as well as Ayurvedic *Vaidya* (traditional Ayurvedic physicians). I noted that these two disciplines differed in their outlooks on the same disease, or symptom. When the situation arose where a senior Ayurveda *Vaidya* would assert that double-blind, placebo-controlled studies are not suitable for studying the effects of Ayurvedic treatments, this was looked upon by the scientific, medical establishment as an excuse and as an example of the Ayurvedic community's reluctance to embrace modern methodology. The *Vaidya* explained that they treat every individual differently; they maintained that cardiac diseases can be prevented, and even reversed. Yoga experts knew that performing simple breathing techniques and asanas can remedy insomnia. When the *Vaidya* and Yoga experts offered this knowledge, they were often ridiculed. Later, the field of pharmacogenomics evolved as basis of person-specific, drug responses; possibility of reversing cardiac diseases was accepted; and top medical centers like the Cleveland and Mayo Clinic integrated Yoga into their treatment regimen. Now after 30 years, when the importance of epistemology is being recognized and whole systems clinical trials are gaining wider acceptance, integrative medicine has already become a reality.

Through my participation in several laboratory experiments, I learned and experienced the value of traditional wisdom. It was hard to believe our own results. I witnessed how traditionally prepared medicines were safe and effective than solvent extracts of the same botanicals; how simple formulation could bring about serological conversion of rheumatoid factor, and how the concept of *Prakriti* has a genetic basis. In 2010, I was invited to lead a new institute in Bangalore, the Institute of Ayurveda and Integrative Medicine. Here, we would endeavor to demonstrate, using scientific methodology, how integrative approaches have enormous potential to benefit humanity.

As a biomedical researcher, I have published over 100 research articles and reviews in scientific journals. In my mind, publishing in this way sufficed to spread our concepts and approaches among peers. At this stage, Dr Gururaj Mutalik, who has a family background in the Ayurveda tradition and rich experience as an eminent professor of modern medicine, became my mentor, and was instrumental in changing my mind-set. At his urging, I decided to take on the project of writing a book on integrative approaches to medicine. My student Dr Girish Tillu is an MD in Ayurveda. He has in-depth knowledge of Indian philosophy and Yoga. He has been working with me as a research scholar and editorial associate for over 12 years. We, the authors of this book, bring a unique blend of relevant expertise from our respective domains. The first author is a biomedical researcher, the second author is a modern medicine physician scientist, and the third author is an Ayurveda and Yoga expert. We prepared a concept note, and decided to approach the best publisher in biomedical sciences—Elsevier.

This book is being published when the world is facing major human health challenges. There is a sharp increase in noncommunicable diseases, difficult-to-treat, chronic conditions, and lifestyle disorders. The current global focus is more on medical care than health care. Current biomedical research is focused more on diseases, drugs, and therapies. The cost of medical treatment is becoming unaffordable. The pharmaceutical industry is facing a severe innovation deficit. The magic bullet approach of the twentieth century is no longer relevant. Modern medicine is facing unprecedented challenges, and people are exploring new ways to meet these challenges.

In this book, we advocate a fundamental shift in mind-set from illness/disease/drug-centric, curative therapies, to person/health/wellness-centric, integrative modes of the future. The experience, knowledge, and wisdom from traditional and complementary systems like Ayurveda and Yoga have much to offer. We are convinced that high quality, biomedical research which integrates modern medicine, Ayurveda, and Yoga is the best prescription for the future of global health care.

Through scientific inquiry, we are now discovering the processes and principles which underlie the healing potential and reality of the ancient wisdom-based practices, such as Ayurveda and Yoga. It is our hope that this book, with its transdisciplinary approach, will bring a new perspective into focus, and serve as an inspiration and impetus for the further integration of these two powerful modes of healing. Humanity has only to benefit from such an exploration.

Bhushan Patwardhan
Pune, India

Acknowledgments

Many friends, philosophers, and guides have helped us during our research career, which has led to development of our thoughts and themes for this book. It is impossible to thank each one enough. We wish to specially thank a few individuals and institutions for support, encouragement, guidance, and help.

We wish to put on record consistent guidance and support from the Elsevier team especially Stacy Masucci, Shannon Stanton, and Julia Haynes which helped us to complete the task in time. Anonymous reviewers of Elsevier have immensely helped us to refine preliminary ideas about the book. Efficient copy editing by Davanna Cimino substantially helped to improve semantics.

Many internal reviewers gave excellent critic and constructive suggestions, which helped in revisions: Darshan Shankar, Sharda Bapat, Vinod Deshmukh, Prabhakar Lavakare, Padmaja Shastri, Girish Welankar, Kalpana Joshi, Kishor Patwardhan, Asmita Wele, Upendra Dixit, Madhav Khanwelkar, and Madan Thangawelu.

We thank Neelmudra Creative Carnival artists Ajey Jhankar, Shalaka Kalkar, and Yogini Gavai for excellent cover designs. We thank Principal, Bharati Vidyapeeth College of Arts and Mangesh Tambe for excellent graphic illustrations and figures.

BP thanks Savitribai Phule Pune University's Vice Chancellor W.N. Gade, Jayakar Library, University and colleagues at the Interdisciplinary School of Health Sciences, Anita Kar, Aarati Nagarkar, Angeline Jeykumar, and Uma Chandran for continuous support.

BP also thanks Indian Institute of Advanced Studies, Shimla, India for awarding visiting professorship, which facilitated book writing and gave opportunity to interact with many scholars, historians, and philosophers especially professors: D.N. Dhanagare, Sharad and Medha Deshpande, Udayan Misra, Radha Vallabh Tripathi, and Director Dinesh Singh. Special thanks to Gerard Bodeker for consistent interactions and advice.

GM thanks B.V.K. Sastry for his assistance and appreciates inputs from Pradeep, Meenakshi, Madhav, Sujata, Shruti, Madhuri, Meredith, Roy, and Nikita—members of his family.

GT thanks Serum Institute of India for award fellowship.

We are grateful to Jijnyasa foundation and in particular to its president Anil Deshpande, Orlando, FL, for his kind support to our undertaking.

We thankfully acknowledge various knowledge sources, which have helped us to get authentic and latest information on various subjects. We wish to specially mention Wiki, PubMed, Google Scholar, and Cochrane Library.

About the Cover

The cover portrays the concept of emerging integration between biomedical sciences, modern medicine, Ayurveda, and Yoga for the future of global health. A Yogi in deep meditation sitting in traditional *lotus posture* with perfect harmony and balance at all levels of his *being*—body, mind, and spirit, symbolizes *swastha*, as a true healthy state. Six circular nodes inside the body depict the ascending levels of comprehension, knowledge, awareness, intuition, elemental power, and consciousness. The line drawing at the background is traditional Indian earthen lamp surrounded by lotus petals. In Buddhist, Indian and other Asian traditions lotus symbolizes purity, beauty, and spontaneous generation. The lotus is rooted in the mud but the flower floats on the water without becoming wet or muddy indicating nonattachment. The cephalic glow depicts fully functional neural networks, emanating power of enlightenment. The mortar and pestle containing basil leaves symbolize Ayurveda while the double helical DNA with Asclepius rod on the top represent biomedical research, modern science and technology.

The back cover portrays twelve steps of *sun salutations* practiced as easiest and most effective exercise based on intelligently designed series of *Yogasana*. Twelve salutations performed every morning maintain health and energize body and mind.

The harmonious blend of collage with dark and light shades embodies the exhortation from Upanishad *"Lead me from unreal to real, from darkness to light and from fear of death towards knowledge of immortality!"*

CHAPTER 1

Advocacy for Integration

He who does not trust enough, will not be trusted

Lao Tzu

BACKGROUND

Before we begin this chapter "Advocacy for Integration," we recognize that, on a global level, the importance of an integrative approach for health and medicine, in general, has been accepted. Indeed, there is no need to convince professionals, governments, and people. However, there is not enough clarity and consensus about what constitutes *integration*. To build advocacy, a critical review of ostensibly different knowledge domains that broadly represent modern and traditional knowledge systems, biomedical sciences, Ayurveda, and Yoga is needed. Interestingly, although these knowledge systems differ epistemologically, their underlying philosophies are similar.

The domain of modern knowledge systems is represented by biomedical science, which integrates biology and medicine. Biological sciences are represented by physiology, biochemistry, biophysics, microbiology, and similar basic sciences. Medicine is represented by pharmacology, pathology, therapeutics, and similar applied sciences. We prefer to call the *medical* component as *modern medicine*, which is also known as conventional medicine, orthodox medicine, Western biomedicine, or more popularly as allopathy. New variants of modern medicine such as molecular medicine emphasize cellular- and molecular-based interventions beyond the conceptual and observational focus on patients and their organs. Personalized medicine brings understanding of individual genomic variations to design specific treatments most suitable for a particular person.

All systems, practices, or forms, and therapies other than modern medicine are contained in the traditional and complementary medicines (T&CM) category. The domain of traditional knowledge systems is represented by Ayurveda and Yoga. Ayurveda deals with health promotion, disease prevention, and personalized treatments with the help of natural medicines. Yoga is a nonpharmacological approach that focuses on mind/body/spirit through exercise, relaxation, and meditation.

1

Integrative Approaches for Health. http://dx.doi.org/10.1016/B978-0-12-801282-6.00001-2

In general, modern science is often regarded as reductionist, while Ayurveda and Yoga are considered to be holistic. We feel that modern and traditional knowledge domains share some similar basic principles. Historically, while there have been some detours, the basic philosophy, doctrines, ethos, and approaches remain the same, even today. In the post-Aristotelian period, modern science has rigorously followed the reductionist approach—imparting a much deeper understanding to the parts. After William Harvey, the understanding of physiology became more precise. After Robert Koch, Louis Pasteur, and Joseph Lister, the understanding of diseases and causative factors became clearer. After Ronald Ross and Alexander Fleming, the magic bullet era began. After Paul Erlich and Linus Pauling, medicinal chemistry became much more precise. Frederick Sanger, James Watson, and Francis Crick advanced knowledge of genetics and genomics. Now the high-throughput technologies are helping us to move toward personalized treatments.

The emergence of systems biology is an indication that even though biomedical sciences have taken the reductionist path, there is a growing recognition of the importance of a holistic approach. New fields like quantum mechanics, robotics, and artificial intelligence are indicative of a convergence toward holistic approaches—going back to basics, but with much more clarity and strength. We feel that both modern medicine, and Ayurveda and Yoga have complementary strengths to evolve into integrative medicine of tomorrow. It is true that in practice, they do not seem to be so. The present biomedical science has compartmentalized itself into several superspecialities, and has become more linear, mechanical, and protocol-driven. Ayurveda and Yoga have remained frozen traditions. However, at the core, both in their own rights are logical, evidence-based, patient-centered, and personalized.

Today, hoping to attain health and wellness, the world is facing incurable diseases and illnesses. Despite powerful drugs, diagnostics, surgical advances, and sophisticated technologies, people continue to suffer. To note, this is not to undermine the power of modern medicine; on the contrary, modern medicine has done wonders to eradicate many infectious diseases, and improve quality of life and longevity. However, limitations of modern medicine exist—especially in the treatment of chronic, psychosomatic, and lifestyle diseases.

In this day and age, medical science knows more about disease and illness than health and wellness. Biomedical sciences may have understood complexity of body and mind to some extent, but a holistic view that takes into account body/mind/spirit is still missing. Until now, the fact of the importance of positive health and wellness has been neglected by the scientific community. The curative approach—with an excessive focus on diseases, diagnostics, and drugs—has overshadowed any holistic methodology of health promotion and prevention. The patient as a *person* is usually undermined in the process of clinical diagnosis and treatment. The clinical process has become more scientific and evidence-based, but at the same time it has become protocol-driven, algorithmic, and mechanical. Experience-based systems like Ayurveda and Yoga might be useful in overcoming the present limitations of medical therapy. We recommend the integration of the best resources and practices from biomedical sciences, Ayurveda, and Yoga as a smart strategy and practical approach.

In our advocacy, we do not take any extreme positions regarding either Western reductive modern medicine or Eastern traditional holistic medicine. Both these systems have strengths and weaknesses. We hypothesize that a synergistic confluence of these seemingly diverse systems may be possible through appropriate integrative approaches. In this book, we try to build a case in support of our hypothesis. In order to provide a philosophical underpinning for our case, we also discuss the evolution of medicine, and concepts of health and disease. We critically review the importance of holistic approaches and their similarities with the emerging science of systems biology. We review the advances in biomedical sciences leading to the development of modern pathology, diagnostics, and drug discovery; and the journey of modern medicine from clinical acumen to scientific evidence-based medicine. We discuss the importance of food, nutrition, and lifestyle management for health, wellness, healthy aging, and longevity. Finally, we attempt convergence of personalized and integrative approaches to imagine the contours of future medicine.

GLOBAL HEALTH SCENARIO

In attempting an advocacy for integrative approaches, it is necessary to review the global health scenario. Of interest is the juxtaposition of the evolution of the health care system against the backdrop of advances in biomedical sciences, including the unmet needs and unresolved challenges in this sector. In the following sections, we sketch this background before discussing the need for integrative approaches.

Historically, health has been a natural and integral part in most of the cultures and traditions. Now, however, peoples' active participation in seeking and maintaining health is overshadowed by their passive dependence on treatments. The basic objective of any health care system is to provide better physical and mental health across the communities through effective interventions. Good health care systems also attempt to improve individuals' satisfaction by respecting their dignity. Health is an element of common good, and a central part of economic and social development. Health is now considered an important human right, crucial to individual dignity, and should be accessible and affordable for all individuals.

As this book is being written, the world has reached a population of approximately 7.2 billion. Out of these, about 80% of the population live in the developing countries. According to the World Health Organization (WHO), about 36 million people have died of HIV so far. Globally, 35.3 million people were living with HIV at the end of 2012. Sub-Saharan Africa remains most severely affected, with nearly 1 in every 20 adults living with HIV and accounting for 71% of the people living with HIV worldwide. Globally, there are at least 300 million acute cases of malaria each year, resulting in more than a million deaths. Around 90% of these deaths occur in Africa, and are mostly young children; malaria is virtually nonexistent in the developed world. The number of people who suffer from tuberculosis is over 532 per 100,000 in Africa and Southeast Asia. Spending on pharmaceuticals accounts for about 15% of the total amount spent on health worldwide; the average per capita spending on pharmaceuticals in high-income countries is over 400 dollars, and barely over 4 dollars in low-income countries. A WHO statistics showed that the density of modern medicine physicians per one hundred thousand persons in some countries is very low: Rwanda, 1.87; Ethiopia, 2.85; Uganda, 4.70; India, 51.25; and China, 164.24. In stark contrast is the numbers for Australia and the United States: 249.13 and 548.91, respectively.

Over 1 billion people exist on less than 1 dollar a day. Over 2.5 billion people lack sanitation. Over 1.5 billion people do not have safe drinking water, and some 3 million people—mostly women and children—die every year from diarrheal diseases directly related to these deficiencies. There is little cause for optimism in these areas as available evidence suggests that the drinking water resource in the poor world is likely to diminish over the next several decades.

According to the WHO, in 2012, there were 56 million deaths worldwide from all causes. The epidemiological transition from communicable to noncommunicable diseases (NCDs) is apparent in the second decade of the twenty-first century. NCDs were found to be responsible for over 68% of all deaths globally; a near 10% increase from 2000. The four main NCDs are diabetes, cardiovascular disease, cancer, and chronic lung disease. Among these, cardiovascular disease tops the list with 30% of deaths. On a global level, collectively, communicable diseases, poor prenatal, natal, postnatal and neonatal care, and poor nutrition, were responsible for 23% of deaths. Globally, 9% of all deaths were the result of injury. Earlier in the twentieth century, the rich and developed countries ranked higher in NCDs, while poor countries were contending with communicable diseases. However, over the past few decades, the epidemiological transition is clearly visible where poor and developing countries are confronted with the double burden of both communicable diseases and NCDs.

Successful elimination of many infectious diseases, substantial reduction in childhood mortality, and the improvement of life expectancy and longevity has occurred in the rich and developed world. Ironically, the contrasting picture is seen in the poor and developing world. Diseases like diphtheria, malaria, and tuberculosis have reemerged in many regions; newly emerged diseases such as HIV/AIDS have become pandemic. Emerging deadly diseases such as Ebola are

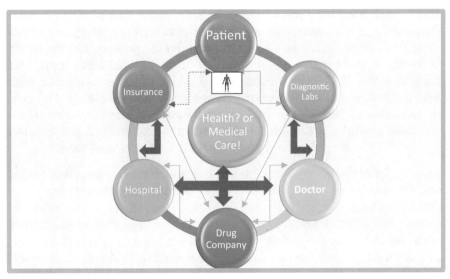

FIGURE 1.1
Vicious cycle of health care indicating the nexus between doctors, drug companies, and hospitals supported by diagnostic laboratories and insurance companies. In the whole process, the interests of the patient are compromised. Under the name of health care, the patient receives medical care many times at high costs because insurance cover is not universal in many countries.

immediate threats. Due to the adverse effects of drugs, many new iatrogenic diseases are on rise. Lifestyle diseases like diabetes are becoming pandemic. Sadly, the poor world continues to suffer the ill effects of malnutrition, and killer infections like malaria and tuberculosis.

At present, health systems focus more on *sick care* than *health care* [1]. Health seems to be dependent upon medical care. The medical care is dependent upon hospitals. Hospitals are dependent upon doctors. Patients are dependent upon doctors and insurance. Doctors are dependent upon diagnostic tests and products marketed by pharmaceutical companies. Diagnostics are dependent upon pathological laboratories. Pharmaceutical companies are dependent upon the marketing of drugs. Many times, in this vicious cycle of dependency, the interests of people and patients are ignored (Figure 1.1). This situation needs to be considered while discussing the relevance, rationale, and need for integrative approaches to health.

THE EVOLUTION OF HEALTH CARE SYSTEM

A good health care system is supposed to deliver quality services to people based upon need. Public health is defined as the science and art of preventing disease, prolonging life, and promoting health through the organized efforts, and informed choices of society and people. The WHO's Alma-Ata Declaration is a milestone in the field of public health; it identified primary health care (PHC) as the key to attaining the goal of "health for all" by 2000. While this goal was too ambitious on a global scale, it did help popularize the PHC. Besides primary medical care, PHC is focused on basic health determinants such as nutrition, safe water, and sanitation. The drastic reduction of infant and child mortality rates up until the turn of the twenty-first century, is certainly attributable to the commendable work done by the WHO at the policy level and by UNICEF at the ground level.

Broadly, the preventive aspects of health care are covered under public health, and the curative aspects are covered under medical care. Health care is provided at various levels. Primary care is the first point where patient enters the system. This involves preventive care, first aid, and immediate medical assistance. Secondary care is mainly hospital and emergency services, and tertiary care is specialized services. Medical care is limited to professional treatment of illness or injury by providing diagnostic, therapeutic, and surgical services.

Earlier, medical care was more holistic where the patient was at the center. Clinicians used to treat patients using knowledge, experience, and acumen. In the United States, the Flexner Report of 1910 was a landmark event that gave a big boost to biomedical research through intensive programs aimed at developing high-quality physician scientists. This opened new avenues for development of specialties and super specialties. However, in the process, medical practice became compartmentalized. The earlier, clinical practice based on experience and judgment, was transformed into a scientific evidence-based, protocol-driven medical system. As a result, clinical acumen, which is crucial for holistic health, was compromised. Now, the patient, who should be at the center of medical care, is overshadowed by expensive diagnostics; diseases predominate, rather than health. The absence of disease is no guarantee of health. Unfortunately, now health care is more focused on drugs, pharmaceuticals, and medical interventions. Optimal health care can and should not be restricted only to drugs or material medicine.

The effective control of diseases like cholera and typhoid was possible not only because of drugs, but mainly because of techniques like pasteurization and disinfections contributed by Pasteur, Lister, and others. Vaccination, and not antibiotics, was responsible for eradicating diseases such as smallpox. Thus, the focus on prevention is not new. However, now when the society is dominated by drugs, treatments, and medicines, the focus should again be on the prevention of disease. This does not mean treatments are not important. Prevention strategies alone are not enough to eradicate or ameliorate the threat of disease. We should intensify research and put more effort into discovering new treatments. However, priority must be given for early diagnosis and prevention of diseases.

The health care model based on treatments and therapeutics was more relevant in the twentieth century when the world was facing major threats due to infectious disease. Discovery of antibiotics and sulfa drugs laid a strong foundation for effective treatment of infectious disease. The influence of antibiotics was such, and confidence in scientific discoveries so high, that in 1967 the Surgeon General of the United States made a statement in US Congress that "It is time now to close a book on infectious disease." Of course, this book was never closed, and new emerging and reemerging infectious diseases are a continuing threat. Therefore, the value and need of new treatments and discoveries cannot be underestimated.

In the twenty-first century, besides newly emerged diseases like HIV/AIDS and Ebola, lifestyle disorders have emerged as major threats. The aging population is increasing. The domination of noncommunicable, chronic, difficult-to-treat diseases is increasing. These diseases are slow progressing, and take a longer time to manifest before leading to life-threatening conditions like cancer. Therefore, at the present time, prevention strategies make much better sense than curative approaches. More emphasis on prevention strategies can reduce disease complications and the duration of morbidity before death [2].

Historically, the medical profession was a noble and service-oriented profession. Providing health care to people was considered the responsibility of the king or the government. Universal health care coverage was the generally accepted norm. In ancient India, Ayurveda doctors known as *Vaidya* did not practice in return for fees. Take fees or money for medical services was not common. This custom changed, and private health care providers began charging professional service fees. Over a period of time this fee-for-service culture became deeply entrenched, and slowly the medical profession became commercial.

According to eminent Indian physician and cardiologist, Dr B.M. Hegde, "Modern medicine has become a costly chaos with no end in sight. We now have significant problems that beg urgent solutions." Dr Hegde highlights the need for a new integrated approach in medical science to come out of "present human-made and drug-industry-protected health problems of society can only be physiology-based" [3].

We feel that root causes of many problems in our health care systems of today are related to the commercialization of this sector. The pharmaceutical and diagnostic companies have started defining and dictating health policies. The private service providers began giving better quality health care than the government, but at hefty charges. Slowly, the increasing costs of health care necessitated insurance coverage. In reality, the cost of health care became impossible to sustain without insurance coverage. The nexus between insurance companies and private health care providers has taken costs beyond the capacity of common people. This has led to commodification of health care and a situation known as the *medicalization of society*.

In 1988, Arthur Barsky from Harvard Medical School expressed concerns about the medicalization of society. He cautioned medical doctors on the paradoxical consequences of public dissatisfaction with medical care [4]. After a quarter century, what Barsky predicted is now manifest to an even greater extent than he predicted. Many experts and voluntary organizations are now demanding the *socialization of health care* as a policy.

The health care systems in developed countries like the United Kingdom and the United States are facing much criticism, and experts are demanding innovative reforms. The Patient Protection and Affordable Care Act, also known as Obamacare, is a recent example of health care reform. This law is an attempt to provide a mandated health insurance system to eliminate some of the bad practices of the insurance companies. Obamacare attempts to provide more subsidies to the poor to help them to buy insurance. However, many experts have pointed out the shortfalls of Obamacare, which according to them ignores health care costs and the quality of care [5].

During the last two decades, Brazil, the Russian Federation, India, China, and South Africa (the BRICS countries) have decided to move toward universal health coverage (UHC). The goal of UHC is to ensure that all people obtain required health services without suffering financial hardship when paying for them. However, good intentions are not enough. It will be necessary to invent innovative practical measures, which can make health care affordable and accessible to needy people.

DIVERSITY OF CHALLENGES

The increase in life expectancy during the twentieth century was mainly due to significant improvements in public health services. This includes better sanitation, cleaner water, mass vaccinations, and improved workplace safety. The problems related to health in developed and developing countries may seem different, but they are interconnected and interdependent. Poor countries continue to face a severe burden of preventable infectious or communicable diseases. In addition, the poorer countries now are witnessing increased incidence of NCDs due to aging population, malnutrition, and unhealthy lifestyles. The richer countries are plagued by NCDs due to sedentary lifestyles and overnutrition; these countries also face the re-emergence of infectious diseases due to environmental and socioeconomic changes.

The health care quality and delivery have become a matter of great concern for world polity. It is of interest to note representative examples of key health indicators from developed and developing countries, in order to understand some commonalities and disparities related to health care.

CASE OF THE UNITED STATES

Many of the problems of health care in developed countries can be attributed to the abundance of facilities and technologies; this leads to an overdependence on drugs and hospitals. The example of the United States illustrates a situation where health care may be available, but is unaffordable for the masses. For instance, a new drug known as *sofosbuvir* can effectively cure Hepatitis C in 90% of patients without use of interferon [6]. This is certainly a breakthrough, however, its treatment cost is approximately US $80,000 per patient, which is just unaffordable. Probably due to the high cost of modern treatments, at present, the US health care remains dominated by the insurance coverage. While various indicators suggest a decline in the prevalence of several diseases, the health status of over 10% of the US population is rated as poor. This population may not even get required prescription drugs. Almost 15% of the population is reported not to receive required health care in a timely fashion. Unintentional injuries and emergencies are on the rise, resulting in complications of care and adverse effects. Nearly half the US population takes at least one prescription drug—half of those take three or more drugs [7]. This is certainly not a healthy picture for a rich country.

According to the Agency for Health care Research and Quality, over 770,000 injuries and deaths are reported in the United States due to adverse drug events. The situation is nearly identical in the United Kingdom where nearly 10% of the patients suffer adverse drug events [8]. Tobacco use followed by poor diet and low physical activity remain the major causes of death in a rich country like the United States [9]. A report by the Institute of Medicine suggests that about one-third of health costs are wasted due to unnecessary, and harmful, early elective deliveries. This amounts to nearly $1 billion every year. A sharp rise in lifestyle disorders like obesity, cancer, medical emergencies, and Alzheimer's disease add to the already burdened health care system. According to the U.S. Department of Human Health and Services, nearly 70% of deaths are due to cardiovascular disease, diabetes, and cancer, which together account for nearly 75% of all health care expenditures. According to the American College of Lifestyle Medicine, almost 80% of chronic diseases are preventable and manageable through a systematic application of *lifestyle* as medicine. These studies strengthen our argument that present models of treatment-centric curative approaches in health care need to be refocused on health promotion and disease prevention.

CASE OF INDIA

Emerging economies like India and developing countries like Nigeria or Bangladesh face a double burden of diseases. The health care problems in India vary drastically even within the country. A broad picture of the health status of India can give a representative picture of such countries. The infant mortality rate, which is one of the important health indicators, is four times lower in the state of Kerala than the Indian average, and five times lower than that of the state of Uttar Pradesh. Communicable diseases continue to pose a major problem in India. Almost 1.5 million cases of malaria have been reported over the past 5 years. The number of dengue fever cases has doubled from 12,317 in 2006 to 27,247 in 2009. More than 10 million cases of diarrhea, 1 million cases of typhoid fever, and almost 2.5 million cases of acute respiratory infections were reported in 2010. India continues to have the largest number of tuberculosis patients in the world.

The number of diabetes patients in India is projected to rise from 31.7 million in 2000 to 79.4 million by 2030. The prevalence of hypertension in India is about 16%; over 13 million people are estimated to suffer from it, out of which an estimated 62% are males. Thus, developing countries like India are bearing a double burden of disease; India has experienced a significant increase in the incidence of NCDs, and at the same time is plagued by most of the communicable diseases.

The nutritional status of countries like India poses a paradoxical situation: persistent undernutrition, as well as the increasing problem of obesity. Low body mass index levels were observed in 35.6% of women and 34.2% of men. On the other hand, 12.6% of women and 9.3% of men were either overweight or obese. Further, it has been estimated that 69.5% of children age 6–59 months were anemic, while 55.3% of women aged 15–49 were anemic.

Since the challenges in the developing and developed countries are different, the health systems models are bound to be different. It is important to learn from each, based on experience, so as to avoid making the same mistakes. Developed countries have built large infrastructures with the trained human resource of medical and paramedical professionals. However, it is expensive to replicate such health care infrastructure in developed countries. For example, Nigeria currently has barely 14% of the doctors per capita as that of the Organization for Economic Co-operation and Development (OECD) countries. To catch up to the level of richer countries, Nigeria would need approximately 12 times more doctors by 2030. Under current training models, about 51 billion USD in investment would be needed, which is about 10 times the current annual public health spending in Nigeria. Clearly, such strategies are unaffordable and unsustainable for any poor countries [10].

We feel that the integration and mainstreaming of traditional medicine can be a smart strategy for developing countries and emerging economies such as India. As a human resource, medical doctors are not the monopoly of modern medicine. In a country like India, separate registries for medical practitioners other than modern medicine do exist. India has a large number of T&CM

doctors from Ayurveda, Yoga, Unani, Siddha, Sowa Rigpa, and Homeopathy (AYUSH) systems. They come with different kinds of skills, and can make contributions to the health care system. AYUSH doctors can offer valuable advice about diet and lifestyle along with time-tested herbal medicines—especially for chronic diseases. Yoga offers physical exercises, and breathing, meditation, and relaxation techniques, which can play important roles in the prevention of several chronic diseases. Most of these treatments are easy, accessible, and affordable. For acute conditions and quick symptomatic relief, AYUSH doctors can be trained to use modern medicines at least for primary care. Such an integration of modern medicine and AYUSH doctors can be complementary.

There are increased attempts to involve AYUSH practitioners in the public health system to enhance UHC. A recent cross-sectional qualitative implementation research study conducted in three states of India has shown that individual and interpersonal efforts are generally in favor of integration. However, the system processes and policies are not favorable enough. Researchers of this study have emphasized the need for high-level political will to facilitate system-level integrative efforts [11].

In India, an experiment is under consideration where the mainstreaming of AYUSH in health care is supported by the government. If this happens, the present doctor to patient ratio of 1:1600 will be improved to 1:700, and the cost of health care might be brought down considerably. Of course, for this to happen, many imaginative steps will have to be taken by the respective governments [12].

Admittedly, spectacular advances in biomedical and pharmaceutical sciences are responsible for the control and treatment of several diseases. It is the result of systematic progress consisting of major discoveries, and advances in the knowledge of human anatomy and physiology by scientists like Vesalius, Harvey, and others. Many powerful tools such as the microscopy, X-ray imaging, electrocardiography, electroencephalography, computerized tomography, and magnetic resonance imaging (MRI) are available today. Advances in molecular biology and genomics have strengthened diagnostic sciences. Of interest, and for background, is a brief review of a few landmarks in biomedical sciences.

ADVANCES IN BIOMEDICAL SCIENCES

The past century witnessed the scientific progress, wherein Western medicine was transmuted into modern medicine. The understanding of molecular, genetic, cellular, and biological processes and their application in the fields of biotechnology has advanced considerably during the last few decades. This has led to the emergence of modern medicine, which is now accepted and practiced globally as conventional or mainstream medicine.

The recently emerged disciplines like genomics, proteomics, and the entire class of them together known as Omics technologies and emerging field of epigenetics have revolutionized our understanding of cellular and molecular biology; as well as significantly contributing to a better understanding of the scientific basis of individual variations and in the development of personalized approaches in health and medicine. In 2014, Japanese scientists reported that simple methods like inducing stress through lowering pH may transform mature somatic cells into pluripotent stem cells [13]. These findings are likely to have significant impact on regenerative medicine and may lead to new ways of effective treatments of several diseases and injuries.

Computational biology, bioinformatics, and supercomputers provide more power to analyze vast, complex biological data for diagnostic and therapeutic decisions. New high-throughput screening technologies, and next generation sequencers, confocal laser scanning microscopes, transmission electron microscopes, nuclear magnetic resonance, matrix assisted laser desorption ionization, liquid chromatography, mass spectroscopy, and hyphenated techniques are revolutionizing disciplines like genomics, proteomics, metabolomics, and molecular medicine, and increasing the precision and sensitivity of biochemical and pathophysiological investigations. Sophisticated imaging technologies such as positron-emission tomography, functional MRI, and single photon-emission computed tomography are offering insights in neuroimaging and whole-body scanning. Nanotechnology, robotic and minimally invasive surgery, organ transplants, advanced yet biocompatible materials for bones, joints, and tissues have revolutionized treatments and have helped improve the duration and quality of life substantially.

Drug discovery and development also progressed rapidly during the past century. Starting with simple, good old-fashioned drugs like aspirin and penicillin, and the discovery of sulfa drugs in the 1930s, drug discovery and development has come a long way. During the last few decades many broad-spectrum, and powerful and specific antibiotics; nonsteroidal, antiinflammatory drugs; antacids; selective serotonin reuptake inhibitors; antidepressants; antiasthmatic drugs; beta-blockers; antihypertensive medications; statins; cancer chemotherapeutic agents; and therapeutic insulin have been discovered. The pharmaceutical industry created blockbusters like Prozac, Lipitor, and Viagra resulting in record sales of these drugs. During the last two decades, many new biological drugs such as therapeutic proteins and monoclonal antibodies, as well as botanical drugs like artemisinin and taxol, have emerged to challenge the monopoly of synthetic chemical drugs.

The advances in cellular and molecular biology, pathophysiology, genomics, and pharmaceutical and biomedical sciences have helped modern medicine earn global acceptance. Undoubtedly, modern medicine can give quick relief

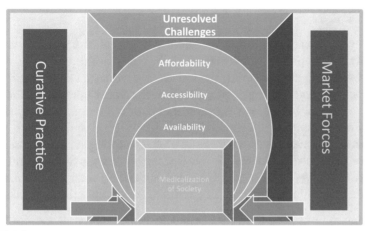

FIGURE 1.2
Health care crisis is the result of focus on curative care driven by market forces.

from acute symptoms like headache, inflammation, fever; it can effectively control many infections, and deficiencies, and offer life-saving, surgical procedures. However, despite powerful drugs, precision technologies, diagnostics, and surgeries, today, health care sectors seem to be struggling with many unresolved challenges. In addition, we are witnessing many new emerging and reemerging diseases. People need easy access to PHC, and affordable medical care to deal with their day-to-day problems.

We feel that the current health care crisis has three main causes (Figure 1.2).

1. Intensely compromised affordability, availability, and accessibility of quality health care
2. The medicalization of society, where market forces are driving the curative medical practice
3. Unresolved challenges, which have remained despite biomedical advances

The traditional knowledge systems like Ayurveda and Yoga may be able to address these causes and help to resolve the health care crisis in the best interest of people. Already, the Commission on Intellectual Property, Innovation and Public Health of the WHO, Geneva has recognized importance of traditional medicine for affordable, global health [14]. If traditional medicine are effectively used and home health care through simple remedies is strengthened, over 30% of expenditure on pharmaceuticals and medical care can be saved.

At this point, it is important to briefly review some major challenges of the biomedical sciences.

UNRESOLVED CHALLENGES

As discussed in earlier sections, advances in biomedical sciences have led to longevity and improved quality of life. Still, many health and medical problems have remained unresolved. The last century witnessed many environmental and socioeconomic transitions, which have adversely impacted the health of the ecosystem. In the West, improvements in sanitation and hygiene were responsible to prevent infectious diseases like cholera and typhoid. On the other hand, polluted environments and global warming are becoming leading causes of new diseases like chronic obstructive pulmonary diseases and cancer. Newer diseases like HIV-AIDS, bird flu, and Ebola have emerged, and many diseases like tuberculosis, malaria, dengue fever, and diphtheria are reemerging.

Aging populations pose formidable challenges. Devising strategies for healthy aging, rejuvenation, and for geriatric care are necessary to preserve the health of, and support senior citizens in achieving a productive, rich life experience. This is one of the great challenges of this century.

The developed countries have relatively more hospitalizations due to injuries, suicides, and complications of care; developing countries have disease categories such as multidrug-resistant tuberculosis. In recent years, a range of initiatives have targeted improving the prevention and treatment of major diseases, especially tuberculosis, HIV/AIDS, and malaria. It is becoming increasingly clear that biomedical research must no longer be confined to drugs, devices, and vaccines. The Mexico Ministerial Summit has publicly asserted that the biomedical model of health research is wholly inadequate to tackle disease alleviation in the less-developed world [15].

In modern times, human behaviors and eating habits have drastically changed—giving rise to many new lifestyle diseases like diabetes, psychological disorders, and more complex conditions like metabolic syndrome. Many deficiency disorders and lifestyle diseases can be easily treated with help of a wholesome diet consisting of fresh fruits, vegetables, and whole grain. However, a new trend in health supplements has emerged in the last two decades. Multivitamins, supplements derived from microbial, botanical, and animal sources, antioxidants, probiotics, and fibers are abundantly available. Many of these do not have sufficient scientific evidence to support their putative claims and health benefits. These products are known as nutritional or dietary supplements, functional foods, or nutraceuticals; their attempted goal is to provide nutrition through tablets and capsules, instead of inculcating healthy habits of natural food consumption.

Drugs may be necessary to manage diseases, but they do not give us health; they treat diseases. Overuse and dependence on drugs must be controlled. During the last six decades, many drugs have been withdrawn from the market due to untoward effects. Pharmaceutical toxicities, autoimmune conditions, iatrogenic disorders, microbial resistance, and multiple organ failure are becoming serious issues of concern. While many modern drugs are effective in acute conditions to treat certain symptoms, chronic degenerative conditions are still waiting for effective therapeutic regimens.

New drug discovery is facing a severe innovation deficit. While billions are being spent by pharmaceutical companies every year for research on life-saving medicines, the rate of new drugs being introduced into the market is still falling. Pharmaceutical companies continue to aspire to create blockbusters, but do not give due enough attention to discovering drugs to treat neglected diseases, which are responsible for millions of deaths in developing countries.

The grand challenges of this century are not only about discovering new drugs, therapies, and surgical techniques, but they include addressing basic determinants of health such as nutrition, lifestyle, environmental factors, genetics, and access to affordable health care. The challenges are about how to improve low socioeconomic conditions, which are responsible for the growing disparities in quality of life and health in different regions of the globe.

The grand challenges also include the need to change the mind-set from hospital, drug-based health care, to new strategies, which focus on disease prevention and health protection. The challenges seem to be more in the areas of ecological sustenance, malnutrition, education, psychological and mental conditions, unhealthy lifestyles, inadequate resources, and poor access to medical care. These grand challenges may appear different in the rich and the poor worlds, yet the foundational causes behind these challenges remain common in this borderless world and interconnected societies.

At present, there is a global health paradox. Leaders, policy-makers, and scholars the world over are grappling with possible solutions to these complex problems. But there is no magic bullet or no single solution [14]. Present situation poses many questions about our approach toward health and medicine:

Have we really understood the concept of health?
Could we define it comprehensively?
Can we really achieve health as a complete state of well-being at physical, mental, and social levels?
Can we assume to be healthy when we do not have visible or measurable disease?
Is today's health care really *health*-centric or *disease*-centric?
In reality, is the present approach of health care, medical care?
Is there any role for traditional and complementary systems?
What is the need for integrative approaches in health?

Answers to some of these questions may be found in collaboration with the experiential wisdom available in traditional systems.

TRADITIONAL AND COMPLEMENTARY MEDICINE

Today, by and large, the onus of medical therapy is on modern medicine, which is scientific and evidence-based. Modern medical therapy is precise, efficient, and technologically intensive. However, it is focused on treating symptoms and diseases rather than on the health and well-being of the person. Modern medical therapy is also unaffordable and inaccessible to the majority of people around the world. As a result, people are exploring various options available from traditional and complementary medicine (T&CM).

Many studies investigating health-seeking behavior have found that people from developed and developing countries are looking for various options. The reasons underlying this search might be different. Many people in developing countries, particularly those in rural areas, have more access to traditional, rather than modern medicines, and take the traditional route more frequently [16]. They often turn to traditional healers to treat sickness. For example, an estimated 90% of people in Ethiopia use traditional medicine to meet their PHC needs, as do 70% of people in Benin, India, Rwanda, and Tanzania. One reason for this is that the traditional systems are deeply rooted in their respective societies and cultures, so their acceptance is very natural.

Even in the developed countries, the use of T&CM therapies is increasing. However, the extent of this use, and the economic implications were not well known until Eisenberg et al. conducted a national survey in the United States [17]. This study determined the prevalence, costs, and patterns of use of unconventional therapies such as acupuncture and chiropractic. One in three respondents reported using at least one unconventional therapy in the past year, and one-third of these saw providers for unconventional therapy. In 1990, expenditures associated with the use of unconventional therapy amounted to approximately $13.7 billion of which almost three-fourths was paid out of pocket. Many other studies have also reported that about one-half the population of developed countries now uses T&CM in one form or other. The national survey data suggest that use of T&CM therapy in the United States is not necessarily due to dissatisfaction with conventional care. People use and value both the conventional modern medicine, and complementary and alternative medicine systems [18]. Despite this situation, the real role and utility of T&CM in mainstream health care is still not clear. Many critics and scientists have very strong opinions about traditional medicine, many still view them more as placebo. In fact, because of the absence of sufficient scientific evidence they continue to be known as *alternative* or *complementary*, and have not become part of mainstream health care.

Doctors trained in modern medicine need not be the only choice for medical and health care. The health system needs to be properly segregated into primary, secondary, and tertiary levels with clear roles for medical and paramedical professionals. In fact, before accessing PHC, people should be empowered to take care of minor illnesses through home care. With help of pre-PHC many common illnesses may be easily treated with traditional medicine, home remedies, and by simple interventions of diet and lifestyle. Many countries have pluralistic systems of medicine as part of their culture.

The WHO policy on traditional medicine supports facilitating integration of T&CM into national health systems by helping member countries to develop their own national policies [19]. Unquestionably, scientific research on many T&CM therapies is inadequate. In many cases, various claims of T&CM therapies are not supported by sufficient, contemporary scientific evidence. Use of such diverse alternative or complementary, or traditional practices should not be mistaken for integrative medicine; at best these practices can be called medical pluralism. The concept of integrative medicine is much more profound; a discussion of such is the focus of this book.

T&CM is a comprehensive term used to refer to traditional systems such as Indian Ayurveda, Traditional Chinese Medicine, Japanese Kampo, Korean Sasang, Greco-Arabic Unani, and various forms of indigenous medicines and complementary therapies like Osteopathy, Homeopathy, and Naturopathy. T&CM practices also include Yoga, Acupuncture, manual and spiritual therapies.

In countries where the dominant health care system is based on modern medicine, or where T&CM has not been incorporated into the national health care system, T&CM is often termed *complementary*, *alternative*, or *nonconventional* medicine. However, in recent years the term *alternative* is not much used, because it is very difficult at times to decide what is alternative. For instance, Yoga is an integral part of Ayurveda, and Acupuncture is part of Traditional Chinese Medicine, but Europeans and North Americans regard them as alternative because they do not form part of their own health care traditions. Similarly, Homeopathy and Chiropractic systems were developed in Europe in the eighteenth century, and after the introduction of allopathic medicine, yet they are not incorporated into the dominant modes of health care in Europe. In this book, we have used T&CM as a common term to represent medical practices other than modern medicine.

The WHO's traditional medicine strategy rightly emphasizes accessibility of T&CM, which is hardwired into the belief systems of some communities. It should come as no surprise that the majority of people from poor and developing countries use T&CM as a means to fill their PHC needs. The low-income, developing countries do not receive the optimal benefits of modern medicine because they cannot afford it; the high-income, developed world does not benefit from the holistic wisdom implicit in T&CM, which is not present in these societies.

The T&CM systems are largely unregulated in most countries. The WHO believes that information and education will help consumers to seek out appropriate types of self-care and as a result, help them to obtain more benefits from T&CM and reduce unnecessary risks. Studies show that many patients use T&CM therapies concurrently with conventional medicine, often without informing their health care provider. This may pose a risk of herb–drug and food–drug interactions. Efforts are needed to improve communication with patients and health care providers to ensure that consumers are better to minimize the risks and maximize the benefits of T&CM use [20].

Although there are very limited systematic studies to compare cost-effectiveness of T&CM, some reports have substantiated the general belief that T&CM is affordable as compared to modern medicine [21]. In a recent randomized controlled study on 401 patients with chronic headache, use of Acupuncture significantly improved health-related quality of life at a small additional cost compared with a number of other available interventions in the modern medicine [22]. Similarly, interventions through Yoga and meditation especially in cardiovascular, psychosomatic, musculoskeletal, and mental disorders have been beneficial at a considerably low cost and risks [23]. Saper et al. have reported that an estimated 15 million American adults had used Yoga at least once in their lifetime and 7.4 million during the previous year. Yoga was used and often perceived to be helpful and cost-effective both for wellness and specific health conditions [24].

Thus, some of the T&CM therapies offer safer, better, and cost-effective alternatives. However, more systematic research is needed to show their usefulness in specific conditions in different populations. Available data, mainly from the developed countries such as the United States make a case with reasonable evidence that T&CM can be a better alternative to modern medicine especially in many chronic, difficult-to-treat lifestyle diseases. Possibility of improved affordability is another good reason for considering possible integration especially when most people in low-income countries pay for medicines out of their own pockets.

The governments of China and India, among others, provide state support to strengthen training, research, and the use of traditional medicine in their national health care strategies.

A number of African countries are considering how to integrate traditional medicine into "mainstream" health care [25]. Apart from medical use, the production, sale and export of traditional medicines is an important component in some economies. China, for instance, exports over $600 million of traditional medicine products annually. Chinese health authorities have launched a nationwide program to build a large number of Traditional Chinese Medicine hospitals, each specializing in the treatment of a particular condition, such as different types of cancers, cardiovascular diseases, and hepatitis.

T&CM can provide novel inputs into the drug development process. For instance, a consortium led by the Medical Research Council of South Africa is carrying out a study on traditional medicines used by communities for the self-treatment of fevers, with the aim of discovering active compounds to treat malaria more effectively. Various institutions, including the Council for Scientific and Industrial Research (CSIR) in India, are exploring alternative paths to modern pharmaceutical research presented by traditional medicine. These approaches known as *reverse pharmacology* could be cheaper, faster, and more effective. Potential of such innovative approaches needs to be explored for developing cost-effective and safe medical products where knowledge of traditional medicine is used for modern drug discovery.

NONPHARMACOLOGICAL APPROACHES

In the present situation where drugs and hospitals are dominating health care, several other nonpharmacological approaches may play a vital role in disease prevention and health promotion. Traditional knowledge systems like Ayurveda and Yoga may give us a more holistic understanding of health and may provide newer insights to disease prevention and health promotion. Ayurveda offers an excellent nutritional guidance based on life cycle approach and Yoga offers several relaxation techniques. If these ancient concepts are studied and properly validated using the twenty-first-century science and technology, we may be able to bring them in practice through proper education, especially of mother and child. We feel such integration of knowledge systems may improve present health care system.

For instance, Ayurveda has valuable advice about ways and means to stay healthy. This is known as *Swasthavritta*. Yoga offers several physical, physiological and mental ways for relaxation, health promotion, and disease prevention. Ayurveda offers personalized advice on nutrition and lifestyle according to seasonal variations. Life cycle approaches to nutrition and developmental origin of adult diseases have emerged as major issues in recent years [26]. Ayurveda also describes ways and means for healthy progeny and it will be interesting to study these concepts using modern developmental biology. Food and nutrition plays significant roles in health care, however, just providing food does not necessarily mean providing nutrition to person.

The modern approach based on mathematical calculations of calories and supply of proximate principles like proteins, fats, carbohydrates, minerals, etc. does not consider the "person" who is consuming them for nourishment. New research is building the optimism toward achieving health through biobehavioral and lifestyle modifications. Physical fitness is not only about building muscles and shape. A physically fit person may not necessarily be healthy. However, physical fitness certainly helps in improving endurance and maintaining health. For example, physical exercise training has demonstrated improvement in cardiac function through formation of new cardiomyocytes [27].

Interestingly, most of the T&CMs across the board consider human beings and their health in the context of their environment. This criterion defines T&CM apart from modern medicine and provides a reason for their reemergence. Thus, T&CM can be used as an input for development of modern medicine to "postmodern" integrative health care. We hope that involving the ancient wisdom from Yoga and Ayurveda will be studied with the help of biomedical science and technology. This may show new ways to understand complex holistic relationships of body-mind-spirit to shape future integrative health care.

EXPERIENCE AND EVIDENCE: LEARNING TOGETHER

The modern medicine is often described as scientific *Evidence*-Based Medicine. We feel T&CM may be described as *Experience*-Based Medicine. Actually both the "*E*"s are important and should not be treated as mutually exclusive. Most of the T&CM systems or therapies have emerged from traditions or through experiential, observational, or cultural routes. In reality, medicine as such, modern or otherwise, evolves and progresses through traditions. Few of these numerous traditions, especially from the Europium region like Greek and Latin followed the scientific rigor of experimentation and evolved as modern medicine and received global acceptability. Whereas, others like Ayurveda and Yoga, hardly crossed the threshold of geographical and cultural locality, remained untested for centuries neither proved nor disproved.

To address present crises in medical therapy, we need an inclusive system involving the drugs, pharmaceuticals, and surgeries, which can be drawn from modern medicine while diet, nutrition, lifestyle modifications, Yoga, meditation, acupuncture and such interventions can be drawn from T&CM.

INTEGRATIVE HEALTH AND MEDICINE

We differentiate terms "integrative medicine" and "integrative health." They should not be used as synonyms. Integrative health is about holistic understanding and recognition of the importance to consider all the vital dimensions of health as body, mind, and spirit. Integrative medicine is more about synergistic convergence of modern medicine and T&CM systems. These integrative approaches cannot be considered in isolation. At the cost of repetition, we wish to reiterate that *health cannot be achieved just by giving medicines or by treatments provided by doctors or merely by discovering effective drugs*. Health requires proactive involvement of individuals or communities. Present models of health care are based mostly on symptomatic treatments, which are good and required but are not good enough.

While few examples in the book are drawn from India, the situation related to health is no different in many other developing countries. In reality, in an interconnected world practically every country whether rich or poor,

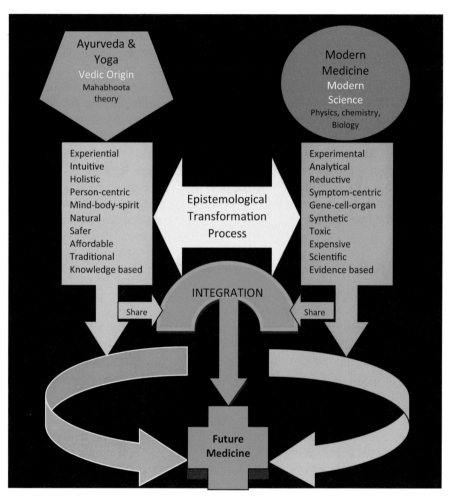

FIGURE 1.3
Integration of Ayurveda, Yoga and modern medicine for future medicine.

underdeveloped, developing or developed, western or eastern is facing a challenging scenario related to health care. Present models of disease- and treatment-centric health care are becoming increasingly unsustainable.

Clearly, modern medicine needs an inclusive and integrative approach for a viable and sustainable solution. Therefore, many experts and thought leaders are advocating drastic changes and disruptive innovations in the basic strategies, concepts, and approaches to health care. The mutual sharing of knowledge may create the necessary environment for integrative approaches where experiential and experimental epistemologies transform each other and make a way for the future medicine (Figure 1.3).

Systematically designed "integrative" approaches are now required. There is lot to learn from each other both from the T&CM represented by Ayurveda and Yoga, and the modern medicine. There is no need to rediscover the wheel. But there is need to take smart steps by avoiding redundant developmental stages, which may have been unavoidable in earlier times. This strategy is known as leapfrogging [28]. The techniques and structures created to meet previous developmental challenges have tended to remain embedded in health systems even after the circumstances have drastically changed. Better methods and approaches are being ignored and the medicine and health care seem to be trapped. The leapfrogging concept helps to avoid such traps and innovate. We suggest that mutually beneficial integration by bridging Ayurveda and Yoga with evidence-based scientific approaches of modern medicine to build a future health care system [29] may be one of the disruptive innovation and leapfrogging strategies.

MODERN MEDICINE, AYURVEDA, AND YOGA

Ayurveda and Yoga deal with healthy life style, health promotion and sustenance, disease prevention, diagnosis, and treatment. Prolonged use of Ayurveda by people has also led to the development and use of time-tested home remedies for common ailments, which continue as local health traditions. Ayurvedic medicines also contain sophisticated therapeutic formulations. Yoga as a way of life is part of Indian tradition and involves systematic physical exercises, breathing and meditation, and relaxation techniques.

We have supplemented a primer on Ayurveda and Yoga in the Annexure so that without losing the flow, readers are able to get a simple understanding of the basic principles, epistemologies, strengths, and limitations of these systems in the context of this book. We have discussed evolution of medicine from primitive shamanism to the twenty-first-century modern medicine. While introducing the need for convergence among various approaches in health and medicine, we have discussed appropriate models for effective integrative health and medicine. With representative examples, mainly drawn from Ayurveda and Yoga, we have tried to portray the interesting evolution and convergence of diverse philosophies, principles, techniques, procedures, practices.

We have taken a life cycle approach drawn from Ayurveda and Yoga to discuss about concepts of healthy lifestyle, food, nutrition as well as concepts of mental and spiritual health. We have shown how Ayurveda-knowledge-inspired efforts may expedite safer and efficient drug discovery. We have discussed relevance of Ayurvedic concepts of aging and rejuvenation in the rapidly evolving cutting edge sciences like stem cells and regenerative biology.

We wish to clarify that we have briefly reviewed a few projects related to biomedical research in Ayurveda and Yoga. However, this book is mainly about how strengths of modern medicine, Ayurveda, and Yoga can play an important

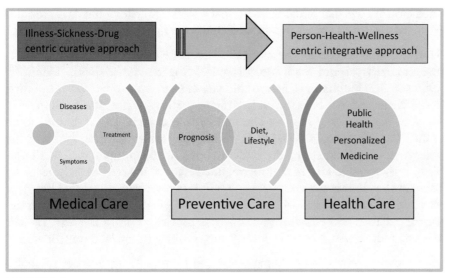

FIGURE 1.4
Changing mind-set from medical care to health care.

role in advancing biomedical research for protecting and promoting health. We have highlighted the need for a fundamental shift in mind-set that needs to change from "sick care" and "medical care" to "preventive care" and "health care." We have advocated necessity for moving from illness-sickness-drug-centric curative therapies to person-health-wellness-centric integrative approaches of the future. This route will transit through the preventive care where diet and lifestyle modification will be involved. Delivery of high quality health care to the public and effective personalized medicine for individuals will be possible through the integrative approaches (Figure 1.4).

While making the advocacy for integrative medicine, we wish to emphasis that the integrative movement is already popular in many countries. Top medical centers like Mayo, Cleveland, and Universities such as Pennsylvania, California, Boston, Arizona, and many from the European region are successfully integrating Yoga and T&CM in clinical practice.

We are aware that fairly good global consensus exists in favor of integration. However, at present the integration is happening not through medical doctors but mainly because of patients, who are seeking various options. In very few countries like China, integration is taken seriously at policy levels. Therefore, an advocacy is needed for rediscovering and developing appropriate integrative models for health and medicine. We trust that the best option to disrupt present impasse in medical therapy and health care may be a judicial integration of modern science and technology led modern medicine with the age-old traditional experiential wisdom of holistic systems like Ayurveda and Yoga for envisaging the future medicine. We feel that bridging traditional knowledge systems like Ayurveda and Yoga with scientific approaches in medicine is the right path for the future [29].

However, true success and test of such "integrative" approach will remain in the mutual trust and ability to recognize, respect, and maintain identities, philosophies, foundations, methodologies, and strengths intact while building a sufficient evidence base for integrating respective systems. It is important to remember: *"absence of evidence is not evidence of absence"* [30]. We need to be a little humble before criticizing or praising any system or monopolizing the science or blindly accepting the evidence-based medicine approach. In all fairness, it is crucial to ensure that the required rigor of the science is achieved by avoiding any hubris of technology power. It is necessary to ensure that the spirit of integration is achieved without compromising the foundations of the traditional practices and knowledge. The words of wisdom from *Lao Tzu "He who does not trust enough, will not be trusted"* may probably show us a way forward.

Integrative approaches are being embraced by a growing number of mainstream medical centers. However, it is still not finding expected acceptance at the practitioner levels. This may be due to a general perception that T&CM is not evidence-based and should only be used with caution. More intense multidisciplinary research efforts are needed so that the advocacy flows as thought leadership from the top medical centers to the practitioner levels through scientific publications, continuing education programs, and public education.

We conclude this advocacy for integration by emphasizing that the foundations of humanity and interests of the sufferers should be the hallmark. The interests of sufferers must override any rigidities, boundaries, or compartments of any system—be it modern medicine or Ayurveda and Yoga. This is possible if the silo approach and ego-related specific system is set aside to make optimal use of plural systems of medicine in the best interest of people. The words of wisdom by *Lao Tzu* should inspire us to create a trusted integrative system based on mutual trust.

REFERENCES

[1] Fani Marvasti F, Stafford RS. From sick care to health care — reengineering prevention into the U.S. System. N Engl J Med 2012;367(10):889–91.

[2] Fries JF. Aging, natural death, and the compression of morbidity. Bull World Health Organ 2002;80(3):245–50.

[3] Hegde BM. Integrated medical care system (complementary systems of medicine – are they scientific?) Journal of Indian Academy of Clinical Medicine 2011;12(4):260–2.

[4] Barsky AJ. The health paradox. N Engl J Med 1988;318:414–8.

[5] Wilensky GR. The shortfalls of "Obamacare." N Engl J Med 2012;367(16):1479–81.

[6] Reardon S. United States to approve potent oral drugs for hepatitis C. (News). Nature, 30th October 2013. Available from: http://www.nature.com/news/united-states-to-approve-potent-oral-drugs-for-hepatitis-c-1.14059. (accessed on January 29th, 2015).

[7] CDC. Health, United States, 2012. With special feature on emergency care; 2013.

[8] Pirmohamed M, James S, Meakin S, Green C, Scott AK, Walley TJ, et al. Adverse drug reactions as cause of admission to hospital: prospective analysis of 18 820 patients. BMJ 2004;329(7456):15–9.

[9] Mokdad AH, Marks JS, Stroup DF, Gerberding JL. Actual causes of death in the United States, 2000. JAMA 2004;291(10):1238–45.

[10] The global economic burden of non-communicable diseases. World Economic Forum and Harvard School of Public Health; September 2011.

[11] Nambiar D, Narayan VV, Josyula LK, Porter JDH, Sathyanarayana TN, Sheikh K. Experiences and meanings of integration of TCAM (traditional, complementary and alternative medical) providers in three Indian states: results from a cross-sectional, qualitative implementation research study. BMJ Open January 2014;4(11):e005203.

[12] Patwardhan B. Planned progress for health. J Ayurveda Integr Med 2011;2(4):161–2.

[13] Obokata H, Wakayama T, Sasai Y, Kojima K, Vacanti MP, Niwa H, et al. Stimulus-triggered fate conversion of somatic cells into pluripotency. Nature 2014;505(7485):641–7.

[14] Patwardhan B. Traditional medicine for affordable global health. CIPIH WHO Report 2005. Available from: http://www.who.int/intellectualproperty/studies/B.Patwardhan2.pdf.

[15] Mexico, 2004: Global health needs a new research agenda, Lancet 2004;364(9445):1555–6.

[16] Traditional Medicine – Growing Needs and Potential, WHO Policy Perspectives on Medicine. No.2. Geneva: World Health Organization; 2002.

[17] Eisenberg DM, Kessler RC, Foster C, Norlock FE, Calkins DR, Delbanco TL. Unconventional medicine in the United States. Prevalence, costs, and patterns of use. N Engl J Med 1993;328(4):246–52.

[18] Eisenberg DM, Kessler RC, Van Rompay MI, Kaptchuk TJ, Wilkey SA, Appel S, et al. Perceptions about complementary therapies relative to conventional therapies among adults who use both: results from a national survey. Ann Intern Med 2001;135(5):344–51.

[19] WHO strategy on traditional medicine 2014–2024. Geneva: WHO; 2013.

[20] Zhang X. Department of essential drugs and medicines policy, who guidelines on developing consumer information on proper use of traditional, complementary and alternative medicine. Geneva: World Health Organization; 2004.

[21] White AR, Ernst E. Economic analysis of complementary medicine: a systematic review. Complement Ther Med 2000;8(2):111–8.

[22] Wonderling D, Vickers AJ, Grieve R, McCarney R. Cost effectiveness analysis of a randomised trial of acupuncture for chronic headache in primary care. BMJ 2004;328(7442):747–9.

[23] Oz M. Emerging role of integrative medicine in cardiovascular disease. Cardiol Rev 2004;12(2):120–3.

[24] Saper RB, Eisenberg DM, Davis RB, Culpepper L, Phillips RS. Prevalence and patterns of adult yoga use in the United States: results of a national survey. Altern Ther Health Med 2004;10(2):44–9.

[25] Kimani C. Kenya to develop traditional medicine action plan, SciDevNet, 29 June, 2004.

[26] Barker DJP. Developmental origins of adult health and disease. J Epidemiol Community Health 2004;58(2):114–5.

[27] Ellison GM, Waring CD, Vicinanza C, Torella D. Physiological cardiac remodelling in response to endurance exercise training: cellular and molecular mechanisms. Heart 2012;98(1):5–10.

[28] World Economic Forum and Boston Consulting Group. Health systems leapfrogging in emerging economies - Project paper. Geneva: World Economic Forum; 2014.

[29] Patwardhan B. Bridging Ayurveda with evidence-based scientific approaches in medicine. EPMA J 2014;5(1):19. http://dx.doi.org/10.1186/1878-5085-5-19.

[30] Patwardhan B. The quest for evidence-based Ayurveda: lessons learned. Curr Sci 2012;102 (10):1406–17.

CHAPTER 2

Evolution of Medicine

The doctor of the future will give no medicine, but will interest patients in the care of the human frame, in diet, and in the cause and prevention of disease.

Thomas Edison

SHAMANIC AND ETHNOMEDICINE

The history of medicine dates back to the period of human existence. The understanding of health and disease has evolved over thousands of years. Most of the great civilizations—Egyptian, Assyrian, Babylonian, Hebrew, African, Arabic, Chinese, and Indian—had their own traditional medicine.

Archaeological evidence of knowledge of diseases and treatments during pre-historic periods is debatable. The customs and traditions among several tribes in Africa and aboriginal people in Australia, and the Americas provide a few glimpses. During ancient times, the priests or shamans were responsible for diagnosing and treating diseases.

During prehistoric times, medicine men, also known as witch-doctors, were responsible for the maintenance of health, and treatment of ill health in many tribal and ethnic communities. Ethnomedicine used medicinal herbs as well as supernatural, ceremonial, and ritualistic treatments such as magic formulas, charms, jujus, prayers, and drumming; these were commonly used to remove evil spirits. These god-men were believed to be able to communicate with gods or spirits, and cure the patient through supernatural powers. Medicine men used to be central figures in the tribal system because of their medical knowledge, and people's faith in their ability to contact the gods. Most of the knowledge about such traditions—religious and medical training—was passed on orally. This unwritten and noncodified knowledge must have undergone several transitions, and transformations, intentional changes and unintentional errors over centuries. Therefore, in the present context, the relevance of the processes and materials of the shamanistic traditions is highly questionable, unless it is subjected to rigorous study and validation.

27

Prehistoric people probably knew about medicinal plants and other substances to treat disease. The *Ebers Papyrus* dating to 1550 BC is considered the oldest and most important medical papyri of ancient Egypt. Imhotep, who was supposed to be the chief minister to the Egyptian King in the twenty-seventh century BC, is considered to be the first doctor; he is identified with the Greek god, Asclepius. Asclepius is the Greek god of medicine, and his daughters represent important branches of medicine. The snake-entwined staff of Asclepius is a symbol of medicine today. Those physicians and attendants who served the god Asclepius were known as the *Therapeutae* of Asclepius. The use of these terms in modern medicine today shows the influence of these civilizations. There is some archeological evidence that clay tablets bearing specific signs were used by physicians of ancient Mesopotamia. One of the oldest, deciphered stone carvings, the Code of Hammurabi, was promulgated by a Babylonian king during the eighteenth century BC, and contains a few laws relating to the practice of medicine.

Quite surprisingly, the widespread practice of the preservation of dead bodies as mummies by the ancient Egyptians did not lead to a deeper understanding of anatomy. However, they indicated evidence of arthritis, tuberculosis of the bone, gout, tooth decay, bladder stones, gallstones, and the parasitic disease, schistosomiasis—maladies which are seen even today. Modern medicine was nurtured in the European region, and draws substantially from the knowledge and approaches found in papyri of Egypt, Greek, and Hebrew origin.

While the Harappan Civilization also existed during the same time as the ancient Egyptians in the Indian subcontinent, reliable archeological evidence is not available; hence the nature of medical practices prevalent in the Indus Valley Civilization is unknown. The nature of the art and science of medicine, which developed during this period still remains a mystery. *Charaka Samhita*, the classic of Ayurveda, is supposed to have been written sometime during the fourth century BC. Hippocrates' contributions to medicine were made around this same time. While medicine traditions existed during Buddhist and Jain periods sometime during the fourth and sixth century BC. However, for all practical purposes, the reference point for cause and effect based, rational medicine in India remains the *Charaka*.

PHILOSOPHY OF MEDICINE

Although, this book is about integrative approaches for *health*, we begin with a discussion of *medicine*. We find much confusion between these two terms. Many times they are used as synonyms. During the last two centuries, the focus on health was slowly shifting to disease and treatments. The possible reasons for this shift are rooted in the concepts and philosophy of medicine. Therefore, it is important to clearly understand the term medicine. It is necessary to explain a few similar sounding, but often interchangeably used terms related to health and medicine.

The contemporary and dynamic dictionary/search engine, Google, gives two definitions of medicine. The first definition is quite broad, and describes medicine as "the science or practice of the diagnosis, treatment, and prevention of disease." The second definition is specific, and relates to a substance or a product as "a drug or other preparation for the treatment or prevention of disease." Thus the term *medicine* incorporates science and clinical practice, together with substance or drug components. This word is supposed to originate from the Latin word, *medicina*, meaning drug, and *medicus* meaning physician. The term *clinician* is often used as a synonym for physician, or a medical doctor. The term clinician has its origin in the Greek term *clinike*, which actually means bedside. The terms physician, clinician, and doctor generally refer to medical doctors who treat sick people. Thus, even today, the notion continues that medicine and doctors are for sick people. However, this is not so. Rather this should not be so. Doctors should actually help people remain healthy. Their role should not be restricted only to the treatment of diseases. To get a little better insight into this approach, it is advantageous to delve into the philosophy, ontology, and history of medicine.

There are diverse views regarding what constitutes a philosophy of medicine. One of the authorities in this field Arthur Caplan, in 1992, raised the question "Does the philosophy of medicine exist?" According to Caplan, the philosophy of medicine is a subdiscipline of the philosophy of science [1]. The philosophy of medicine includes medical ethics and bioethics, but its scope is much wider, involving epistemological, metaphysical, and methodological dimensions of medicine, which include therapeutic, experimental, diagnostic, and palliative aspects. If we consider medicine only as a branch of the basic sciences like anatomy, physiology, and biochemistry, then the philosophy of medicine will be a branch of philosophy of science.

However, medicine is not limited to pathophysiology, pharmacology, psychology, microbiology, or genetics. The main purpose of the science of medicine is to understand human body functions and dysfunctions through sound hypothesis, robust experimentations, observations, analysis, and a continued search for truth. However, the real purpose of medicine is more than all this. It is indeed a search for truth, with the definite objective to attain health and healing. Probably this is the reason why religious places like churches, temples, monasteries, and mosques were involved in medical care before the evolution of clinics and hospitals. These sacred places protected traditional knowledge during invasions and dark periods in their respective regions. In many ancient civilizations like Egypt, priests worked as doctors and healers. Even now in some parts of Africa and India, such practices exist. During the ancient time, medicine was not restricted to material drugs; rather, the approach was holistic—involving body, mind, and spirit.

Medicine has a much larger role than that of science. It is not merely a sum of philosophies of science, biology, or the humanities. Medicine has a very strong ethical aspect that relates to the good of patients [2]. Its present commercial nature as a fee-based service goes against the true spirit of this noble profession. In any case, the philosophy of science can, and probably will lead to a better understanding of the philosophy of medicine. Philosophers have been trying to define the term science for a long time.

PHILOSOPHY AND SCIENCE

During ancient times, science and philosophy were considered as one. Therefore, ancient Egyptian, Roman, and Greek thought actually exhibits a dominance of holistic philosophy over reductionist science. These Western traditions slowly were transmuted into modern science as a result of sharp-eyed observations, and rigorous experimentation and analysis by several philosophers and thinkers such as Aristotle, Archimedes, Galileo, Newton, and Einstein. On the other hand, many of the ancient Eastern traditional knowledge systems became frozen parts of history. They remained static for a long time, and lost the dynamism that is necessary for the progression of science. As a result, many of the Eastern traditions, like that of Tibet, remained more a philosophical bulwark against the adaptation of the culture of science. The ancient knowledge systems like Ayurveda and Yoga seem to fall somewhere in between.

The discipline of natural philosophy gradually became natural science, which is now known as modern science. Socrates first tried to apply philosophy to the study of human nature and human knowledge. At that time, he actually criticized the way in which physicists were thinking. In the process, he challenged the mythology and dominion of traditions. As a result, Socrates was viewed as a rebel, and was executed in 399 BC. Later, Aristotle (384–322 BC) proposed a less-controversial systematic program; he maintained the sharp distinction between evolving experimental science, and the traditional, practical knowledge of artisans. He advocated that theoretical studies are of higher intellectual activities than skills of artisans. In a way, he started differentiating levels of intelligence, and hierarchical, professional classes. Aristotle emphasized importance of the theoretical steps in deducing universal rules from raw data. According to him, mere experience, and the collection of raw data is not part of science. For the first time, Aristotle proposed that the earth was round, and not flat. He promoted theories of reductionism, and encouraged deeper studies on parts. This Aristotelian reductive logic and quest for new knowledge has led to today's science of analytics.

A paradigm shift happened when philosophy and science were separated, and the seeds of reductionism were sowed. Quite interestingly, almost up to the seventeenth century AD, the words *science* and *philosophy* were used interchangeably [3]. Probably during Newton's time natural philosophy branched out, and evolved as natural science [4]. As philosophy and science were separated, the division between East and West became sharper. With the help of the Aristotelian logic, modern science rapidly progressed in the West, and grew into many branches such as physical, chemical, and biological sciences; this led to new technological applications. Aristotelian deductive reasoning used cause and effect, premises, and analysis to draw conclusions. With valid and true premises, and strict observance of deductive logic it is possible to derive accurate conclusions. These approaches subsequently led to hypotheses building, and theories such as generalization, causal relations, statistical syllogism, detachment, validity soundness, predictions, and finally, empiricism.

Empiricism holds that any knowledge comes primarily from experimentally measurable or experiential, sensory experience. This was further supported by rationalist criteria where proof is drawn from intellectual reasoning and deductive methods. From antiquity through to the Middle Ages, the Aristotelian reductionist approach was rigorously used in understanding several natural phenomena; this gradually prepared the way for the evolution of the modern science. However, in the process of scientific advancement, some valuable insights and ancient knowledge were either lost, or remained in obscurity. This background will be useful when we discuss the meaning of evidence, and its application to medicine.

SCIENCE IN THE WEST

The term *science* normally denotes modern science and involves a broad spectrum of human activities—from understanding the universe and nature through understanding of bosons, electrons, atoms and particles; lasers, microscopes and telescopes; animals, plants, microbes and species; tissues, cells and genes; and an array of activities from space odysseys to decision analysis. Generally, science can be considered as a body of knowledge, and the processes of its acquisition, retrieval, and applications. It is not only about the collection of information or laboratory investigations, or statistical analysis; it is a process of logical, consistent discovery leading to the enhancement of holistic understandings. The knowledge generated by science is reliable, reproducible, and can be applied to develop new technologies.

Science is very dynamic in nature; concepts, hypothesis, assumptions, relevance, and interpretations may keep on changing as the body of knowledge grows. The serious study and practice of science normally lead to new questions for future investigations. Therefore, theoretically, true science never will reach finality. Several traditional knowledge systems existed in the premodern era. These are categorized as the ancient sciences. Today's modern medicine essentially developed through the theories and principles of modern science. Modern science relies more on rational, experimental, and reductionist approaches; traditional practices like Ayurveda rely more on holistic, empirical, pragmatic, experiential, and intuitive strengths.

In the *Oxford English Dictionary*, science is defined as "an intellectual and practical activity encompassing the systematic study of the structure and behavior of the physical and natural world through observation and experimentation." Thus, modern science is restricted to the physical and natural world. Modern science has revolutionized our life by generating knowledge that provided us with powerful diagnostics, new medicines, vaccines, and many technological marvels like robotic surgery. Modern science helps us to understand many important problems, and provides answers to questions: Who were our evolutionary ancestors? Why and how do humans differ from monkeys? How do microbes cause disease?

While modern science can address questions related to physical, chemical, and biological phenomena or processes, it also has its limitations. Typical, searching, philosophical questions: "Who am I? Where did I come from? What is the goal of my life?" cannot be answered in the context of our present understanding of science. Modern scientists are able to arrive at different theories concerning the origin of life, but they do not have an answer to the question: What is the purpose of life? Modern science does not deal with moral or aesthetic judgments. These issues typically are the domain of philosophers. It is interesting to see how science and philosophy are related to each other.

SCIENCE IN THE EAST

While modern science took shape in the West, the East faced invasions and instability. The geopolitical situation in the East was not conducive to research or development. For a long time, the Indian subcontinent was under the rule of Mughal emperors, followed by the British. Although, significant knowledge, wisdom, and competence existed in this region, hardly any scientific advancement was possible; Indian communities were more concern about preserving knowledge. This led to the compartmentalization and monopolization of knowledge. This was a period of great stagnancy. Some examples of notable contributions, which emerged from Eastern regions include Charaka (400 BC), who is considered the father of medicine. He compiled a comprehensive encyclopedia of medicine, the *Charaka Samhita*, which elaborates principals, diagnoses, and treatments. This classic text gives detailed knowledge of human anatomy, embryology, pharmacology, blood circulation, and diseases like diabetes, tuberculosis, heart disease, and details of thousands of medicinal plants used in Ayurveda.

Other notable contributions from Indian subcontinent include work of Aryabhatta (476 AD), a master astronomer and mathematician. In his classic text known as *Aryabhatiyam*, he described the process of measuring the motion of planets and eclipses. Aryabhatta proclaimed that the earth was round, it rotates on its axis, orbits the sun, and is suspended in space. Aryabhatta's most spectacular contribution is the concept of zero. Varahamihira (499–587 AD) in a classic text, *Panchsiddhanta*, notes that the moon and the planets are lustrous not because of their own light, but due to sunlight. In the *Bruhad Samhita*, he detailed many discoveries in the domains of geography, botany, and animal science. Another great scholar, and master of arithmetic and astronomy is known as Bhaskaracharya (1114–1183 AD). In his classic text, *Surya Siddhanta*, he made a note on the force of gravity. Nagarjuna (800 AD), in the classic text *Rasa Ratnakara*, outlined many interesting experiments in metallurgy, alchemy, and chemistry. He also used metal-based ingredients in medicinal preparations. Many other spectacular contributions have come from the Indian subcontinent, however, due to limitations of space, only a few representative ones are mentioned here.

Indian Ayurveda and Traditional Chinese Medicine (TCM) are part of the Eastern or Oriental traditional medicines. India and China have rich traditions, spanning thousands of years, in the arts, humanities, sciences, and medicine. From 400 BC, Indian knowledge became available to European and Arabic regions through invasions by Alexander the Great, and others. There is evidence of some communication between Greece and India even before Alexander the Great's invasion, possibly through the Achaemenid Persians [5].

A closer look at the history of Western science, uncovers the influence of Vedic and ancient Indian knowledge systems on many philosophers and thinkers from the West. Evidence of this influence is also found in the works of pioneers in the field of medicine such as Hippocrates, Avicenna, Galen, and Osler; the influence of Vedic knowledge on Greek science and medicine is apparent. Yet hardly any mention or recognition appears in the documents including the most cited book of Osler on history of modern medicine.

In contrast to the lack of attribution in the Western scientific, philosophical and medical texts, Persian scholars duly acknowledged their sources of knowledge. For instance, Abdulla-bin-Ali, Manka, and Ibun-Dhan translated several Ayurveda texts into Persian and Arabic. *Charaka Samhita* (Sharaka), *Sushruta Samhita* (Kitab-E-Sushrud), *Ashtanga Hridaya* (Astankar), and *Siddhayoga* (Sindhashtaq) reached the Arabic and Persian physicians in the sixth to eighth centuries. In 850 AD, Ali-bin-Raban-al-Tabri authored a textbook of medicine, *Firdausu'l-Hikmat*. He appended a chapter on the Indian system of medicine in this book; this was much earlier than Avicenna wrote his famous *Canon of Medicine*.

The Eastern process of dialogue, dialectical method of enquiry, and hypothesis building also form the foundation of Western philosophy. Many concepts have roots in Vedic literature. The Delphi method of participatory, consensus building, which is used widely today, might have roots in a tradition of Ayurveda known as *Tadvidya Sambhasha*. Detailed rules and procedures for conducting consultations, debates, and discussions are given in Ayurveda. The Greek theory of macrocosm and microcosm, the concept of the four elements (fire, air, water, and earth), the four qualities (hot, dry, wet, and cold), and the four humors (yellow bile, blood, phlegm, and black bile) have striking similarities to the theory of *Loka–Purusha*, the five *Mahabhoota*, the three *Dosha*, and six *Rasa*—which form the basis of Ayurveda.

The Indus Valley Civilization, and the ancient Chinese, Mesopotamian, and Egyptian civilizations had several similarities. In the beginning of the first millennium AD, three principal systems of medicine—Ayurveda, Greek, and Chinese—were active. The fundamental attitude to the relationship of man and nature was similar in all the three ancient civilizations. However, their explanations of the human body, physiology, pathology, and therapy were different in some respects [6]. Their approach to health was holistic.

Later, Greek medicine followed the modern science path, and gradually evolved as modern medicine. Ayurveda emerged as a remarkable, holistic system both in its foundational ideas, and therapeutic measures. Later, during the period of invasions in India, Ayurveda was suppressed and stagnated for several centuries—a severe setback for systematic progress and development of Ayurveda. Moreover, because of the dominance of modern medicine during the colonial period, Ayurveda was further fragmented, deteriorated, and was marginalized.

In illustrating a few examples from India and Ayurveda, it is not the authors' intention to raise a debate about the supremacy of any civilization or approach. We simply wish to highlight the tradition of exchanges, dialogue, and similarities between Eastern and Western thought—especially during the time of ancient civilizations. Similarity in approaches is evident in the advice given by Charaka and Hippocrates. It is unfortunate that due to geopolitical, environmental, and socioeconomic changes, a healthy dialogue and a free exchange of knowledge between East and West were significantly reduced. Western societies, especially during the Renaissance, embarked upon the dynamic, reductionist path of modern science and technology. On the other hand Eastern societies remained isolated, and struggled to protect and continue their own traditions.

Before the Renaissance, the division of the world into East and West was not very sharp. The Renaissance of the fourteenth to the seventeenth centuries AD, started in Italy and spread throughout Europe. This triggered a new era of creation, transformation, and progress of civilization. In real sense, this is an early example of disruptive innovation. With the availability of paper and early printing tools, the dissemination of ideas was facilitated. This period witnessed spectacular revolutions in humanism, and social, political, and intellectual pursuits. Some of the world's greatest artistic achievements were made during this period. Remarkable artists like Leonardo da Vinci and Michelangelo made timeless impacts during this period. The benefits of the Renaissance to various disciplines including art, culture, science, medicine, and humanism were mainly seen in the West and Middle East including European, English, Arabic, and Russian regions. Sadly, these developments of science, technology completely bypassed the East, including the Indian subcontinent, and other parts of Asia. As a result, today's modern science is known as Western science, and modern medicine is also known as Western biomedicine.

The Western ideas fueled by the Aristotelian reductive approach shaped the modern world due to robust developments in sciences and technologies. The modern world became rational, logical, mechanistic, materialistic, and self-centered. The Eastern world continued with traditions, remained suppressed, pragmatic, intuitive, and spiritual. If an epistemological bridge between these two worlds would have existed, the face of Ayurveda as well as modern medicine would have been much different.

The Western mind and philosophy seems more restrictive, reductive, quantitative, calculative, and analytical—relying more on empirical, measurable parameters within the understanding or experiences of human sensory organs, or from instruments as their extensions. In a way, Western philosophy is dominated by the left brain.

The Eastern mind and philosophy seem more open, inclusive, holistic, qualitative, creative, and intuitive in approach, and not restricted to sensory experiences, but often marked by transcendence involving body/mind, and spiritual insights. In a way, Eastern philosophy is dominated by the right brain. The earth can be visualized as a brain: the East and South dominated by the right brain, and the West and North by the left. Interestingly, the difference in Eastern and Western approach is seen in the respective regions (Figure 2.1).

We are aware that while psychologists agree with the left and right brain theory, brain scientists differ. We feel that psychology as a systemic science focusing on mind, represent right side, while the neurology as a structural science focusing on brain, might represent the left side. According to Jeff Anderson, neuroscientist from the University of Utah, who studied over 7000 regions of brain with help of magnetic resonance imaging (MRI), "It's absolutely true that some brain functions occur in one or the other side of the brain. Language tends to be on the left, attention more on the right. But people don't tend to have a stronger left- or right-sided brain network" [7]. Agreeably, at a physical level, the division between left and right brain may not be seen by MRI. However, we feel that current science still has many limitations to know the real functioning of the

Left Brain
West/North World

Right Brain
East/South World

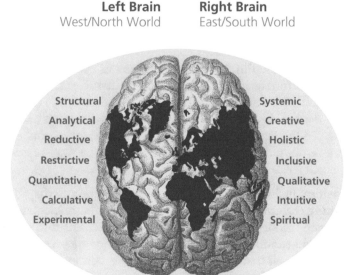

Structural

Analytical

Reductive

Restrictive

Quantitative

Calculative

Experimental

Systemic

Creative

Holistic

Inclusive

Qualitative

Intuitive

Spiritual

FIGURE 2.1

East and West mind-set depicting left and right brain. The earth can be visualized as a brain: the east and south dominated by the right brain, and the west and north by the left brain. Western philosophies dominated by the left brain. Eastern philosophy is dominated by the right brain.

brain leave apart its relation with an entity called *mind*. The analysis of the brain by a neuroscientist is typical feature of Western analytical mind, while an artist depicting the brain as a globe to show the right and left divide is a typical feature of Eastern creative mind. We are considering the division of the East and West, not merely in geographic sense, but more in terms of approaches and mind-sets.

Both the Eastern and the Western approaches are important, and should not be seen as mutually exclusive. Both the left brain and the right brain approaches are essential, and complementary. A judicious convergence of Western and Eastern philosophies is vital for the evolution of future medicine, and for constructing integrative approaches for health.

EASTERN TRADITIONAL MEDICINE

Eastern medicine, also known as Oriental or Asian medicine, comprises mainly health systems and traditional practices from India, China, Japan, Korea, and Tibet. In addition, there are traditional medical systems in most of the Asia–Pacific region including Myanmar, Thailand, Malaysia, Vietnam, Cambodia, and Indonesia. Ayurveda and Chinese medicine remain the largest systems in the East. Other traditional medical practices from this region include Japanese

Kampo, Arabic Unani, Tamil Siddha, and Tibetan Sowa Rigpa. Acupuncture is a specialized therapeutic technique predominantly practiced in TCM. Yoga remains a stand-alone knowledge system from India that focuses on health of mind and body, using relaxation, meditation, mindfulness, and consciousness techniques.

A brief review of a few important Eastern medical traditions will help to understand the interplay between science and traditions.

CHINA

The organized form of TCM appears to have originated sometime during the third century BC when the Emperor Huang Ti, also known as the Yellow Emperor, wrote the *Canon of Internal Medicine*, also known as the *Nei Ching*. Other classics of TCM include the *Mo Ching* (known as the *Pulse Classic*) of the Han dynasty (202 BC–220 AD), and the *Golden Mirror*, which were compiled sometime during 1700 AD. From the nineteenth century onward, Western medicine started becoming popular in China; however, TCM continues to be popular, and remains in practice even today.

TCM is based on the dualistic cosmic energy theory comprised of yin and yang. The yang is the male principle. It is believed to be active and light, and represents the heavens. The yin is considered as the female principle. It is passive and dark, and represents the earth. Both principles and their properties are quite similar to the *Prakriti* and *Purusha* principles described by Indian philosophy. Quite similar to Ayurveda, TCM also holds that the human body comprises five phases or elements: wood, fire, earth, metal, and water. These phases are associated with other groups including the five planets, the five conditions of the atmosphere, the five colors, and the five tones.

TCM concepts of anatomy are based on presence of 12 channels, the 3 burning spaces, 5 storage organs (heart, lungs, liver, spleen, and kidneys), and 5 viscera organs (stomach, intestines, gallbladder, and bladder). Each organ is associated with one of the planets, colors, tones, smells, and tastes. According to TCM, there are 365 bones and 365 joints in the body. In TCM physiology, the blood vessels contain blood and air as per proportions of the yin and the yang; the blood and air circulate in the 12 channels, and control the pulse. In TCM pathology and diagnosis, the physician makes use of a detailed history of the illness, and the patient's taste, smell, dreams, voice, color of the face, tongue, and most important—the pulse.

The Chinese materia medica is exhaustive, consisting of over 1000 vegetable, animal, and mineral remedies. *The Great Pharmacopoeia* prepared in the sixteenth century AD has 52 volumes, and is still authoritative. Drugs are used mainly to restore the harmony of the yin and the yang, and are prescribed in relation to the five organs, the five planets, and the five colors. The art of diagnosis and prescribing in TCM is very complex and personalized, and quite similar to Ayurveda.

Many herbal drugs from TCM have entered the modern pharmacopoeia: rhubarb, castor oil, kaolin, aconite, and camphor. The most frequently used plant from Chinese sources is ginseng, which is used as a tonic and aphrodisiac. The herb *Ephedra vulgaris* has been used in China for at least 4000 years; the alkaloid, ephedrine, isolated from *Ephedra vulgaris*, is used in the West to treat asthma. Artemisinin is another successful modern drug, which originated in TCM. Artemisinin is isolated from *Artemisia annua*, and has been used for thousands of years in China for the treatment of malaria.

Another popular TCM treatment, moxibustion, consists of small, moistened cones of the powdered leaves of mugwort or wormwood for external application. This practice is often associated with acupuncture. Acupuncture technique uses special needles to modulate the distribution of the yin and the yang in the meridians, or hypothetical channels and burning spaces of the body. It is claimed that acupuncture has been practiced since almost 2500 BC. Recently, there has been much research into the effects of acupuncture on neurotransmitters, and on its usefulness for relief from pain, and other symptoms. It is believed that acupuncture might trigger the brain to release analgesic substances like endorphins.

JAPAN

Japanese traditional medicine, which uses herbal drugs, is known as Kampo. Its practices are like acupressure and acupuncture; known as shiatsu. The term Kampo is synonymous with Japanese medicine. This system has been in use for nearly 1500 years, and is integrated into Japan's modern health care system. In the sixth century AD, along with the spread of Buddhism, Japanese scholars imported concepts and practices of Chinese medicine. The Chinese medical practices were modified to suit local needs, and became *Kampo*. This term literally means "Chinese style" in Japanese language. At the beginning of the nineteenth century, Kampo was suppressed by legislation that allowed only doctors trained in modern medicine to practice medicine. However, the Japanese people continue to use Kampo. The study of traditional medicine facilitated drug discovery in Japan: the isolation of ephedrine from *Ephedra vulgaris*, rotenone from *Derris ellipticai*, and the synthesis of methamphetamine. Many Japanese people object to calling Kampo *alternative*, or *complementary* [8]. After World War II, Kampo practice was strengthened; in 1967, it was covered under health insurance [9].

The Kampo system is a pragmatic and simplified version of Chinese medicine. Kampo treatment is based on the symptoms of the patients at the moment (*sho*), interpreted into three basic substances (*qui*, blood, and water). It considers hypofunction and hyperfunction, heat and cold, superficies and interior, and yin and yang. There are five parenchymatous visera, and six stages of disease. The selection of the treatment is based on sho: symptoms, systemic conditions, and physical constitution. The formula is prescribed for the specific sho. Kampo has been included in curricula of medical and pharmacy education since 2002. There are no separate courses in Kampo, the result of which is a lack of trained teachers [10]. A survey of patients reports that half of the patients use Kampo in combination with Western medicine. The factors associated with the usage of Kampo in Japan include people's belief in philosophy, holistic approach, and efficacy of Kampo [11].

KOREA

Traditional Korean medicine (TKM) possibly has its roots in Chinese and Japanese medicine [12]. There is evidence of the importation of Chinese literature into Korea around the tenth century. The classification of Korean drugs was known since the twelfth century. In the last 700 years, Korean medicine developed its therapeutic regimens—mainly whole, person-centric Sasang constitutional classification, and Saam acupuncture methods. By the seventeenth century, traditional Chinese and Korean medicine were distinct as Saam acupuncture practices were in use by this time [13].

Sasang constitution types are based on anatomical characteristics, temperament, and other traits of an individual. Every individual can be categorized into any of the four constitutional types: Tae-eum, So-Yang, So-eum, and Tae-Yang. Each constitution type is specific for appearance, personality traits, disease proneness, drug responses, and physiological attributes of the individual [14]. Treatment is based on constitution type, rather than symptoms. The Buddhist monk, Saam propounded acupuncture based on the five elements—earth, metal, water, wood, and fire, and six types of *qi* (energy). He suggested that the five elements have either nourishing or suppressing relationships. The Saam technique is based on 12 meridians, representing all the physiological processes. Each organ has a dominant element, and energy type.

TKM remained progressive in its theory and practice. Recently, a Korean scientist D.W. Kwon proposed eight constitution acupuncture types. The four Sasang constitutions types were further divided by functioning of organs (strong or weak functioning). The Korean doctors tried to integrate modern concepts into their practices. They also developed different acupuncture types based on specific meridians and constitution types. They developed the newer packages as bactericidal formula, and paralysis formula. Korean acupuncture types evolved further as herbal acupuncture, where the practitioners used herbal extracts (or bee venom) at specific acu-points [15]. Though the Korean medicine has roots in TCM, today it is an independent medical system with it's own principles, philosophy, and practices. Another contributing factor to the growth of TKM is the acceptance of paradigms and diversity. TKM doctors have formed various academic societies, which contributed to scientific discussion, and development of clinical disciplines [16].

TIBET

Traditional Tibetan medicine is known as Sowa Rigpa. It is based on Buddhist tradition. Buddha's teachings spread in Tibet after the eleventh century. *Gyud-Zhi* (Four Tantra) is the classical text that describes eight branches, and three principles, similar to Ayurveda. Health is the state of the balance of the three principles: *Kapha*, *Pitta*, and *Vata*. Diagnosis is based on pulse analysis, and urine examination. Treatment consists of behavioral and dietary modifications, and certain herbs. There are physical therapies similar to Ayurveda, such as massage. The Tibetan system also uses acupuncture and moxibustion in treatment. Apart from Tibet, the system is practiced in India, Nepal, and Bhutan, and the Himalayan regions, China, Mongolia, and in some parts of Europe. The Indian government has included Sowa Rigpa in AYUSH systems.

INDIA

The Indus Valley Civilization represented by the Harappan phase seems to be from twenty-seventh century BC to the seventeenth century BC. This was followed by Aryan Vedic civilization which lasted until 1000 BC. The Jain and Buddha period was around 500–200 BC. Alexander the Great's invasions of the fourth century BC were followed by the golden period of Ashoka from the third century BC, until that of Chandragupta, of the fourth century AD. The most authentic record and scholarly works on Ayurveda—*Charaka* and *Sushruta Samhita*—seem to be from the Chandragupta period.

India has codified systems like Ayurveda and Yoga, as well as noncodified, ethnomedicine and local health traditions. Siddha and Sowa Rigpa are mainly practiced in Tamil Nadu and Tibet, respectively. In addition, India has embraced Unani, which has Greek–Arabian origin, and Homeopathy, which is of German origin. All these are bundled together as AYUSH (Ayurveda, Yoga, Unani, Siddha, Sowa Rigpa, and Homeopathy). Yoga is an integral part of the Indian knowledge system. Principles of Yoga existed in the *Vedas*, *Upanishadas*, and ancient Indian texts. Yoga is considered a path toward the ultimate reality. Broadly, it has two parts. The first part relates to meditation, breathing, and relaxation techniques. The second part describes sophisticated, physical postures, and exercises.

More details about the philosophy, epistemology, logic, and foundational principles of Ayurveda and Yoga can be found in the "Primer" in Annexure A.

EVOLUTION OF AYURVEDA AND YOGA

Ayurveda literally means *the knowledge of life*. It considers the harmonious relationships between macrocosm and microcosm as a continuum. The holistic approach of Ayurveda addresses the preventive, as well as the curative aspects of medicine. Ayurveda is not folklore or a herbal tradition. It has its own epistemology and principles of anatomy, physiology, pharmacology, and therapeutics. Its history goes back to the Vedic and post-Vedic periods.

Ayurveda went through several transitions from 800 BC until about 1000 AD, and slowly moved away from a faith-based, ritualistic system of medicine, to a cause and effect-based, rational, and evidence-based system with sophisticated concepts of pathophysiology and therapeutics. Scholars like Charaka, Sushruta, and Vagbhata laid foundations of systematic, cause and effect relationships based on logic and evidence. There is no agreement concerning the time period of Charaka and Sushruta—which ranges from 800 BC to 200 AD. Many scholars believe Sushruta was sometime during 700–800 much before Charaka ~400 BC. Many scholars also believe that *Sushruta Samhita* was compiled during ~100 AD after *Charaka Samhita*. The time of Vagbhata is considered to be ~700 AD. The three classic compilations known as *Charaka Samhita*, *Sushruta Samhita*, and *Ashtang Hridaya* of Vagbhata are still considered authoritative.

Charaka gave a very sophisticated account of internal medicine, including information of over 500 medicinal plants. The Indian materia medica had a large list of over 3000 botanical drugs. Ayurveda also advocated animal products such as the milk of various animals, bones, gallstones, shells, horns, minerals, and employed metals like sulfur, arsenic, lead, copper, and gold. The Ayurvedic physicians collected, prepared, and dispensed their own medicines. Many of the original Ayurvedic plants and spices are listed in Western pharmacopoeias as well.

Sushruta is considered the father of modern surgery. He described several sophisticated surgical procedures including removal of tumors, incision sections, draining of abscesses, punctures to release fluid in the abdomen, extraction of foreign bodies, repair of anal fistulas, splinting of fractures, amputations, caesarean sections, and stitching of wounds. According to Sushruta, the surgeon should be equipped with 20 sharp, and 101 blunt instruments of various descriptions. The instruments were largely of steel. Alcohol seems to have been used as an anesthetic during operations, and bleeding was stopped by hot oils and cautery. Sushruta listed over 760 medicinal plants.

The earliest plastic surgery is believed to have been performed by Sushruta. Amputation of the nose was one of the prescribed punishments for adultery, and reconstructive surgery like repair was carried out by cutting from the patient's cheek or forehead a piece of tissue of the required size and shape, and applying it to the stump of the nose. The results appear to have been tolerably satisfactory, and the modern operation is certainly derived indirectly from this ancient source. Indian surgeons also operated on cataracts by displacing the lens to improve vision. Recent reports suggest strong evidence that sometime during 800–600 BC an extraocular expulsion of lens material through a limbal puncture was described by Sushruta [17]. This recognition was made in 1940 by John Davis in the presidential address at the annual meeting of the American Surgical Association [18].

Sometime during the eighth century, Nagarjuna developed *Rasashatra* (Ayurvedic alchemy). This was a major development in history of Ayurveda. Rasashatra literally mean "science of mercury," which also includes metals and minerals. Hundreds of new substances were added as Ayurvedic medicine. Apart from medicine, newer technologies and processes were added to Ayurveda pharmaceutics.

Madhava Nidana, a classic on clinical diagnostics was compiled in the eighth century. This comprised pathology and diagnosis of prevalent diseases. Many diseases that were not mentioned in earlier texts were compiled in the Madhava. This text also describes several additional risk factors, and variations in manifestations of diseases. It also describes prodromal symptoms, disease types, and possible complications. The trend of revision, review, and analysis leading to gradual, progressive development of this ancient science of life continued at least until the tenth century.

After the Indian subcontinent became the target of Persian and Greek warriors like Alexander, much knowledge from Vedic sources was acquired by the Persians and Greeks. Followed by the collapse of Gupta Empire in the sixth century BC, the Indian subcontinent remained almost continuously volatile, and a larger portion of India began to be ruled by invaders; the tenth century onward saw invasions by Gazani, Ghori, and several others. This was the beginning of a dark period for Ayurveda, which continued until India's independence from British rule in 1947. Ironically, even after independence, due to lack of proper vision and strategy, Ayurveda and Yoga remained stagnant and could not be brought into the mainstream health care.

The development of Ayurveda took an interesting trajectory from mythology to logical, rational, and evidence-based practices. However, due to hostile environments as discussed earlier, the Indian traditions seem to have taken the ritualistic path to preserve their knowledge and identity. As a result, progress of scientific culture in India was arrested. The epistemology of Indian science was compromised. Due to unfavorable environments like colonial rule, and a wider acceptance of modern medicine, the growth and development of Ayurveda suffered for many decades. As a result, the rich, living tradition of Ayurveda was frozen for almost 10 centuries, and its growth was arrested; its tradition of science gradually became ritualistic.

It is interesting to note that, like Ayurveda, TCM, TKM, and Kampo were also suppressed and marginalized. However, they were revised and reestablished in line with their respective cultural backgrounds through systematic efforts. These traditional systems of medicine have been integrated into the national health programs in their respective countries [19].

SCIENCE AND TRADITIONS

Any science is dynamic in nature. It is about the continuous quest for knowledge. Many traditions have given important direction to the development of science. Ancient systems like Ayurveda and Yoga are considered science, as well as part of living traditions. Traditions refer to knowledge systems and practices in various domains of human experience. Such traditions contribute to the history of every society. They might concern science, philosophy, culture, language, and religion, and can be reasonably classified as healthy, or unhealthy. Healthy traditions change and adapt to contemporary needs. Much of the fabric of modernity has threads from evolving, living traditions. Traditions continue to evolve from generation to generation to become a heritage; the heritage shapes the religious, philosophical, and cultural practices of society.

Traditions can be living and dynamic. Living traditions allow debates and self-critique. A living tradition rejects uncritical approach and dogmatism of any kind. With a rational approach, living traditions can evolve into contemporary science. In the Western world, the Greek traditions followed a rationalist path, and progressed during the Renaissance, the age of Enlightenment, and Industrial Revolution. Many scientific discoveries and inventions occurred during this period. The emergence of modern science and modern medicine is a result of traditions adopting the rationalist and progressive approach.

Whereas traditions can be continuously evolving, rituals tend to be static. Traditions can remain frozen. Such frozen traditions become rigid and ritualistic. When this happens, adherence to traditional knowledge and practice degenerates into mere ritualism. As a result, tradition loses its contemporary relevance. Further, with the ritualistic approach, traditions can lead to blind faith and superstitions.

Indian health traditions have gone through these phases.

When the core ideology and purpose behind the tradition is lost, it becomes a relic in the form of rituals. Rituals can lead to customary practices, mysticism, and stereotyped, mechanical behavior. Many customs of yesterday may seem outdated, irrelevant, and redundant today. The process of transferring knowledge from generation to generation is adversely affected due to foreign invasions, dictatorial rules, and rigid social systems. Traditions can become irrelevant or outdated if they do not adapt to a changing environment. The ritualistic approach encourages worship instead of critical inquiry. Many times, ritualistic behavior is egoistic and aggressive. For instance, the concept of fasting has a scientific basis but today it has become a mere ritual. Healthy, living traditions like fasting, and *Ritucharya* can have a tendency to be reduced to worship, festivals, and food recipes. The Eastern world holds several healthy traditions. However, because of historical and cultural reasons, many living traditions are have tended to become ritualistic [20].

In the modern world physical, chemical, biological, and computational sciences are converging on sophisticated technologies like genomics, biotechnology, nanotechnology, scientific computing, artificial intelligence, and robotics. New theories based on nonlinear, holistic, and systems approaches are evolving. Discussions about science and spirituality, consciousness, mind–body interactions, ecological sensitivities, value systems, and the searching questions about purpose are finding new grounds. We seem to be completing the circle by creating new bridges between science, philosophy, and spirituality.

Some concepts of Ayurveda have contributed to the development of modern medicine, surgery, and various specialties. Sushruta's contributions to anatomy, surgery, urology, ophthalmology, and plastic surgery are well-acknowledged [21–23]. A few concepts like *Prakriti* are now appearing in modern scientific literature. However, several contributions have remained unrecognized in the development of modern medicine [24]. The reasons may be related to a language barrier, ontological divergence, different health care models, as well as bias against traditional knowledge.

MODERN MEDICINE

Following the Greek revolution in philosophy and science during the era of Copernicus, Pythagoras, Socrates, Plato, and Aristotle is the notable period, the Galen era. This is recognition to the outstanding work by Claudius Galenus during the second century AD. Galen contributed significantly in several scientific disciplines such as pharmacology, pathology, physiology, and anatomy. Galen's understanding of medicine was greatly influenced by Hippocrates' theory of humorism. Galen also showed that the brain controls muscle movements through the cranial and peripheral nervous systems. He started systematic use of vegetables, animal products, and plant extracts. Even now, these dosage forms are known as Galenicals. Galen was a physician, and a philosopher interested in the debate between the rationalist and empiricist streams.

He used direct observation, and dissection and vivisection to bridge these extreme viewpoints. According to Galen, the arterial and venous systems were separate; blood circulates between the ventricles through invisible pores. His anatomical observations and reports based on the dissection of monkeys and pigs remained uncontested practically until the sixteenth century. The Galen era remains historically important because the world witnessed the influence of Galen's ideas over 1000 years after his death.

Paracelsus (1493–1541) was a medical doctor, alchemist, and philosopher who believed that a human being has two bodies: a visible body that belongs to the earth, and an invisible body of heaven. The invisible one is closely attuned to imagination, and the spiritual aspect of the individual. Paracelsus understood the psychosomatic abilities of people, as he stressed the importance of suggestion, and using signs and amulets to help a patient form mental images, which translated into profound, physical cures. He was the first to challenge theories of Galen.

In 1543, Andreas Vesalius also challenged Galen's theories on anatomy and physiology with help of over 600 anatomical drawings in his classic book the *Fabric of the Human Body*. In 1616, William Harvey proved Galen wrong by showing that the heart, which acts as a pump, is responsible for blood circulation [25].

Typifying the advancement of the Renaissance, and the Enlightenment, during the seventeenth and eighteenth centuries, Bacon and Descartes stimulated robust advancements of natural science based on mathematical, method-driven, experimental processes. During the nineteenth century, John Herschel and William Whewell systematized methodology and coined the term *scientist*. Charles Darwin proposed the theory of natural selection and provided scientific explanation of origin of species. Many other spectacular discoveries marked the last two centuries where atomic theory, the laws of thermodynamics, electromagnetic theory, relativity theory, and quantum theory were established—raising new questions which could not be easily answered using Aristotelian logic or Newtonian approaches. Einstein's theory of relativity and the development of quantum mechanics led to the replacement of Newtonian physics with a new physics containing two parts which describe different types of events in nature.

Scientific innovations during the two world wars of the last century also led to several discoveries, space missions, and to the endorsement of the value of modern science. In the process, the humanity also witnessed the misuse of science and technology through changes in socioeconomic fabric, growing violence, ill health, new diseases, industrial pollution, and ecological degradation. History is full of bloody violence, however, weapons of mass destruction are a gift of modern technology!

Sir William Osler gave a lucid overview of the evolution of modern medicine in series of lectures delivered on "Evolution of Modern Medicine" at the Yale

University. The lectures give an exhaustive account of the history of ancient medicine, and the emergence of modern medicine. This lecture series has been published as a 233-page book, which provides an excellent overview of primitive and traditional medicine including Egyptian, Assyrian, Babylonian, Hebrew, Chinese, and Japanese medicines, with landmark contributions from Hippocrates, Galen, and Harvey. Surprisingly, Osler did not mention India or Ayurveda anywhere in the Yale lecture series, or even in the book [26].

Modern Medicine grew from the firm foundations of modern science. However, during its evolution some of the important theories were left behind, and escaped serious attention. Those omitted include the theory of *conditional probability* proposed by Thomas Bayes during the mid-seventeenth century, and the theory of falsification proposed by Karl Popper in early 1930s. Among different schools of thoughts in the philosophy of science, the most popular position is empiricism, which claims that knowledge is created by a process involving observations. Empiricism also holds that scientific theories are the result of generalizations from such observations. Karl Popper challenged the empirical approach in science. Popper rejected the connection between theory and observation. He claimed that theories are not generated by observation; rather, observation is made in the light of theories. Popper coined the term *critical rationalism* to describe his philosophy. Both these theories are gaining the attention of scientists, because they seem to be more relevant to holistic approaches in health and medicine.

Consistent and magnificent progress based on rationalism, reductionism, and inductive logic have led to the spectacular growth of modern science and technology. Modern biology and modern medicine also followed the same path.

FUTURE APPROACHES

Modern science and modern medicine are about testable and repeatable experiments and predictions. Modern science has always been dynamic, and continuously evolving. In this journey, many path-breaking inventions were either considered as idiotic, irrelevant, unwanted, or even impossible. New technologies and industries like aviation, computers, telecommunication, and the Internet are good examples. These were in real-sense disruptive innovations, which changed many paradigms of their times. Today, cloud computing has become a reality, while many other innovations like Google Glass, sixth and seventh sense, brain-to-brain hypercommunications, and brain–computer interfaces are on the way to becoming a reality.

An interesting example of future trends is *The 2045 Initiative*. This was founded by Russian specialists in the field of neural interfaces, robotics, artificial organs, and systems regarding life extension possibilities. Its main goal is "to create technologies enabling the transfer of an individual's personality to a more advanced nonbiological carrier, and extend life, including to the point of immortality." More interestingly, this initiative is facilitating dialogue between the world's major spiritual traditions, science, and society to discuss the possibilities of human development on the path of cybernetic immortality (http://2045.com/). Another ambitious program related to biological technology of the Defense Advanced Research Projects Agency of the United States portray astonishing picture of future developments in biology and medicine (http://www.darpa.mil/Our_Work/BTO/Programs/).

Science fiction books like *Coma* and *Nano* by Robin Cook, the *Robot* series by Isaac Asimov, the *Terminator* film series, James Cameron's film, *Avatar*, and the Wachowskis' *Matrix* film series, give us glimpses of where future technology is headed. The progress of modern science has reduced the gap between imagination and reality, but more importantly, the progress of science has also raised the issue of its relation with tradition, ethics, and humanity. The brain-to-brain communication through direct information transmission from one human brain to another using noninvasive means between six participants has been reported by scientists from University of Washington [27]. Scientists are already envisioning that such hyperinteraction technologies might have a profound impact on the social fabric and raise crucial ethical issues [28].

As we were making final improvements to this book, in December 2014, renowned cosmologist Stephen Hawking publicly expressed worries about artificial intelligence and machines that can outsmart humanity. Dr Hawking in an interview to the BBC said, "The development of full artificial intelligence could spell the end of the human race. It would take off on its own, and re-design itself at an ever-increasing rate. Humans, who are limited by slow biological evolution, couldn't compete, and would be superseded."

We feel that in the thrilling yet terrifying journey of modern science, traditional knowledge and wisdom may be useful in several ways so that the future can be more humane and sustainable.

Modern medicine became what it is today because of the positive influence and advantages from developments in science and technology. This transformed the Greek, Roman, and Arabic traditions into modern medicine. Eastern medicine including Ayurveda mostly remained isolated, got suppressed during invasions, and was bypassed by the whole scientific progress and the industrial revolution.

In a process to become scientific, modern medicine lost sight of the person behind the patient. Eastern medicine preserved the holistic nature of healing involving body, mind, and spirit but lost an understanding of physics, chemistry, and biology. Eastern traditional medicine remained philosophic, nonlinear, creative, and intuitive; Western biomedicine became scientific, linear, logical, analytical, and evidence-based.

Arguably, today's medical practice in most general clinics is not necessarily based on scientific evidence, or ethical practice. Many treatments, drugs, and diagnostics seem to have survived without rigorous, intellectual challenges. In the race to become evidence-based, modern medicine is relying more on pathology reports, and rigid protocols. The hegemony of large medical centers, high-impact journals, and iconic physicians is influencing the nature of clinical practice. Many drugs like statins, steroids, analgesics, psychotropic agents, vitamins, and antioxidants are being prescribed in blind acceptance of the protocols of professional bodies, and pharmaceutical companies.

The protocols, standards, and checklists are necessary requirements. However, their relevance and context must be kept in mind when making therapeutic judgments. Clinical acumen cannot be blindly replaced by laboratory reports. Because of an overdependence on rigid protocols, measurement tools, and techniques, contemporary, evidence-based, medical practice is going more the ritualistic way [20]. How health, physiology, and pathology are defined is driven by commercial interests, rather than by scientific evidence.

Modern medicine has become more scientific, rational, and evidence-rich. Unfortunately, in the process, it is becoming mechanical, impersonal, and ethics-poor. Going against the original ethos of this noble profession, both systems of medicine—Eastern and Western—have become commercial, where fees are charged for services rendered. This shift in attitude of medical doctors has led to the commercial interests overpowering the interest of the good of the patient. Slow, progressive deterioration of value systems, ethics, and morality, coupled with a distortion of the nature of science is the main reason for the contemporary challenges. Today's medicine needs to be liberated from unnecessary medicalization, and need to refocus on health and wellness. Today's medicine must maintain the interests of the public, and the patient as the primary concern. Present medicine needs a balanced integration of both sides of the brain—the creative *and* analytical. What is needed is a scientific and holistic mind-set where modernity and traditions go hand in hand.

It is interesting to witness that the European region, which was the epicenter of modern medicine evolution, is now rediscovering the importance of the holistic approach; the focus is shifting from illness to wellness. The new movement of integrative medicine is based on personalized approaches through diet and lifestyle modification. Thought leaders are proposing that the appropriate changes in lifestyle where increased physical activities can reduce use of machines and save natural resources and energy. The consumption of wholesome natural food instead of processed foods and a daily routine in relatively stress-free environments can actually have dual benefits where public health is improved and the global environment is saved.

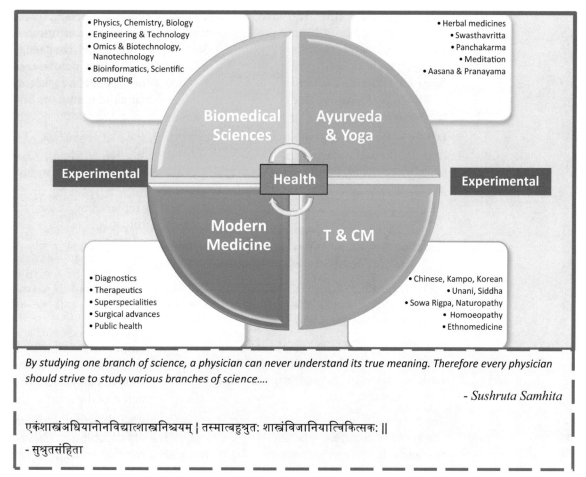

By studying one branch of science, a physician can never understand its true meaning. Therefore every physician should strive to study various branches of science....

- Sushruta Samhita

एकंशास्त्रंअधियानोनविद्यात्शास्त्रनिश्चयम् | तस्मात्बहुश्रुत: शास्त्रंविजानियात्चिकित्सक: ||

- सुश्रुतसंहिता

FIGURE 2.2

Integrative approaches for health through a confluence of experiential traditional knowledge systems like Ayurveda and Yoga, and the power of experimental biomedical sciences with the advances in modern medicine. A future physician scientist must have the knowledge from multiple disciplines.

There is growing realization that the ultimate purpose of science should not be only for economic development, or creation of wealth, but also should contribute toward humanity, human values, and nature. The global consensus is building in favor of using science for environmental sustainability and not just for human comforts and economies [29].

Perhaps this is a new beginning in creating integrative approaches for health through a confluence of traditional knowledge systems like Ayurveda and Yoga, the advances in modern medicine and the power of biomedical sciences (Figure 2.2). Indeed, tomorrow's physicians will have knowledge of multiple disciplines and medicine will probably be more personalized, and will move beyond mere drugs and cures. Future medical practitioners might embody the essentials of an ideal doctor. New initiatives like predictive, preventive, participatory, and personalized medicine, and person-centered medicine and whole-person medicine, and other integrative variants emerging from Europe and other parts of the world seem like modern avatars of Ayurveda.

In concluding this chapter of the evolution of medicine, we wish to emphasize that the present approach to medicine and therapeutics has to change. The principle of the evolution of species is also applicable to medicine. History and science tell us:

> It is not the strongest of the species that survives, nor the most intelligent that survives. It is the one that is most adaptable to change. In the struggle for survival, the fittest win out at the expense of their rivals because they succeed in adapting themselves best to their environment.

Whether or not to assign this statement to Darwin is debatable, but its soundness is largely accepted. The modern medicine must adapt itself to suit present environment and meet expectation of people.

The history and evolution of medicine tells us how the faith-based, ritualistic, mumbo jumbo practices gradually waned giving way to scientific and evidence-based medicine. However, this is not the end, and the process of evolution must continue.

The complexity and expanse of biomedical sciences are touching new heights. The speed of technology seems to be far ahead of human capabilities. How to use available knowledge, resources, drugs, diagnostics, medical procedures and surgical technologies is emerging as a major challenge. The ability to recognize the limits of current science and more importantly limitations of human capabilities; identifying and reducing the areas of ignorance and keeping an open mind to learn from experiences will be the key attributes while moving to the future medicine. The hubristic and monopolistic mind-set need to be changed to more humble, inquisitive, and open approach.

The next phase of evolution of medicine seems to be more holistic, personalized, and integrative. In the following chapters, we will try to visualize the shape of future medicine. The role of the future doctor might change from prescriber of medicine, to advisor for health. Tomorrow's health care may indeed be what Thomas Edison visualized where doctors will give no medicine but will interest people by advising how to protect health through diet and lifestyle and how to avoid causes of disease. This will be the desired revolution in the evolution of future medicine.

REFERENCES

[1] Caplan A. Does the philosophy of medicine exist? Theor Med 1992;13(1):67–77.

[2] Velanovich V. Does the philosophy of medicine exist? A commentary on caplan. Theor Med 1994;15(1):77–81.

[3] Lindberg David C. The beginnings of Western science: the European scientific tradition in philosophical, religious, and institutional context. 2nd ed. Chicago: University of Chicago Press; 2007.

[4] Newton I. Philosophiae naturalis principia mathematica; 1687.

[5] Filliozat J, Chanana Dev Raj MRML. The classical doctrines of Indian medicine Delhi; 1964.

[6] Subbarayappa BV. A perspective. History of science, philosophy and culture in indian civilization. Part 2. In: CD P, editor. Medicine and life sciences in India, vol. IV. New Delhi, India: Center for Studies in Civilization; 2001.

[7] Nielsen JA, Zielinski BA, Ferguson MA, Lainhart JEAJ. Right-brain hypothesis with resting state functional connectivity magnetic resonance imaging. PLoS One 2013;8(8):e71275.

[8] Terasawa K. Evidence-based reconstruction of Kampo medicine: part I—Is Kampo CAM? Evid Based Complement Alternat Med 2004;1(1):11–6.

[9] Yu F, Takahashi T, Moriya J, Kawaura K, Yamakawa J, Kusaka K, et al. Traditional Chinese medicine and Kampo: a review from the distant past for the future. J Int Med Res 2006;34(3):231–9.

[10] Motoo Y, Seki T, Tsutani K. Traditional Japanese medicine, Kampo: its history and current status. Chin J Integr Med 2011;17(2):85–7.

[11] Hottenbacher L, Weißhuhn TE, Watanabe K, Seki T, Ostermann J, Witt CM. Opinions on Kampo and reasons for using it–results from a cross-sectional survey in three Japanese clinics. BMC Complement Altern Med 2013;13:108.

[12] Seong-GyuKo CSY. Introduction to the history and current status of evidence-based Korean medicine: a unique integrated system of allopathic and holistic medicine. Evidence-Based Complement Altern Med 2014;2014:740515. http://dx.doi.org/10.1155/2014/740515.

[13] Cha W-S, Oh J-H, Park H-J, Ahn S-W, Hong S-Y, Kim N-I. Historical difference between traditional Korean medicine and traditional Chinese medicine. Neurol Res 2007;29(Suppl. 1):S5–9.

[14] Lee SW, Jang ES, Lee J, Kim JY. Current researches on the methods of diagnosing sasang constitution: an overview. Evidence-Based Complement Altern Med 2009;6:43–9.

[15] Yin C, Park H-J, Chae Y, Ha E, Park H-K, Lee H-S, et al. Korean acupuncture: the individualized and practical acupuncture. Neurol Res 2007;29(Suppl. 1):S10–5.

[16] http://tkmedicine.blogspot.in.

[17] Grzybowski AAFJ. Sushruta in 600 BC introduced extraocular expulsion of lens material. Acta Ophthalmol 2014;92(2):194–7.

[18] Davis JS. Address of the president: the story of plastic surgery. Ann Surg 1941;113(5):641–56.

[19] Park HL, Lee HS, Shin BC, Liu JP, Shang Q, Yamashita H, et al. Traditional medicine in China, Korea, and Japan: a brief introduction and comparison. Evidence-Based Complement Altern Med 2012;2012:429103. http://dx.doi.org/10.1155/2012/429103.

[20] Patwardhan B. Traditions, rituals and science of Ayurveda. J Ayurveda Integr Med 2014;5(3): 131–3.

[21] Bhattacharya S. Sushrutha – our proud heritage. Indian J Plast Surg 2009;42(2):223–5.

[22] Das S. Susruta, the pioneer urologist of antiquity. J Urol 2001;165(5):1405–8.

[23] Kansupada KB, Sassani JW. Sushruta: the father of Indian surgery and ophthalmology. Doc Ophthalmol 1997;93(1–2):159–67.

[24] Patwardhan K. The history of the discovery of blood circulation: unrecognized contributions of Ayurveda masters. Adv Physiol Educ 2012;36(2):77–82.

[25] Aird WC. Discovery of the cardiovascular system: from Galen to William Harvey. J Thromb Haemost 2011;9(Suppl. 1):118–29.

[26] Osler W. The evolution of modern medicine. In: Garrison FH, editor. Silliman memorial lectures. New Haven: Yale University Press; 1921. p. 233.

[27] Rao RPN, Stocco A, Bryan M, Sarma D, Youngquist TM, Wu J, et al. A direct brain-to-brain interface in humans. PLoS One 2014;9(11):e111332.

[28] Grau C, Ginhoux R, Riera A, Nguyen TL, Chauvat H, Berg M, et al. Conscious brain-to-brain communication in humans using non-invasive technologies. PLoS One 2014;9(8):e105225.

[29] Anon. Late lessons from early warnings: science, precaution, innovation. Europian Environment Agency Report; Luxembourg: Publications Office of the European Union; 2013.

CHAPTER 3

Concepts of Health and Disease

Disease is a lack of health. Health is not a lack of disease

Unknown

HEALTH AND DISEASE

In view of the focus of this book on health, it is necessary to review its historical evolution, and revisit prevailing concepts of health. Often, the concept of health revolves around a few key words like well-being, wellness, and happiness. The etymology of term *health* is very interesting. *Health* in Old English actually means "wholeness, sound, or well." It seems to have its origin in the proto-Germanic word *hailitho* meaning "whole, uninjured, of good omen." An Old Norse term *heill* means "healthy," and *hælan* means "to heal." Health also denotes prosperity, happiness, welfare, preservation, and safety. Often the terms *health and wellness* are used together in Western culture. In Oriental language, the Sanskrit word for health is *aarogya*, which is indicative of absence of disease. A more appropriate term for health used in Ayurveda is *Swasthya*, which is very profound, and will be discussed in later sections.

In many cultures and traditions the proverb "health is wealth" is popular. For every individual, health is something very important and crucial for life and existence. Everyone hopes and aspires to health, naturally. However, the concept of health is not only limited to individuals, but also relates to community health, public health, human health, animal health, plant health, and environmental health. The concept of health even extends to specific ecosystems like oceans, rivers, habitats, towns, cities, and nations. Health is an equilibrium, a balance, and a state of harmony. When we lose it, we know that it existed. It is easier to know illness, but very difficult know wellness. Probably due to this, mostly health is described and discussed in relation to disease/illness/sickness [1].

Many times the term *health* is equated with the term *medicine*. The erroneous perception of health is based on disease models, which might have led to the current disease focus. Today, health is considered as a commodity, which can be acquired through medicines and health care. In reality, health is a positive and dynamic concept. It can be attained, acquired, and maintained only

53

Integrative Approaches for Health. http://dx.doi.org/10.1016/B978-0-12-801282-6.00003-6

through the active participatory efforts of individuals, and all interdependent stakeholders. Health is one of the fundamental human rights. We must struggle to provide this right to unfortunates who are born with avoidable birth defects.

Health is a natural phenomenon; disease is mostly an invited trouble. Our present system has moved away from nature; right from pregnancy, the newborn is exposed to risks of disease. This could be due to deficient maternal nutrition, the psychological condition of the mother, or environmental pollutants and toxins present in food and water. It could be due to inadequate and unhygienic facilities in the poorer parts of the world. Artificial environments and procedures in modern nursing homes might also be to blame. Knowingly or unknowingly, the seeds of future diseases are sown right from the conception. Diseases of developmental origin are an important, and emerging branch of biomedical sciences.

Diseases like metabolic syndrome, and type 2 diabetes are thought to be due to poor nutrition in early life, which produces permanent changes in metabolism. This is also known as the thrifty phenotype hypothesis, or Barker hypothesis, which suggests an association of retarded fetal growth to various metabolic diseases in adulthood [2]. This is true for many other serious diseases and conditions of adulthood, such as hypertension, coronary heart disease, stroke, obesity, cancers, and many more. These diseases may be a result of poor nutrition and inadequate care during pregnancy. These are avoidable, undesirable birth gifts, which result in disease, illness, and sickness; there follows an unavoidable sequence of interventions in which medical care and medicines take over the birthright of natural health.

CONCEPT OF DISEASES

The question "Why do we fall ill, or get diseases?" has long remained a puzzle. The human body is an autoregulated, complex system where trillions of cells and molecules continuously interact and grow from conception to death. Different tissues, organs, and systems work in harmony during various stages of development. In a state of health and well-being, we are not aware of the existence of any of these complex, dynamic, and continuous processes. Organs and body parts announce their presence only when something goes wrong. It could be a mechanical injury, infection, infestation, or any of a number of other types of assault. These assaults can produce certain symptoms like pain, inflammation, spasm, nausea, vomiting, and diarrhea. This is a natural mechanism, which draws our attention to the respective regions of the body and body organs. Certain sets of signs and symptoms are associated with specific diseases. A cluster of specific signs and symptoms can be seen in different diseases.

Knowledge of diseases is not new. Many diseases are mentioned in ancient cultures and civilizations. The knowledge of diseases, and the understanding of reasons why the physiological process becomes pathological has evolved through sharp-eyed observation, and documentation. This accurate knowledge has evolved over several centuries. In earlier times, diseases were thought to be due to evil spirits and supernatural forces. Prehistoric people practiced rituals, made sacrifices, and uttered prayers to satisfy the spirits, and get rid of diseases. Many traditional healers and priests still propagate this superstition-based theory. In the past, another theory, *miasma*, from Greek mythology, dominated the thoughts concerning disease causation. The miasma theory suggested that diseases are produced due to unhealthy or polluted vapors rising from the ground, or from decomposed material. However, as science advanced, faith-based, supernatural theories of diseases were challenged. A quest for cause and effect relationships related to signs, symptoms, and diseases gradually developed as the science of pathogenesis and led to diagnostic science.

Pasteur's germ theory, and Koch's postulates pinpointed microbial infection as the cause of disease. However, these postulates cannot sufficiently explain etiopathology of noncommunicable diseases. Max von Pettenkofer (1818–1901) proposed the concept of multifactorial causation. He linked medicine with physics, chemistry, statistics, and the environment, and proposed that diseases may have many causative factors [3]. Later, epidemiology emerged as a new branch of science, which addresses the involvement of multiple causative factors.

Epidemiology as a discipline probably started in 1854, with the study of a cholera epidemic in London. After careful investigations, British physician John Snow reported an association between the patterns of cholera cases and the water supply. He found that a public water pump was responsible for contaminating drinking water with cholera bacteria. Another landmark epidemiological investigation was made by Hungarian physician, Dr Ignaz Semmelweis. He studied puerperal fever, which was very common in obstetric hospitals, and caused almost 30% of the mortality in the obstetric wards. In 1846, Dr Semmelweis discovered that the use of hand disinfection with chlorine substantially reduced the incidence of puerperal fever. He published a book *Etiology, Concept and Prophylaxis of Childbed Fever*. This work highlighted the importance of hygienic practices like washing hands with germicidal lotions before surgical procedures.

The science of epidemiology studies patterns and causes of diseases in a given population. It describes transmission and control of diseases. Better knowledge of causation, symptomatology, and disease progression has led to the evolution of epidemiology as an important branch of biomedical sciences. Epidemiology deals with the control of diseases and other health problems by studying distribution and the determinants of health and disease. In general, epidemiology

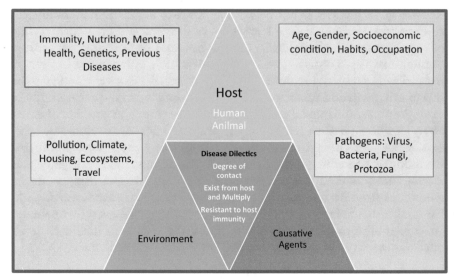

FIGURE 3.1
Epidemiological triad showing host, pathogen, and environment relationships.

explains the relation between host, environment, and the causative agent. This is known as the *epidemiological triad* (Figure 3.1). A new approach in epidemiology takes onboard a complex, interdependent model known as the *web of causation* [4]. For example, a high fat diet, and sedentary lifestyle are risk factors for coronary heart disease and obesity. Obesity itself is known as a risk factor for hypertension and coronary heart disease. Hypertension is known to cause atherosclerotic changes leading to myocardial ischemia (Figure 3.2).

British epidemiologist and statistician, Sir Austin Bradford Hill, established convincing evidence about the relation between use of tobacco and the incidence of lung cancer [5]. Hill discussed the association between the use of tobacco as the cause, and lung cancer as the effect. Hill provided sound logic and statistical backing to the concepts of epidemiology. He discussed the fundamental difference between association and causation, which provided a framework for assessing the cause and effect relationships involved in various diseases [6].

Advances in genetics and molecular biology have provided newer tools for epidemiology. Genome-wide association studies (GWAS) integrate epidemiology and molecular biology, and consider macro and microlevel of biological organization. The GWAS explain the association between specific genetic traits and specific diseases. The GWAS typically compare the single nucleotide polymorphism of diseased people to healthy controls. One of the largest GWAS involved 14,000 patients, and suggested genetic susceptibility for various

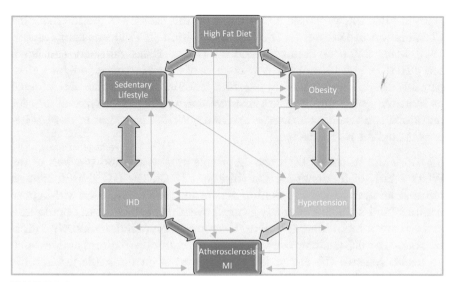

FIGURE 3.2
The web of causation.

diseases, including bipolar disorder, coronary artery disease, Crohn's disease, rheumatoid arthritis, and diabetes [7]. With this additional power of genetic and molecular epidemiology, and more precise and sensitive medical devices, our contemporary understanding of pathophysiology has been considerably enhanced. This has helped in the endeavors of systematic disease classification and therapeutics.

DISEASE CLASSIFICATION

Efforts to classify diseases have been ongoing for the past four centuries. An Australian statistician, George Knibbs, first systematically attempted to classify diseases (www.who.int/classifications/icd/en/HistoryOfICD.pdf). William Farr, a medical statistician, initiated efforts to standardize nomenclature and disease classification. At the International Statistical Congress at Brussels in 1853, William Farr and Swiss nosologist Marc D'Espine were tasked with preparing a uniform classification of causes of death, which could be applicable internationally. Farr classified diseases into different categories: communicable diseases, general diseases, sporadic diseases, developmental diseases, and others. D'Espine classified diseases according to their nature: gouty, herpetic, hematic, and other categories. There was no consensus for a long time. Later, French statistician Jacques Bertillon revised the classification of diseases according to anatomical structures and the nature and frequency of diseases. The American Public Health Association endorsed Bertillon's classification in 1897, and suggested that the classification should be revised every 10 years. The International

Statistical Institute proposed a new name for this exercise as the International Classification of Diseases (ICD). The first revision (ICD-1) was completed in 1900. Many European countries, and the United States, Australia, and Japan accepted ICD-1 by 1909. A few new diseases of the endocrine glands, as well as parasitic diseases were added in the third revision, ICD-3, which was adopted by more than 40 countries. The fourth revision added various types of cancers, nutritional, and rheumatic diseases. Since the WHO's inception in 1948, it has coordinated the ICD initiative.

ICD-10 is the current ICD. It has 22 groups of diseases, and is a part of the WHO Family of International Classifications (FIC). The FIC aims at storage, retrieval, analysis, and interpretation of data and their comparison within populations (nations, regions, or any groups) over time, and between populations at the same point in time, as well as the compilation of internationally consistent data. The FIC facilitates development of a common platform and research on health systems. The FIC makes comparisons of data possible through the common use of terminologies. The reference classification also includes International Classification of Functioning, Disability and Health, and International Classification of Health Interventions, along with the ICD.

The eleventh revision of the ICD is due by 2017. The WHO has furthered its development through a participatory, Internet-based process. The beta draft was published in 2012, and is still available for comments and revision http:// apps.who.int/classifications/icd11. The latest version of the ICD includes many emerging lifestyle diseases, disorders of the fetus, drug-induced or autoimmunity-induced disorders, and drug-induced or iatrogenic diseases due to treatment complications. In the latest version, many idiopathic diseases of unknown origin or cause are also included. This inclusion of diseases of unknown origin or cause illustrates the limitations in knowing causative factors for these diseases, despite advances in modern diagnostics. The fact of the existence of many diseases and conditions of unknown origin or cause can provide the impetus to examine causative factors in light of traditional knowledge systems. It is important to note that the ICD-11, for the first time during a decade-long process of revision, has considered inclusion of traditional medicine. As of now, it covers traditional Chinese and Korean medicine-based disease patterns, for example: principles-based patterns (yang, yin, heat, cold), body constituents patterns (blood, *Qi*), meridian patterns, and Sasang constitution medicine patterns. However, Ayurveda and Yoga do not find a place in this important database, yet.

Interestingly, new trends are emerging in the understanding of disease, its classification and distribution. For example, the ICD has started using new terms like *drug-induced diabetes*, *sleep–wake disorders*, and *restless leg syndrome* [9]. This terminology illustrates that we do understand the real causes of many diseases, and we prefer to classify them into the ready-made categories, which perhaps do not accurately reflect the nature of the disorder, disease, or condition. Many new diseases are the result of the failure of current drug treatments, or of socio-economic and environmental changes. For instance, drug-induced diabetes is actually not a disease, rather, it is a condition created by therapeutic interventions. Thus, the concept of disease has changed during last few decades.

DEFINITION OF HEALTH

The WHO defines health as "a state of complete physical, mental and social well-being and not merely the absence of disease or infirmity" [10]. However, in practice, there is no robust and reliable way to measure *well-being*. Generally, the absence of disease or infirmity is considered to be *health*. In many quarters, the terms *health* and medicine are used interchangeably, without an appreciation of the difference in meaning of the two words. Today's health care revolves predominantly around medicine—primarily dealing with disease diagnosis and treatments, and not as much on prevention and health promotion. This mind-set has its roots in the misunderstanding of health.

Many experts feel that the WHO definition of health is not complete without the inclusion of the spirituality component. The WHO definition highlights something like well-being that is difficult to measure, and hence health care decisions are being made based upon the apparent absence of disease. Moreover, modern definitions of health are mostly restricted to the body, and to some extent, the mind. However, the holistic picture of body/mind/spirit is completely missing in modern descriptions of health. The WHO have been discussing the concept of spiritual health, and several recent studies have endorsed its importance. In 1997, the WHO's executive board resolved to recommend to the General Assembly of the United Nations, a new definition of health: "Health is a dynamic state of complete physical, mental, spiritual and social well-being and not merely the absence of disease or infirmity." A WHO's spokesperson stated that "if we can find ways to approach practically the spiritual dimension of health, we may be able to alleviate and combat some of the pressures of modern life."

While advances in science and technology have helped us to understand etiology, pathophysiology, and treatments they have hardly helped us to understand *health*. For example, nearly 30,000 genes are known. Many genes responsible for causing diseases are also known. Oncogenes are responsible for cancers; ob genes are responsible for obesity. In general, we know which genes are responsible for many disorders or conditions. However, genes that are responsible for health are hardly known. While powerful drugs like antibiotics are available to treat infectious diseases, or to alleviate symptoms like fever, pain, or inflammation, there is no medicine to achieve health. Doctors can treat diseases but cannot give health. Seeking health on the part of the individual is not a passive process like taking an administered medicine, or undergoing surgery. Achieving health requires active participation of individuals and communities.

Today, health care practices are dominated by pharmaceutical or surgical interventions. This trend is based on the erroneous notion that newer and more powerful drugs may increase our means to achieve better health. Therefore, there exists an incorrect notion that discovering more and more new drugs will increase our chances of achieving health. This mind-set seems to be a global phenomenon, which cuts across East and West, rich and poor, and developed and developing countries. In a consumerist society, this seems to be an attractive business proposition for various industries including pharmaceutical, diagnostics, and in particular, insurance. Therefore, from policy makers to practitioners, all seem to be comfortable with keeping real, public health at bay, and continuing to focus on curative aspects and medical care.

A closer look at budgets of ministries or departments of health make it clear that most of the expenditures are made on medicines, equipment, diagnostics, surgeries, and salaries. In general, efforts toward health promotion or disease prevention has taken a back seat. The disease-centric curative approach has strong roots in education as well. For instance, many states in India have established universities of health sciences. A critical review makes it amply clear that these universities are actually conglomerations of medical and paramedical colleges, with emphases on medicine, nursing, and physiotherapy, among others. Thus, in health science universities there is a domination of medicine, diagnosis, therapeutics, and surgery; the actual *health* component is largely ignored.

The confusion between health and disease is deep, and the domination of disease over health is pronounced. For instance, in the United States, the largest and most advanced scientific network of 27 institutes and centers is known as National Institutes of Health (NIH). Even a casual glance at the NIH Web site reveals the dominance of disease-, diagnosis-, and treatment-related activities in contracts for the prevention and promotion of health. The first page of this Web site has a heading, *Health Information*—with a list of diseases like diabetes, obesity, cancer, asthma, cholesterol, depression, hypertension, and Alzheimer disease. Except for its information on nutrition and environment, the NIH Web site gives more information about *disease*, than it does regarding *health*.

In contrast, the National Library of Medicine (NOT Health) gives more current terminologies of *health* under Medical Subject Headings. It includes 16 key words, which describe various aspects of health, such as family health, public health, women's health, and holistic health. This confusion between concepts of health and medicine, visible at the highest levels, is of great concern.

It is necessary to state again: *Health* is not a synonym for *medicine*. *Health care* is not equivalent to *sick care*, and as such, it cannot be restricted to encompass only *medical care*.

HEALTH CARE AND MEDICAL CARE

At the outset, it is necessary to differentiate between the concepts of health care and medical care. There seems to be confusion among our conceptions of health care, medical care, public health, and global health. As rightly stated by Dr Atul Gawande from Harvard Medical School "global health is about the idea of making care better everywhere." Advantages of health care must reach to all the needy people and to the poorest of the poor.

Most of the poor and developing countries fail to provide effectively for basic public health determinants such as safe drinking water, environmental sanitation, and nutrition. This is part of health care. However, the Western notion of *health care*, in actual fact is *medical care*; in most Western countries, basic public health needs like safe drinking water are universally provided, and defecation is confined to sanitary latrines. Therefore, the Western public expects to receive *medical care*, which they call *health care*. The Third World is far from providing for even the basic, minimum, public health needs. Therefore, a proper understanding of the concept of health care is fundamental to the realization of worldwide acceptance and practice of primary health care as a developmental concept—as opposed to *first contact care* in medical terms, as it is in the United States.

When effective health care is not provided, then diseases may appear and the community will obviously require *medical care*. Health care is not equivalent to medical care. Today, global health care is dominated by modern medicine, and compromised policies have given way to pharmaceutical companies who dictate health care strategies. The cost of medical care in most societies is rising to a prohibitive range. Visionary experts advocate a need for new approaches to medical care: the best of corrective surgery, and emergency measures of modern medicine, integrated with holistic concepts from traditional knowledge systems. The situation can be frustrating for sensitive professionals who raise the question: Are medical science and drugs artificially created entities like a golem? [11]. The medicalization of society has indeed become a global concern. Actually, people need simple ways to achieve health and prevent disease. People need a healthy diet and healthy lifestyle. People need mental peace and relaxation.

Major limitations and challenges are faced by the health care models of established systems such as the National Health Scheme of the United Kingdom [12]. These models were appreciated at one point in time. The experience of the implementation of these models has raised many questions:

Are we searching the solutions without acknowledging real problems?
Are we expecting health with help of only medicines?
Do we need to change our strategies?

Well-known innovation guru, Clayton Christensen, of Harvard Business School, has commented on the United States' health care system, and has offered an innovator's prescription. He has proposed "disruptive innovation": a process initiated at the grass roots level, relentlessly moving up to displace existing systems. For example, after the introduction of email, communication became much easier, and postal services were disrupted. We really need to disrupt present systems to evolve a health protection-centric, health care system over the disease-, treatment-, and cure-centric, medical care system [13].

DETERMINANTS OF HEALTH

Determinants of health may be biological, behavioral, sociocultural, economic, and ecological. Broadly, the determinants of health can be divided into four, core categories: nutrition, lifestyle, environment, and genetics, which are like four pillars of the foundation. When any one of the pillars of health determinants becomes weak, a support system is needed. This is considered the fifth determinant of health and involves medical care (Figure 3.3). A brief review of these core determinants of health will provide more insight.

Interestingly, two determinants, nutrition and lifestyle, are totally in our hands, and hence are called modifiable factors. Many diseases are caused by bad practices of nutrition and lifestyle. The degraded ecosystem, and environmental pollution are the causes of several disorders and diseases. With the help of powerful technology and screening methods, many disorders of genetic origin can be prevented. If one or more core determinants become weak, then only the support of medical care is needed.

Over 75% or more of the resources allocated in health care budgets, especially from rich countries, are used for the treatment of lifestyle-related conditions. There is a growing consensus that lifestyle modifications should be the foundation of any health care system. According to the American College of Lifestyle Medicine, nearly 80% of all chronic diseases are preventable by readily available means—lifestyle modification as medicine.

People should be empowered to "take their health into their own hands" through lifestyle modifications. This will drastically reduce dependence on

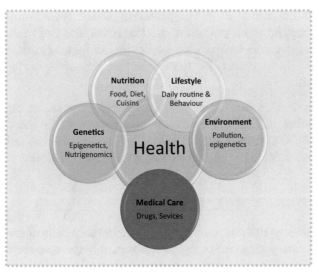

FIGURE 3.3

Determinants of health: Nutrition, lifestyle, environment, and genetics are considered as core determinants and four pillars of health. When any one or more of these is compromised, health is at risk and medical care is required as a support system.

doctors. Traditional knowledge can be immensely useful to design appropriate lifestyle interventions. For instance, *Swasthavritta*, a branch of Ayurveda is dedicated totally to healthy lifestyle. *Swasthavritta* dictates do's and don'ts for a healthy daily regimen, and outlines diet and lifestyle modifications appropriate to different seasons. *Swasthavritta*, and biobehavioral practices suggested by Yoga are very useful sources for lifestyle medicine.

Nutrition is another important determinant. It has individual, family, and community dimensions. The East/West, and rural/urban regions have remarkably different challenges related to nutrition. Generally, at one end of the spectrum, in Western and/or urban spheres, there is less physical activity, calorie overload, but poor nutrition mainly due to junk food consumption. At the other end of the spectrum, in the East and/or rural spheres, there is calorie deficiency, protein malnutrition, and undernourishment. The lower socioeconomic communities may have a greater incidence of premature and low birth weight babies, higher risk of heart disease, stroke, and some cancers. Poor people living in urban areas may have a diet consisting of cheap energy mainly from sugar-rich foods, with little intake of vegetables, fruits, and whole grains. They have relatively less physical activity. On the other hand, poor communities from rural areas might have intense physical activities, but not sufficient energy and protein.

In general, urban communities face problems related to environmental degradation, and air and water pollution; rural communities face problems related to sanitation, hygiene, insecticides, pesticides, and agrochemicals. Thus, the poor are most likely to suffer because of the interplay of the deranged determinants of health.

In the interconnected, borderless world, determinants of health cannot be considered in isolation. They will always be interdependent. The substantial health inequity in different parts of the world is today's reality. This inequality of health is due to inequalities in income, education, gender, and availability of resources.

DIMENSIONS OF HEALTH

The WHO adopted this definition of health in 1948: "A state of complete physical, mental, and social well-being, and not merely the absence of disease or infirmity." Thought leaders have raised several questions related to this definition. How can one define "a state of *complete* well-being"? How do we measure well-being? Actually, there is no reliable instrument to measure well-being. In the year 2011, *The BMJ* published a debate initiated by Huber et al., on the definition of health [14]. According to Huber et al., the word *complete* in relation to well-being is unqualified. Before arguing more on this point, it is necessary to understand various dimensions of health, and the meaning of wellness.

There are seven prominent dimensions of health and wellness. These include social, emotional, spiritual, environmental, occupational, intellectual, and physical wellness (Figure 3.4). All these dimensions contribute to our quality of life (QoL).

1. The physical dimension relates to the maintenance of a good QoL, allowing the individual to perform daily activities without undue fatigue or physical stress, and the importance of adopting healthy habits like balanced diet and exercise, while avoiding bad habits like tobacco, drugs, and alcohol.
2. The social dimension is about the ability to connect and relate to other people in positive relationships with family, friends, and colleagues.
3. The emotional dimension is the ability to meeting challenges, and to understand own strengths and limitations, while respecting others, and overcoming personal ego. The ability to acknowledge and share feelings of anger, fear, sadness, or stress, and hope, love, joy, and happiness in a productive manner.
4. The environmental dimension is the ability to recognize our responsibility to preserve nature, and protect the ecosystem: air, water, land, forests, homes, and communities, or our planet, thus improving the standard of living, QoL, and status of the environment.

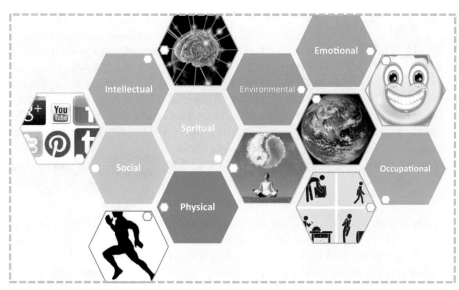

FIGURE 3.4
Dimensions of health.

5. The occupational dimension is about balancing work pressures and job satisfaction, while making positive contributions to enterprise organizations and also to society.
6. The intellectual dimension is the ability to keep an open mind to new ideas and experiences. The desire to learn new concepts, improve skills, and engaging in lifelong learning contributes to the intellectual dimension.
7. Awareness of the spiritual dimension is a recent realization, especially in the West; it is a quest to find our sense of purpose, and apprehend the meaning of life. It is about the ability to think beyond selfish motives, to establish peace and harmony, and understand the importance of values, personal purpose, and a common purpose that binds humanity.

The spiritual dimension of health underlines the vital importance of mind as a major factor in health promotion, disease prevention, and effective recovery from illness. The importance of economic stability in life is not new, but its recognition as dimension of health and wellness is now appreciated even in the East. The financial dimension is about the economic stability of the individual, and the family in the face of financial emergencies.

The dimensions of health and wellness now encompass individual or public health, and involve healthy environments, unpolluted rivers, healthy towns, healthy cities, healthy nations, and planetary health. Global awareness is now moving toward *health*. Growing recognition about the importance of determinants of health and wellness is clearly palpable.

WHAT IS WELL-BEING?

In the WHO's definition, dimensions of health are connected to the term *well-being*. Wellness and well-being are not synonyms, but the differences are very subtle. Well-being as a concept involves total life experience, happiness, and prosperity. Wellness refers mainly to the individual feeling of one's body/mind/spirit. Today, corporations talk about employees' well-being; spas offer tranquility and relaxation, but the concept of wellness is beyond physical health and fitness. Wellness is considered an integration of states of physical, mental, social, and spiritual well-being. Because it is difficult to differentiate between these concepts, it is necessary to discuss the dimensions of health, well-being, and wellness together.

What is a state of *well-being*? Indeed, it is very difficult to answer this question. A state of well-being is very difficult to measure. It is very subjective, consequential, judgmental, and qualitative. According to the principles of management, "what cannot be measured cannot be managed." Can we manage well-being? What is the meaning of a *state*, when we say a *state of well-being*?

Thus, the first part of the WHO definition seems obscure, while the second part is based on a negative connotation. How can the "absence of disease or infirmity" be assumed to be the presence of health? The negative connotation in the definition, and understanding health in the context of disease has several implications. It shifts the focus from *health* (which is difficult to measure) to *disease* (which is easy to measure). Indeed, it is relatively easy to manage disease, rather than managing health. Because health is difficult to measure, the pathological laboratories and pharmaceutical companies have developed tests or drugs only for known diseases.

Arguably, as an obvious outcome, today's health care is dominated by drugs, diagnostics, and disease-curative attempts. Although Ayurveda is considered a holistic medicine, its current practice is focused on curing disease, which closely mimics the approach of modern medicine. This is not in the real spirit of Ayurveda, which first stands for health protection and disease prevention. The *curative* approach, which predominates in most of our societies, might have its roots in our current, narrow understanding of health.

The concept of well-being is slowly transmuting into a broader concept of *wellness*, and is getting closer to the Ayurvedic concept of *Swasthya*. Well-being is about wellness. The term *wellness* was first explained by Halbert L. Dunn, who was chief of the United States' National Office of Vital Statistics. Dunn described a high level of wellness as "performance at full potential in accordance with individual age and makeup." He suggested that health dimensions range from high-level wellness to death. The concept of wellness has a spiritual and social context. Dunn suggests that to "know yourself" is the fundamental principle of wellness. In-depth knowledge of the outside world is possible only with the knowledge of one's inner self. The role of wisdom and maturity possessed by senior citizens, and successful people contribute to the wellness of society. He further explains the connection between biological nature, and the spirit of man as "creative expression." An expression of the self in search of universal truth provides inner satisfaction to humanity on an individual level. Core values like creative expression, trust, love, security, maturity, and altruism bring about wellness at a societal level. Dunn presents an optimistic vision that encompasses the wisdom of the East and West. He also admits the practical difficulties of his proposition, but suggests that dreaming can produce wellness for society [15].

The concept of wellness provides a conscious way of living for individuals and society to reach their fullest potential. However, the term *wellness* is generally used in the context of spa-going, relaxation, and personal recreation. The core concept of wellness is being distorted, and now is perceived as "pseudoscience."

HOLISTIC HEALTH

The term *holistic health* is used many times in literature with a variety of different connotations. Holism also has its origin in the Greek word *holos*, which means "whole." Holism is not about any cult or religion, rather, it is an approach that looks at things in a total perspective. As far as we know, it was first used in 1926 by Jan Smuts, in his book *Holism and Evolution*. The superspecialization of scientific disciplines has created a silos mentality—leading to a myopic understanding of knowledge, and compromising our ability to deal with the most obstinate problems. Although Jan Smuts used evolution as an example to explain his concept of holism, this book actually became a trigger for systems thinking, and complex, interdisciplinary and integrative approaches in science.

Holistic health typically considers the whole person—body, mind, and spirit. A holistic approach to healing goes beyond just eliminating symptoms. Many times, holistic health is considered holistic medicine involving traditional and complementary medicines (T&CM). We feel this is not correct. Holistic health is an approach, which modern medicine also needs to adopt. Our idea of integrative approaches for health is, in a way, holistic—where modern medicine, Ayurveda, Yoga, and T&CM are considered as a whole.

Ayurveda is about the ways and means of restoring and promoting health. The broader goal of Ayurveda is to maintain a dynamic balance between internal and external environments. In a positive and broad manner, Ayurveda defines health as *Swasthya*, which actually means "being contented in ones' natural state of inner harmony." According to Ayurveda, one is considered as healthy when body, mind, and spirit are in the state of equilibrium, comfort, and bliss.

According to Ayurveda, health is not just the absence of disease; much more is needed for an individual to qualify as being in a healthy state. The individual's capacity to adapt to conditions like extreme hot or cold temperatures, and the experience and expression of psychological feelings like joy or sorrow are signs of a healthy state. Modern researchers now accept health as the ability to adapt and to self-manage. Optimum levels of sensory and motor function are the requirements of a healthy state, which the individual must possess in order to be able to conduct social tasks, and follow spiritual aspirations. The Ayurveda health model is based on proactive efforts to monitor, maneuver, and maintain healthy status. Ayurveda considers the surrounding environment, and the society as part of the broad concept of health. Ayurveda stresses the importance of creating an awareness of health. Ayurveda empowers individuals to know their strengths, limitations, inclination, and proneness to disease, and to carry out continuous self-assessment. Thus, Ayurveda helps every individual to take charge of his or her own health. This philosophy of "our health in our hands" is practiced as *Swasthavritta*. In contrast to the contemporary model of outsourced health care, Ayurveda and Yoga actually advance the insourcing of health care.

QUALITY OF LIFE

Health, wellness, or well-being is a subjective perception. Health is multifactorial and QoL has several components. Many approaches have been proposed to describe and measure QoL. Studying QoL is important—especially in regard to health, therapeutics, disabilities, and rehabilitation. It is an indicator of a meaningful and enjoyable life. It is a measure of individual assessment of various aspects of life. QoL is related to economics, politics, culture, and environment. Merely a good standard of living does not necessarily lead to a better QoL. QoL is not only about employment and wealth, but also involves life purpose, and satisfaction. QoL has moral, philosophical, and spiritual dimensions. Several studies like the human development index, and human happiness index involve the use of appropriate instruments to measure QoL.

In 1991, the WHO initiated a project to develop a QoL assessment instrument. QoL measurement involves contextual, individual perceptions related to culture, value systems, personal goals, standards, and concerns. The WHO's QoL instrument comprises 26 attributes, which measure the physical health, psychological health, social relationships, and environment (Figure 3.5). Health-related QoL (HR-QoL) takes account of factors affecting health. These tools consider self-reported effects of diseases, and their risk factors. There are many tools and scales to measure QoL, such as Euro QoL, Sickness Impact Profile, and the Health Utilities Index.

FIGURE 3.5
Health is multifactorial and quality of life has several components.

The Healthy People 2020 program of the Unites States Department of Health and Human Services emphasizes HR-QoL. This is a multidimensional concept, which includes physical, mental, emotional, and social functioning. HR-QoL goes beyond direct measures of population health, life expectancy, and causes of death. It considers the impact of health status on QoL. The concept of well-being takes in positive aspects of a person's life, such as positive emotions, and life satisfaction; it is closely related to HR-QoL. Clinicians and public health officials are using HR-QoL, and well-being to measure the effects of chronic illness, treatments, and short- and long-term disabilities. Several approaches to the measurement of HR-QoL, and well-being are being explored, but definitive and reliable methodologies have yet to evolve.

Well-being is measured by assessing people's perceptions: feeling healthy, satisfied or content with life, the quality of relationships, positive emotions, resilience, and realization of their potential. Assessments of their social participation, education, employment, civic, and social and leisure activities are also considered. The assessment of functional limitations like vision loss, mobility difficulty, or intellectual disability also helps in determining HR-QoL—especially in determining whether an individual's life is productive and of a good quality. These assessments form a global health measure known as Patient Reported Outcomes Measurement Information System (www.nihpromis.org). A close look at these attempts indicate that in the search to know more about QoL and well-being, psycho-social, philosophical, and spiritual dimensions are getting due recognition.

ASSESSMENT OF HEALTH

It is indeed difficult to measure health; yet, it is important to know the extent of health status at an individual and a community level, on a continuing basis.

The assessment of individual health is normally attempted by a routine health checkup, done with a physician's detailed, physical, clinical examination, and laboratory investigations involving pathological tests, and radio imaging procedures. Normal constituents of blood, urine, and stool are assessed. Images of normal organs are part of our knowledge. We know normal images of organs like heart, lungs, kidneys, liver, and other organs. Normal values of various biochemicals in the body have been established over time through observation, research, and scientific studies. The routine health checkup protocols can indicate any abnormality through a series of laboratory investigations. Absence of any detectable abnormality is assumed to be an indicator of a disease-free condition. In practice, this is interpreted as a *healthy* status. Quite in line with the conventional definition of health, the assessment is mainly based on indicators of diseases.

The assessment of public health is normally done with the help of a set of quantifiable indicators of health. The WHO Indicator and Measurement Registry classifies key indicators into eight broad categories: demography, equity, health service coverage, health system resources, mortality, risk factors, socio-economics, and morbidity data. Some of the key indicators include life expectancy at birth, population growth rate, fertility rate, immunization coverage, availability of medicine, doctor population ratio, under-five child mortality, maternal mortality, per capita income, education, and nutrition status.

Both individual and public health indicators are strongly related to the incidence or prevalence of diseases, and delivery of medical care; most of these markers are not direct indicators of health but are assumptive, speculative, and indirect indicators.

Ayurveda puts forward the assessment of health right from the stage of a newborn. The description of the healthy baby contains anthropometric parameters as predictors of health and longevity. Ayurveda propounds that the assessment of the physical component of adult health can be done effectively by using a unique method based on relative, anthropometric measurements. This method uses the dimension of individual fingers as proportionate measures. This is known as *Anguli Parimana*, and is a very good self-help tool to measure the status of individual health [16].

Ayurveda describes a measurable scale for the functional assessment of health status. Several parameters are included: degree of hunger, bowel practices, and efficiency of sensory organs (Figure 3.6). The ultimate objective of Ayurvedic treatment is balancing of the homeostasis of the *Dosha*, and establishing the normal *Prakriti*. Ayurveda gives detailed description and advice about desired endpoints as treatment outcomes (Figure 3.7).

HEALTH PROMOTION

Health promotion as defined by the WHO is "the process of enabling people to increase control over, and to improve their health." The 8th Global Conference on Health Promotion at Helsinki, endorsed the proposition that the assurance and sustenance of health requires peace, shelter, education, food, income, a stable ecosystem, sustainable resources, social justice, and equity as basic requirements. Earlier concepts of health promotion were more about the prevention of lifestyle disorders, and noncommunicable diseases; new approaches to health incorporate communicable disease prevention and involve multiple sectors. This is a proactive and positive step in moving closer to health, and beyond a disease focus.

Few academic institutions are actively involved in propagating the concept of wellness. The University of California, Riverside has a wellness program for its faculty, staff, and retired members. The program includes social, emotional, spiritual, environmental, occupational, and intellectual and physical wellness [17]. The University of Pittsburg starts wellness orientation right from the student admission. Wellness related research, and its translation into national programs are very limited, despite having a sound, philosophical background. We discuss the basics of health and wellness with a focus on preventive, protective, and proactive approaches. We suggest that, in looking at health, an optimistic view is necessary. The domination of a pessimistic view, with the lifelong dependence on disease treatments is leading to the medicalization of society. As a result, people are trapped in the nexus of drugs, doctors, and

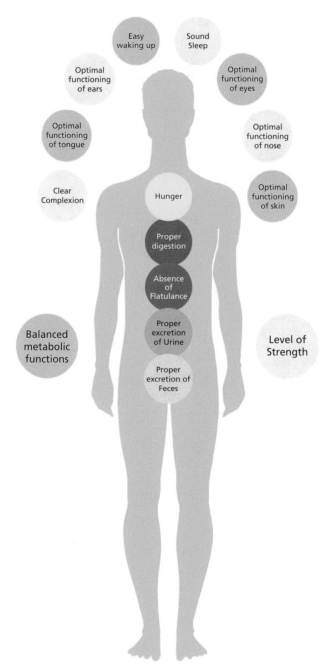

FIGURE 3.6
Health assessment—Ayurveda approach.

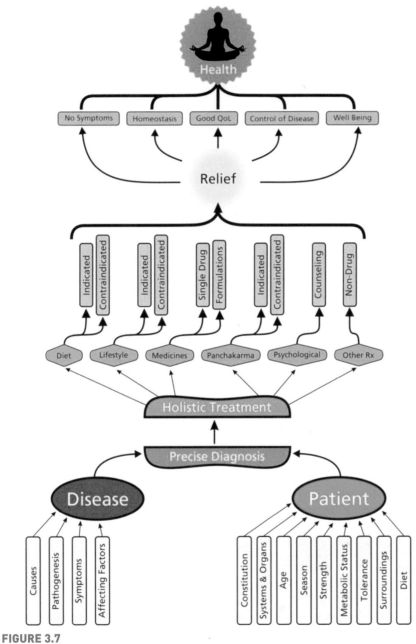

FIGURE 3.7
Ayurveda treatment approach.

diagnostics. We hope that with systematic efforts, and experience-based guidance from Ayurveda and Yoga it will be possible to achieve, maintain, and protect health.

Today's wellness programs based on putative health benefits, along with commercial packages, dietary supplements, and exercise devices have many limitations. The commercialization of wellness as a package is actually against the ethos and might take it down an exploitative path. Wellness is a very individual experience. To achieve a state of wellness, it is important to empower the individual.

Ayurveda and Yoga empower individuals with the knowledge, attitude, and behavior for positive health and wellness; modern medicine provides objective tools for health assessment. The concept of health promotion in Yoga and Ayurveda are quite comprehensive. Ayurveda advocates for health promotion through advice on diet and lifestyle. This branch is known as *Swasthavritta*, which is an Ayurvedic pronouncement for healthy life. Any deviation from *Swasthavritta* may lead to an imbalance in *Dosha*, resulting in ill-health. The *Swasthavritta* incorporates advice pertaining to age, nutritional status, metabolic attributes, and individual tolerance and sensitivity in the context of environmental and seasonal variations. According to Ayurveda, the three *Dosha* undergo chronobiological changes during the day and night cycle. The accompanying illustration of the Ayurvedic clock is very interesting. *Dosha* are also affected by the pattern of daily routine, foods habits, and behaviors. Ayurveda suggests matching daily food and behavior in such a way that the *Dosha* are maintained in the balanced state. For example, one should consume food when *Pitta Dosha* is at the highest peak, exercise when *Kapha* is increased, and rest when *Vata* is aggravated. If daily activities are not synchronized with *Dosha* levels, the result might be imbalance and disease. Hence, the history of patient's diet and lifestyle is important for diagnosis. The treatment includes detailed advice on when and what to eat, and aims to harmonize *Dosha* with the properties of diet and drugs. Sometimes a useful substance may be harmful if consumed at the wrong time. An asthmatic may be advised to drink medicated water at bedtime to avoid the aggravation of *Kapha* during the night.

Ayurveda advises specific changes in diet and lifestyle according to various seasons. This is known as *Ritucharya*. Specific interventions like *Panchakarma* are also advocated in particular seasons. The guidelines for daily routine, seasonal changes, and treatment regime are specific to *Dosha* changes, and aim to achieve a dynamic balance of the *Dosha*. A chronobiology clock based on Ayurveda is very useful to plan various activities during the day for optimal output and health (Figure 3.8(a) and (b)). The accompanying chart *Ritucharya* is useful to understand the specificity and richness of Ayurvedic advice for health protection during the year depending on predominance of *Dosha* in each season.

(a)

Daily routine as advised by Ayurveda (*Dinacharya*)

1. Wake up before sunrise
2. Clean face, mouth and eyes
3. Drink warm water
4. Bowel and Bladder Evacuation
5. Tooth, Tongue and Gum cleaning
6. Light oil massage and Bath
7. Use natural perfumes
8. Exercise, Yoga, Pranayama
9. Meditation
10. Breakfast
11. Work, Lunch, Work
12. Dinner with family before sunset
13. Light reading, socializing
14. Sleep

(b)

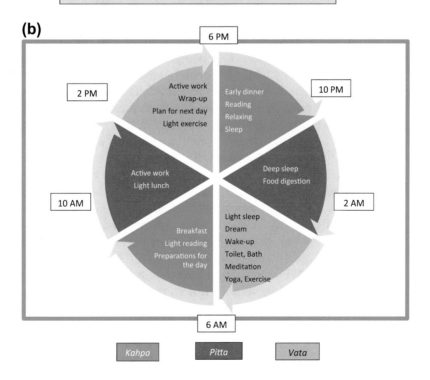

FIGURE 3.8

Daily routine and Ayurveda chronobiology of health. (a) Daily routine as advised by Ayurveda (*Dincharya*) (b) Ayurveda chronobiology clock of a day. The daily routine advised by Ayurveda is based on dominance of *Dosha* during 24 h of the day, which are divided in six spells. Blue color denotes *Kapha*, Red color denotes *Pitta*, and gray color denotes *Vata*. Maintaining specific activities during the respective *Dosha* dominance helps in maintaining their balance. This is one of the fundamental principle of Ayurveda for health protection.

FROM PUBLIC TO PLANETARY HEALTH

The importance of health promotion and disease prevention can no longer be overshadowed by disease treatment and therapeutics. Therefore, while we recognize and promote the idea of integrative medicine, we advocate integrative approaches for health. Public health care encompasses four levels of prevention. It focuses more on perceptions about health than on health-related products like drugs and devices. Health should be considered an *approach*, rather than a *commodity*. Today, there is considerable awareness that research in biomedical sciences should translate into direct benefits for people and society. While some blue-sky research will continue, many countries are focusing on real needs and challenges in public health, and medical care. Such translational research can help to resolve the unresolved problems of health care. A welcome trend that recognizes a broader meaning of health is being witnessed globally. In this context, a recent step taken by the NIH in the United States to change the name, the National Center for Complementary and *Alternative Medicine*, to the National Center for Complementary and *Integrative Health* is representative, and indicative of future trends.

It is heartening to witness new ideas reflected in the manifesto released by the Editor of the medical journal, *The Lancet*. This bold manifesto is a clarion call to consider health beyond mere drugs, pharmaceuticals, science, technology, individuals, or peoples. This manifesto attempts to transform public health by taking it beyond personal, community, national, regional, and global levels—to planetary health. The concept of planetary health recognizes health of a planet itself, which can be achieved only when all ecosystems living on the planet are healthy. This is a very timely, proactive step, which addresses threats to the human health and well-being, sustainability of civilization, and natural and human-made systems. The vision is for a planet that nourishes and sustains the diversity of life with which we coexist, and on which we depend. The goal is to create a movement for planetary health. *The Lancet* manifesto has added a new, philosophical dimension and introspective attitude toward life and living. It emphasizes the importance of people, and not just diseases and treatments. The manifesto seeks to minimize differences in health, according to wealth, education, gender, and place. While accepting knowledge as a major source of social transformation, the manifesto aims to progressively realize attainable levels of health and well-being [18]. The term *planetary health* is not restricted to human health or individual health. In reality, the concept of health can be applied to communities, animals, plants, and various ecosystems like ocean, rivers, forests, mountains, cities, and nations. Such an inclusive, humane approach, in contrast to the restrictive, anthropocentric attitude, certainly should help to expand the vision and highlight the importance of mutual existence in nature.

FIGURE 3.9
Illness can be treated passively but seeking health requires participatory approach.

In short, to move from disease to health and from Illness to Wellness, it is necessary to replace "I" with "We" (Figure 3.9). It is necessary to first protect *well* and then *ill* should be treated. Clearly, it is an active and participatory effort, which cannot be achieved in isolation. Health is something natural, but the attributes of perfect health state are hazy. As against, a perfect disease state is fairly known. Undoubtedly, present science understands diseases better than health. While diseases or illness may occur as a part of life, and their treatment is necessary, the focus on health must be regained and strengthened. We need more emphasis on nutrition, diet, lifestyle, and behavior modification.

As rightly stated by Dr Dean Ornish, a well-known researcher of lifestyle medicine from the University of California, San Francisco: "Poor health is not caused by something you don't have; it's caused by disturbing something that you already have. Healthy is not something that you need to get, it's something you have already if you don't disturb it." This is very much in tune with Ayurveda and Yoga, which propose a greater focus on mind, and strengthening the inner strength.

Finally, as the renowned physician from Edinburgh, Peter Mere Latham, said, "Perfect health is very difficult to attain, but it is easy to diagnose and treat a perfect disease." Today, the importance of health promotion and disease prevention can no longer be overshadowed by disease treatment and therapeutics. Therefore, while an idea of integrative *medicine* is often promoted, this book advocates approaches for integrative *health*.

Health is basis for development of ethical, economic, aesthetic, and spiritual dimensions.

Charaka Samhita

REFERENCES

[1] Filho N de A. For a general theory of health: preliminary epistemological and anthropological notes. Cad Saúde Pública 2001;17(4):753–99.

[2] Hales CN, Barker D. Type 2 (non-insulin-dependent) diabetes mellitus: the thrifty phenotype hypothesis. Diabetologia 1992;35(7):595–601.

[3] Locher WG. Max von Pettenkofer (1818–1901) as a pioneer of modern hygiene and preventive medicine. Environ Health Prev Med 2007;12(6):238–45.

[4] Park K. Park's textbook of preventive and social medicine. Jaipur, India: Banarasidas Bhanot; 2009.

[5] Hill AB, Doll R. Lung cancer and tobacco; the B.M.J.'s questions answered. BMJ 1956;1(4976):1160–3.

[6] Hill AB. The environment and disease: association or causation? Proc R Soc Med 1965;58:295–300.

[7] Wellcome Trust Case Control ConSortium. Genome-wide association study of 14,000 cases of seven common diseases and 3000 shared controls. Nature 2007;447(7145):661–78.

[8] www.who.int/classifications/icd/en/HistoryOfICD.pdf.

[9] http://apps.who.int/classifications/icd11/browse/l-m/en.

[10] World Health Organization. Preamble to the constitution of the world health organization as adopted by the international health conference. Off Rec World Health Organ 1946:100. Available from: http://www.who.int/governance/eb/who_constitution_en.pdf.

[11] Hegde BM. Medical science–a golem? Indian Acad Clin Med 2013;14(1):6–7.

[12] Anon. The NHS: a national health sham. Lancet 2005;366(9493):1239.

[13] Christensen Clayton M, Grossman Jerome H, Hwang J. The innovator's prescription: a disruptive solution for health care. New York: McGraw-Hill; 2009.

[14] Huber M, Knottnerus JA, Green L, van der Horst H, Jadad AR, Kromhout D, et al. How should we define health? BMJ 2011;343:d4163.

[15] Dunn HL. High-level wellness for man and society. Am J Public Health Nations Health 1959;49(6):786–92.

[16] Shirodkar JA, Sayyad MG, Nanal VM, Yajnik C. Anguli Parimana in Ayurveda and its association with adiposity and diabetes. J Ayurveda Integr Med 2014;5(3):177–84.

[17] http://wellness.ucr.edu.

[18] Horton R, Beaglehole R, Bonita R, Raeburn J, McKee M, Wall S. From public to planetary health: a manifesto. Lancet 2014;383(9920):847.

CHAPTER 4

Evidence-Based Medicine and Ayurveda

The absence of evidence is not the evidence of absence.

Carl Sagan, Cosmos

WHAT IS EVIDENCE?

Before addressing various aspects of evidence-based medicine (EBM), it is important to discuss the meaning of evidence. In his famous book *The Idea of History*, R.G. Collingwood rightly states "when we try to define *evidence*, we find it very difficult." Evidence by definition is "the available body of facts or information indicating whether a belief or proposition is true or valid." Scientific evidence relies on observations, and well-planned, controlled, and reproducible experimental results to support, refute, or modify a hypothesis or theory. In philosophy, evidence is closely tied to epistemology. In a way, evidence is *proof*. Methods of obtaining proof are connected to the nature of knowledge. Thus, scientific evidence is just one of the many forms of evidence, such as anecdotal evidence, intuition, personal experience, and testimonial cases.

Thus, the meaning of evidence is not straightforward. Evidentialism is a theory which mandates justification based on evidence. Evidence comes from repeatable and reproducible sets of observations, measurements, and reasoning. Obviously, the capacity to observe and measure, as well as the ways and means of reasoning, have a direct influence on the nature, quality, and reliability of evidence. Therefore, as scientific knowledge advances, and our understanding deepens, the acceptability of evidence also changes. This is how many notions or beliefs have been proved to be wrong in history. For example, a notion that all organisms were designed by an intelligent creator or God—known as design hypothesis—was shunned after the Darwinian hypothesis was scientifically accepted. Now, Darwin's theory of biological evolution and the diversification of life is being challenged by new theories like biocentrism, which hypothesizes that life, and consciousness, are the keys to understanding the true nature of the universe.

Thus, as our knowledge expands, the nature of evidence also changes. Because science is dynamic, scientific evidence cannot be a static entity. Science only states that based on whatever is known, accepted, and reproducible at a given

79

point of time, certain hypothesis, theories, observations, interpretations, analysis, or outcomes have a sufficient evidence base. The expectations and nature of evidence in jurisprudence may be different from those required for clinical practice.

Earlier in our discussion of the evolution of medicine, we went over how faith-dominant, primitive, shamanic treatments transmuted into scientific, modern medicine. The nature of diseases became better known with the development of etiopathology, and an understanding of the underlying biological processes. As a result, therapies also became more precise, dependable, rational, and scientific.

During the past century there was consistent progress in biomedical research. We have seen how several discoveries in medicine and surgery affected the quality of human life for the better. To keep pace of advancements in scientific research, and technological developments, medical education also underwent a huge transformation. Medical education had to move from mere bedside observations to laboratory-based investigations in order to bridge the gap between biomedical research and clinical practice. The transformation of the medical profession from one of clinical practitioners, to physician–scientists required a strong research base, and the experience of medical practice.

In 1910, the American physician and educationist, Abraham Flexner, published a landmark report that revolutionized medical education in the United States and propelled the rapid growth of biomedical research and development. Flexner's report recommended the intensification of biomedical research and the development of a new cadre of medical professionals who were physician–scientists. Thus, excellent clinicians were exposed to systematic research, documentation, analysis, and applications. One of the positive impacts of this change in the medical education system is the substantial increase in the number of scientific publications in the field of modern medicine. For instance, in a scintometric analysis of the term *medicine* in Google Scholar, we get 4,810,000 hits, indicating the number of scientific, peer-reviewed papers, citations, and patents. These results refer to scientific publications or patents related to medicine. In the 100 years from 1810 to 1910, there were hardly 40,600 papers. From 1910 to 1950, there were 184,000 papers; from 1950 to 1970, there were 208,000 papers; from 1970 to 2000, there were 328,000 papers related to medicine; and from 2000 to 2013—just 13 years—there are 416,000 papers. The number of papers related to medicine from 2000 to 2013 is almost 10 times more than the total number of papers in the first 100 years [1] (Figure 4.1). While we grant some degree of overlap, and exclusion errors in the Google search engine, the sharp increase in the volume of scientific literature is very clear. Due to the impetus of advances in science, technology, research, and development there is an information overload of

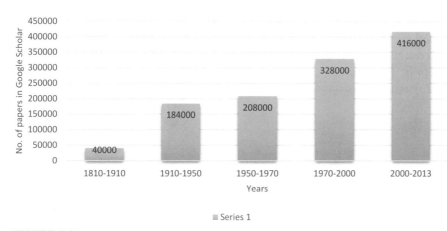

FIGURE 4.1
Sharp rise in scientific publications.

quality, scientific literature. On the one hand, this offers an unprecedented opportunity to update knowledge, and to improve the quality and precision of medicine; on the other, it can create a threat of a selective, biased, or vested interest-driven use of information. Therefore, for the best practice in medicine to be achieved, it is important to critically analyze the scientific information used in decision making.

Earlier, medical practice was mainly based on clinical acumen, and not necessarily on available, and critically analyzed scientific evidence. The cause and effect relationships were known and practiced even before the emergence of EBM. However, there was no uniform system where available information could be scientifically analyzed, and used as a basis for clinical decision making. It is hard to believe that today's gold standard method of randomized controlled trials (RCTs) was not on the horizon until the 1960s. It is surprising and satisfying to witness the fast paradigm shift in the practice of modern medicine, which was powered by the pace of research and technological advances. Many new concepts evolved, new hypothesis replaced old ones, and modern medicine began building a good scientific base. This was probably the beginning of the transformation of EBM from mere experience-based medicine. Interestingly, both terms have the acronym EBM; in this chapter we will use EBM to denote evidence-based medicine.

Medicine is ever evolving, both in its scope and methodology of application. For instance, in the pre-EBM era, the diagnosis of malaria was based on the symptomatology such as fever with rigors, body pain, and profuse sweating on remission of periodic fever. In addition, an experienced physician's acumen, which allowed the clinician to feel the soft, enlarged tip of the spleen, was put to use in distinguishing malaria from other fevers. This diagnostic method was as valid as EBM practice. Later the actual identity of the malarial parasite in a blood film was used as confirmatory criteria. So in a way, the concept of EBM was not new. However, the new EBM has brought robust analysis of available scientific information through systematic reviews, meta-analysis, protocols, and systematic processes.

CAUSE AND EFFECT RELATIONSHIPS

Each disease has a specific natural history, or a predictable path. Each disease is different, and no two patients are the same. Many diseases are self-limiting. Some infectious diseases, especially viral infections, are self-limiting. Many chronic diseases follow a certain pattern of regression. Patients with many skin diseases, or asthma, can improve by avoiding allergens, and just by adopting suitable changes in lifestyle. Patients with rheumatoid arthritis can show temporary phases of regression. Hepatitis C virus can progress slowly for many years, but can be aggravated due to changes in diet, lifestyle, or coinfection [2]. In many diseases such as hepatitis, an understanding of host and pathogen genotypes and patient stratification is necessary when looking for evidence [3].

The quality and reliability of evidence depends on several factors. Systematic reviews and meta-analyses are valuable tools in evidence hierarchy. Today, scientists and professionals have to deal with information overload. In such a situation, systematic reviews come in handy for analysis, and in empowering rational decision making. Systematic reviews indicate if scientific findings are reliable and consistent, and if the outcomes can be generalized or if they differ according to subsets. In the decision-making process, meta-analyses add further value by identifying bias, increasing the power and precision of analysis, identifying possible risks, and improving the reliability and accuracy of conclusions [4]. Therefore, in evidence gathering, appropriate approaches and methods should be used, in accordance with the type of disease, and the nature of the ill patient. Against this background, the validity of clinical observations, case reports, and case studies might have limited value, because data of carefully monitored control groups is not available.

Renowned British epidemiologist and statistician, Sir Austin Bradford Hill, in a president's address of the Royal Society of Medicine meeting on January 14, 1965, presented a detailed argument about the cause and effect hypothesis. He dwelt upon relationships between the observed association and a verdict of causation. He elaborated nine key aspects in relation to association and the most likely interpretation of its causation. These include strength, consistency, specificity, temporal relationship, biological gradient, plausibility, coherence, experimental evidence, and analogy. Thus, experimental evidence is just one of the nine key factors.

No formal statistical tests of significance can answer questions regarding association and causation. Such tests can indicate the chance of specific effects and its likely magnitude but cannot contribute to the *proof* of hypothesis. Tests of significance are valuable tools, but in many situations they are unnecessary because the difference might be clear and obvious, it is negligible, or it may not have any practical importance. When the experimental results need to be compared with standards, placebos, or other variables, statistical hypothesis tests are of great value. Statistics is the study of the collection, organization, analysis, interpretation, and presentation of data. It also deals with methods of data collection, sampling size, and confounding factors in the design of surveys and experiments. It offers probability, confidence limits, and levels of significance. By qualifying the result as statistically significant, statistical analysis rules out the possibility of any unlikely result as having occurred by chance. Statistical analysis is made according to a predetermined threshold of probability, and a certain significance level. Many such tests are used to prove or disprove the null hypothesis in reaching a logical and rational conclusion. The importance of statistical methods in the analysis of scientific data is certainly important. The p value and other tests of statistical significance, like t and x^2 tests, are extremely useful in scientific research. However, the nature of statistical hypotheses testing in clinical research has been questioned. A recent article published in *Nature* examines how p values cannot be always considered the gold standard of reliable measures of statistical analysis. A decision-making frame has been proposed for addressing this critique, and also for transcending ideological debates on probability [5].

We must remember that all observational or experimental scientific work is incomplete. It is always a work in progress. It can change with the advancement of knowledge. However, we cannot ignore the value of existing knowledge in today's decision making. Thus, clinical judgment is a complex process, and has a lot of overlap with subjective and objective variables. The experience, knowledge, and acumen of the clinician drive the clinical judgment. There exists the possibility of bias, and the influence of other confounding factors. Therefore, there exists a need to bring more rationality, objectivity, and scientific support to strengthen clinical judgment.

The foundation of today's EBM was laid in 1967 by the American physician and mathematician, Dr Alvan R. Feinstein, in his groundbreaking work, *Clinical Judgment*. Dr Archie Cochrane drew the attention of the scientific and medical communities to the collective ignorance of the effects of health care in his landmark publication of 1972, *Effectiveness and Efficiency: Random Reflections on Health* [6]. Dr Cochrane was a strong proponent of RCTs, and noted the need for a critical summary of RCTs. This became the foundation for systematic review, and meta-analysis. In honor of his contributions, The Cochrane Foundation was established.

Today, The Cochrane Collaboration is an international network of more than 2,800,000 dedicated people from over 100 countries. The Cochrane Foundation helps clinicians, physician–scientists, health care providers, policy makers, and patients to make well-informed decisions using state-of-the-art Cochrane reviews. Today, over 5000 such reviews have been published online and are available in the Cochrane Library. Such efforts played a major role in empowering physicians, clinicians, and researchers in informed decision making [7].

In 1985, The Cochrane Collaboration published the first classified bibliography of 3500 reports of controlled trials in perinatal medicine published between 1940 and 1984. In 1985, meta-analysis started gaining increasing acceptance as a method of summarizing the results of a number of randomized trials affecting the setting of treatment policy [8]. This led to a very powerful research and analysis tool in the form of *systematic reviews*, which were aggressively promoted, supported, and practiced by the Cochrane Collaboration. Even today, systematic reviews and meta-analyses are considered to be reliable and valuable evidence in guiding and supporting clinical practice.

ACCESSING EVIDENCE

Significant contributions from many physician–scientists reinforced the importance of scientific evidence in clinical diagnosis and therapeutics. As a result, during the late 1980s and the early 1990s, McMaster University in Hamilton, Canada, formed a working group on EBM. This group proposed a new approach, mechanism, and process to bring more rational and analytical evidence for a research-based and research-backed practice of medicine [9]. The EBM process is described as an ability to assess the validity and importance of evidence before applying it to day-to-day clinical problems [10]. The scope of EBM was later expanded to include many professions allied to health and social care. Recently, *The Journal of The American Medical Association* (*JAMA*), together with McGraw-Hill Global Education, launched the initiative, JAMA Evidence. This online education portal provides access to the latest information, books, tools, and other resources for the study of EBM in the context of improving health care (http://jamaevidence.com).

EBM is a complex process; in general, there are five steps.

1. Critical questioning, study design, and levels of evidence
2. Systematic retrieval of the best evidence available
3. Critical appraisal of evidence for internal validity
4. Evidence-informed decisions, and practice
5. Continuous evaluation of practices

Assessment of internal validity involves studying systematic errors of selection bias, information bias, and confounding factors; quantitative aspects of diagnosis and treatment; the effect size, and precision; clinical importance of results; and external validity or generalizability in clinical practice. Evidence generation is not the ultimate goal of EBM. Evidence is a tool to upgrade practice and improve health indicators of society. Evidence can never remain static. It changes according to available data, practices, patients' preferences, and circumstances. Like pharmacovigilance, evidence is also a continuous process of studying ongoing practices, and modifying them based on unbiased decisions. Thus, reducing bias, improving quality of research, disseminating evidence, and facilitating better decision making are processes included in EBM.

In previous times, medical practice was mainly based on observations from clinical experience, the value of diagnostic tests, and the efficacy of treatment. It also involved the study and understanding of the basic mechanisms of disease and pathophysiologic principles as a guide for clinical practice. Later, critically analyzed scientific information from preclinical and clinical studies—especially RCTs—received increased importance as body of evidence. Still, it was believed that a combination of rigorous medical training and experiential clinical acumen would be sufficient to evaluate new tests and treatments. It was also believed that content expertise and clinical experience are sufficient for good clinical practice. Clinicians used to have a number of options for addressing clinical situations. They could use their own experience, understanding of pathophysiology, clinical acumen, and judgment, or refer to textbooks and consult peer experts to decide the best course of clinical diagnosis and treatment. However, many scientists were not satisfied with this system; it was based on personal experiences, expertise, judgments, and opinions of physicians, and so it was prone to biases and errors. Scientists wanted to raise the bar of the evidence base so that it could be used in a systematic fashion, for more accurate and reliable prognosis, diagnosis, prevention, and treatment as part of clinical decision making.

QUALITY AND LEVELS OF EVIDENCE

Systematic reviews of medical literature, meta-analysis, risk–benefit analysis, and RCTs help health care professionals to arrive at conscientious, explicit, and judicious use of current best evidence. The quality of clinical trials and reporting standards are the highest forms of evidence. There are many other guidelines for scientific documentation and reporting to improve the conduct of clinical research. The EQUATOR (Enhancing the Quality and Transparency of Health Research) Network is an important international initiative by methodologists, statisticians, and editors. The network has developed an online resource center (www.equator-network.org) for authors, editors, guideline developers, and teachers for improving reliability and quality of published health research literature.

EBM categorizes different types of clinical evidence, and grades them according to the strength and quality of the evidence. The strongest evidence for therapeutic interventions is provided by the systematic review of randomized, triple-blind, placebo-controlled trials. Patient testimonials, case reports, and even expert opinions are not considered as strong evidence, because of the placebo effect and inherent biases. Most agencies, including the United Kingdom's National Health Service (NHS), and the United States Preventive Services Task Force categorize evidence in four grades or levels. These levels are weighted; properly designed, RCTs have highest value. This is followed sequentially by nonrandomized controlled trials; well-designed cohort, case-control, multicentric studies; and multiple time series. The lowest weight is given to the opinions of authorities, clinical experience, descriptive studies, laboratory studies, and reports of expert committees. Depending on the strength of the evidence, and the balance between benefits and risks, evidence is categorized as good or fair. Any kind of conflicting evidence, where the risk-to-benefit balance cannot be assessed, is considered as insufficient, or poor-quality evidence. In 2000, an informal collaboration of people with an interest in reviewing health care grading systems formed a working group. This initiative is now known as the Grading of Recommendations Assessment, Development and Evaluation (GRADE) working group [11] (www.gradeworkinggroup.org). There are four general levels of quality of evidence according to GRADE guidelines: high, moderate, low, and very low quality of evidence.

Indeed, it is very difficult to reach correct judgments about the quality and reliability of evidence. Making any recommendations based on the evidence is not easy, because such a recommendation depends on several factors, many choices, and situation-specific outcome expectations. For instance, several categories of drugs like β-blockers, calcium channel blockers, diuretics, angiotensin-converting enzyme inhibitors, and angiotensin receptor blockers are available to treat hypertension. The right choice in specific situations will depend on the nature of the supporting evidence. While systematic reviews provide essential evidence, they may not have sufficient information for well-informed decisions. Different reviewers might have different conclusions, and different experts might have different interpretations. The GRADE working group has developed a software application to assess the overall quality of evidence in the facilitation of systematic decision making.

The EBM process, supported by initiatives like GRADE, has provided systematic, critical, consolidated, authoritative, unbiased, statistically significant, and reproducible scientific evidence. This helps to get distilled knowledge that can be uniformly applied in clinical practice. However, experts have also warned that *best available evidence* should not be regarded as a mere collection of data, which might be *suitable evidence*. The authoritative aura attached to the collection of information may lead to abuses or rigidity in clinical practice [12].

EBM IN PRACTICE

The practice of medicine is more than merely following theoretical guidelines arising out of EBM. First, the physician has to keep in mind the basic principle of medicine: *do no harm*. Second, he has to observe the nature of the patient and keep track of the patient's response to treatment. Based on these minute, day-to-day observations, subjectively derived from the patients' histories, and objectively seen from clinical examination, the physician has to do midterm corrections during the course of the illness. That is why it is important to realize that while EBM provides a general framework to guide the modern physician's clinical practice, there is more to it—the practice of medicine is an art and science. Our current approach to EBM does not sufficiently take onboard this reality. As a result, if one critically looks at actual clinical practice, it would be quite daring to say that it is evidence based. Today, clinicians follow cause and effect relationships in terms of symptoms of a disease or syndrome. However, in a mechanical process, they tend to ignore the patient who is suffering. For example, for a patient with suspected metabolic syndrome, the prescription invariably contains a combination of drugs for hypertension, hypercholesterolemia, obesity, and diabetes. Under the pretext of preventive cardiology, the patient is likely to get drugs like aspirin, β-blockers, statins, and angiotensin-converting enzyme inhibitor. Now, cocktail polypills are available. Although there is no sufficient evidence for such formulations, our current symptom-dominated clinical approaches are leading to irrational practices which are masquerading as EBM. Especially in developing countries like India, clinical practice is far away from EBM principles. It is no secret that the majority of clinicians receive continuing medical education from drug companies. Obviously, the prescriptions will be influenced by various business tactics. In a few countries, including India, a practice where a certain percentage per patient, and per diagnostic test is paid for referral (also known as kickbacks practice), has been condemned by the judicial system. Thus, while EBM is a very valuable contribution that has changed medical practice for good, in reality the picture of its actual use varies greatly across different landscapes.

T&CM AND THE COCHRANE APPROACH

Cochrane systematic reviews and meta-analysis have substantially contributed to raising the bar of EBM practice. Cochrane reviews attempt to address the specific research question with the help of prespecified eligibility criteria which must be met in order to be included in the analysis. These systematic reviews evaluate the risk of bias through rigorous and methodological assessment. Systematic reviews also attempt quantitative assessments of the heterogeneity of studies. Meta-analyses pool data of related studies and present findings systematically. The Cochrane review data and analysis uses transparent, rigorous methods. The reproducibility of results is ensured through the freeware, Review Manager. This allows anyone to access the review data, and check the validity of methods and the quality of statistics. The Cochrane Reviews' objective is to help policy makers, health professionals, biomedical scientists, and clinicians. Each review also gives a summary in simple language for lay people [13].

In the analysis of the evidence base for traditional and complementary medicine (T&CM) there are limitations as well. The Cochrane reviews are mainly based on available evidence, and data from clinical trials. The Cochrane methods might not have enough emphasis on secondary data analysis. The Cochrane methodology is based on available literature and is more relevant to clinical trials at a specific point of time. Present focus on RCTs as the main measure to provide certain kinds of evidence is myopic. Sometimes observational studies can uncover conclusive cause and effect relationships about an intervention. If there are no RCTs, a conclusive statement about evidence is not possible. This point is well discussed by Dr Gordon Smith, a researcher from Cambridge, and his colleagues in a witty "systematic review." This article asks evidence whether parachutes are effective in preventing trauma or deaths related to gravitational challenge (Figure 4.2). As there are no "clinical trials" published on this problem, they conclude that "individuals who insist that all interventions need to be validated by a randomized controlled trial need to come down to earth with a bump" [14]. Although the example of parachutes and gravity is humorous, the message of this paper is serious. We should not restrict the search for evidence only to selected scientific methods.

The primary emphasis of Cochrane on "high internal validity" demands stringent selection criteria for systematic review; this might undermine the generalizability of the findings, since the selected studies focus on a very narrow range of patients. In this case, *external validity* (ability to generalize findings) goes down, especially when study protocols are different, and variations in clinical practices exist due to different schools of thoughts.

Double-blind RCTs are viewed as the gold standard in modern medicine. However, RCTs are impractical for many therapies, such as surgeries and complex lifestyle changes. They encourage a one-size-fits-all approach to medical treatment that fails to address the huge diversity among individual patients, in terms of their physical and emotional symptoms, social and cultural upbringing, and other factors. An alternate model has been proposed by Stanford-based research scholar, Dr Somik Raha. This model, Rishi principles, is based on Ayurveda, and

Parachutes reduce the risk of injury after gravitational challenge, but their effectiveness has not been proved with randomised controlled trials

FIGURE 4.2
Evidence for parachute. *Reproduced with permission from Ref. [14].*

epistemology, and includes inductive learning, whole systems thinking, and individually optimized therapy. According to this approach, individually optimized therapy can be interpreted using the lens of decision analysis [15]. Dr Raha suggests integrating the use of Ayurveda tactic known as *Yukti* with observational research methods and Bayesian logic for strengthening clinical practice. He proposes mathematical approach for improving clinical decisions by a process of continuous learning and updating practices. Clinicians can formulate a theory and assign expected fraction of desired benefits as prior probability, which can be assessed with the help of well-documented outcome measures. This approach is important for prospective observational studies on T&CM.

Cochrane reviews appear to have limitations in the assessment of T&CM, especially when the treatments are individualized according to the *Prakriti* and geoclimatic seasons. For example, use of *Bhallataka* is indicated mainly in cold seasons, and *Amalaki* in hot seasons. There may be diverse practices, for example, the Kerala tradition of Ayurveda uses medicated oils and decoctions, while the North Indian practice consists of metals and minerals. The current Cochrane methodology is not able to understand these nuances, which are necessary to harmonize these variations.

There are ongoing efforts to develop protocols for T&CM therapies. Cochrane CAM Field is an activity of the Center for Integrative Medicine at the University of Maryland. The center promotes training and development of systematic reviews on T&CM therapies [16]. Protocols based on an Ayurvedic approach to treatment using Cochrane framework have been proposed. For example, Dr Narahari from the Institute of Applied Dermatology published a protocol to study evidence of Ayurvedic interventions for vitiligo. His protocol uses the Cochrane structure, and the Ayurvedic approach of personalized, whole system management [17]. However, a dearth of clinical trials which have proper study designs (to ensure quality data, and a low risk of bias) remains a hurdle. These efforts suggest that it will be beneficial to develop connections between Cochrane methodology and holistic T&CM approaches.

The EBM approach and holistic principles of T&CM practices need to be integrated so that clinical practice moves away from a straight-jacketed, mechanical process, which is aimed mainly at managing symptoms. The traditional practice may bring the patient back to the center of clinical practice. The Cochrane Collaboration is now moving in the direction of a holistic approach, and is exploring *mixed method research*, where learning from qualitative studies may be considered as evidence. This will change the present mind-set from a hierarchal model with RCTs at the top of the evidence pyramid to a circular model. This has provided new opportunities for development of an epistemologically relevant research methodology for T&CM.

The value of evidence for correct decision making is crucial for many academic disciplines, including scientific interpretation and law and judiciary systems. Evidence is extremely crucial to ensure safe and effective medicine. However, evidence should not be restricted to or monopolized as modern medicine. EBM is a systematic and rigorous process, which can be adopted by any system of medicine. However, the predominance of modern medicine in EBM has created an equivalency between modern medicine and EBM. There have been a few attempts by the complementary and alternative systems to enter the domain of EBM, and also possibly to ride the wave of its popularity. Osteopathy, acupuncture, traditional Chinese medicine, and homeopathy underwent huge turmoil in such attempts. While osteopathy made some headway, most of the others were not successful in justifying an EBM tag. A representative example of homeopathy provides a good case study of the T&CM sector.

THE CASE OF HOMEOPATHY

Can homeopathy be considered EBM? Before we address this question, a quick primer on homeopathy, and an understanding of a few terms like placebo and hormesis, will be useful.

In 1796, a German physician, Samuel Hahnemann developed a new branch of medicine now known as homeopathy. Homeopathy is based on a principle proposed by Hahnemann that "like cures like." According to Hahnemann, a substance that causes the symptoms of a disease in healthy people will cure similar symptoms in sick people. Part of homeopathy is the "law of minimum dose": the lower the dose of the medication, the higher the effectiveness. Homeopathic medicines are derived from plants, minerals, or animal sources. Many potentially poisonous substances like nux vomica, arsenic, poison ivy, belladonna, and stinging nettle are used in extremely diluted forms. Homeopathic remedies are mostly prepared as lactose globules, and also as ointments, gels, drops, creams, and tablets. Homeopathic treatments are individualized to suit specific needs of a particular person. Many homeopathic remedies are so diluted that no molecule of the original substance is likely to remain in the diluted product.

The laws of chemistry state that there is a limit to dilution that can be made without losing the original substance altogether. This limit, which is related to Avogadro's number, is roughly equal to homeopathic potencies of 12C or 24X (1 part in 10^{24}). However, Hahnemann advocated 30C dilutions for most purposes, which goes to dilution by a factor of 10^{60}. Chemically, this means that there remains not a single molecule of original substance. Therefore, critics of homeopathy call these medicines placebo. In their landmark study published in *The Lancet* which compared findings of 110 homeopathy verses 110 matched conventional medicine trials, Shang et al. supported the conclusion that the clinical effect of homeopathy are placebo effects [18].

The homeopathic community has been trying to convince the scientific community that homeopathic remedies are not the equivalent of placebos. Many studies were published in journals to defend the proposition that homeopathic dilutions might have a scientific basis. Homeopathic proponents also cite articles published by reputed scientists in mainstream journals. Most of these articles have indicated that solutions of high dilution can have statistically significant effects on organic processes including the growth of grain, histamine release by leukocytes, enzyme reactions, and degranulation of basophiles; this evidence is disputed since attempts to replicate them have failed.

A representative example of such attempts is a paper by French immunologist Jacques Benveniste, and the accompanying editorial, "When to Believe the Unbelievable," published in *Nature* [19]. Benveniste suggested that water molecules might organize themselves as ghost molecules which mimic drug structures and can evoke biological responses—even in the absence of any original drug molecules. Benveniste's paper certainly gave hope to proponents of homeopathy. However, the experiment was found to be ill designed and irreproducible when investigated by an independent investigative team [20].

In this decade, support for homeopathy was found in an interesting concept in toxicology known as *hormesis*. Hormesis does not follow a typical pharmacological dose–response relationship: low dose, low effect; high dose, high effect. In toxicology, hormesis is a dose–response phenomenon characterized by low-dose stimulation and high-dose inhibition. Over the past two decades, hormesis has become an important concept in biomedical disciplines. This is evident from the fact that the number of papers in the scientific database—a mere 10 to 15 per year in the past decade—rose to over 3200 in 2011 [21]. The resemblance of homeopathy to hormesis is so striking that many have been tempted to relate them to each other, while not making the claim for a toxicological relation. The prevalent thought is that Hugo Schulz, the founder of the hormesis concept, made a mistake in thinking that he had discovered the explanatory principle of homeopathy [22]. Taking a lesson from Benveniste, scientists and proponents of homeopathy are more cautious about reaching unsupported conclusions regarding the relationship between hormesis and homeopathy.

Proponents of homeopathy indicate that the epistemological relationship to the evidence is ignored when studying complex interventions. They believe that homeopathy has important lessons to teach with regard to complexity of individualized treatment, patterns of outcomes, and even the nonlinear, dynamic processes of healing in the patient as a whole system. These critics also maintain that the body of scientific evidence on homeopathy extends far beyond the limitations of the Shang et al. study. They point out that a number of curious, and sometimes clinically beneficial, phenomena can occur during homeopathic treatment. It is argued that homeopathic outcome measures are different from those in conventional medicine; homeopaths report global and hierarchically organized, multidimensional changes in the body. Therefore, it is suggested that future studies involving comparisons of homeopathy with modern medicine need to be designed on a level playing field in order to evaluate outcomes, from both a reductionist and a whole systems point of view [23].

One of the strong critics of homeopathy Dr David Shaw from the Institute of Biomedical Ethics, University of Basel, Switzerland, states "It should by now be very clear that homeopathy is a form of faith healing" [24]. In any case, the debate about the scientific rationale and evidence for homeopathy still continues.

PLACEBO AND NOCEBO EFFECTS

To understand what Shang, Ernst, and Shaw are saying about homeopathy, it is important to understand the term *placebo*. Clinical research and studies assessing cause and effect relationships encounter three main challenges in evidence generation: bias, variability, and confounders. Methodologists have identified about 50 types of biases in clinical research. The placebo effect study can minimize only a few types of biases. Variability is another challenge, and is due to individual variations. Confounding factors can affect or mask the cause and effect relationship. Use of placebo is just one of the approaches to minimize investigators' and patients' biases in interventional clinical studies. The limitations of the placebo effect must be kept in mind before taking any stand about the efficacy of homeopathy, and for that matter, the efficacy of any type of T&CM.

Generally, a placebo is an inactive substance or preparation used as a control to determine the effectiveness of a pharmaceutical preparation. In 1955, Henry Beecher first recognized the clinical importance of the placebo effect [25]. Even though placebos do not act on the disease, they seem to affect how people feel. According to the American Cancer Society, the placebo effect can have an impact on the feelings or symptoms of almost one in three patients. Usually, the placebo effect usually lasts only a short time; sometimes the effect can also cause unpleasant symptoms including headaches, nervousness, nausea, or constipation. The untoward effects of a placebo is called the nocebo effect.

It is accepted that the mind can affect the body. In the quest to attain health or treat diseases, many ancient traditions, including Yoga and Ayurveda, are cognizant of the mind–body connection. Some people think that the placebo produces a cure, but placebos do not cure. For example, the role of placebos in tumor reduction has not been demonstrated, although placebos reduce pain. Still, in some people, placebos clearly can help reduce certain symptoms such as pain, anxiety, and insomnia. In earlier times, placebos were sometimes given by doctors if nothing else was available; they are still used today. A 2008 study found that nearly half of the doctors studied used a placebo. However, use of placebos is not considered ethical, and many governmental health agencies are on record as against using placebos. The United Kingdom Parliamentary Committee on Science and Technology stated: "prescribing placebos…usually relies on some degree of patient deception and prescribing pure placebos is bad medicine. Their effect is unreliable and unpredictable and cannot form the sole basis of any treatment on the NHS." Patients may believe that a larger size placebo pill is more powerful than a small pill, an injection can have a stronger placebo effect than a pill, and expensive medicines are sometimes perceived as having more powerful effects.

A recent systematic review published by the Nordic Cochrane Centre in Copenhagen, Denmark, concludes that, in general, placebo interventions have important clinical effects. However, in certain settings placebo interventions can influence patient-reported outcomes, especially in those involving pain and nausea, although it is difficult to distinguish patient-reported effects of a placebo from biased reporting. The effect on pain varied, even among trials with a low risk of bias, from negligible to clinically important. Variations in the effect of placebo were partly explained by variations in how trials were conducted and how patients were informed [26]. This clearly indicates that the trial design, cultures, doctor–patient relations, and many other factors can influence the findings concerning the placebo response.

Another scientific observation that goes against homeopathy, and in favor of its placebo effect, is its ineffectiveness in animals and cell systems. In a study undertaken by scientists in Maryland, it was demonstrated that the highly diluted homeopathic remedies used by homeopathic practitioners for cancer show no measurable effects on cell growth, or gene expression, in vitro [27]. In April 2012, the Veterinary Clinical Research Database in Homeopathy containing 302 data records, including 146 RCT and 57 nonrandomized control trials, also did not show that homeopathic medical intervention was better than a placebo [28]. This strengthens the homeopathy opponent's view that homeopathy is actually a treatment based on placebos.

The debate on the status of homeopathy as EBM, or placebo has gone through several phases—up and down, like a roller coaster ride. Sometimes homeopathy was riding high in popularity, sometimes it reached a low point. The final blow came from a systematic review based on the Cochrane database, published in 2010 by a professor from the University of Exeter, Dr Edzard Ernst. He concluded that the most reliable evidence fails to demonstrate that homeopathic medicines have effects beyond that of a placebo [29].

The skeptics of homeopathy proposed an interesting warning label that is based on what researcher Dylan Evans suggested in his book, *Placebo: The Belief Effect* [30]. In his book *Trick or Treatment*, Ernst suggested a warning label on every homeopathy product: "This product is a placebo. It will work only if you believe in homeopathy, and only for certain conditions such as pain and depression. Even then, it is not likely to be as powerful as orthodox drugs. You may get fewer side effects from this treatment than from a drug, but you will probably also get less benefits." He also suggested extending the warning to other treatments like acupuncture, chiropractic, and herbal medicines [31].

Placebos work because of several factors related to mind and body interactions. Just the presence of a brain is not enough, as shown in animal studies. For a placebo to have an effect, a thinking mind and cognitive capabilities to create awareness of the desired effect is needed. Modern science's understanding of the relationship between brain and mind is still evolving. Candace Pert discovered opiate receptors, where opiumlike drugs and morphine bind to nerve cells in the human brain [32]. This breakthrough led to the discovery of endorphins, which are analgesic, opiatelike chemicals naturally present in the brain, and responsible for feelings of happiness and bliss. We feel that it is possible for a conscious mind to actually order the brain to produce substances like endorphins.

It is known that even a placebo, or nocebo, effect can exert strong therapeutic responses, especially in cases of pain, inflammation, and psychosomatic disorders [33]. Therefore, while critics continue to believe that the putative effects of homeopathy primarily are a placebo response, many who benefit continue to put their trust in it. As stated by eminent cardiologist Dr B.M Hegde "the placebo doctor can provoke the human immune system much more powerfully than all medicines put together." Many may feel that this is an exaggeration; it might indeed be so. However, we cannot ignore the fact that our present knowledge about mind–body relationships and therefore about the placebo is inadequate to reach to any definitive conclusions.

EVIDENCE FOR YOGA

Research on Yoga provides an interesting slant on mind and body background. Yoga and Ayurveda propose lifestyle interventions, postural changes, meditation, and few therapeutic procedures. In recent years, Yoga therapies have been examined in rigorous research, and it is now more generally accepted that there is scientific backing to claims that Yoga can be used to achieve better health [34]. Research institutions in the United States, Europe, Canada, Australia, and China are actively involved in researching and teaching the practice of Yoga. The Cochrane Library cites 21 systematic reviews, and the PubMed cites 172 reviews on Yoga—much more than those on T&CM. While several institutes which teach Yoga have thrived, the contribution to biomedical research in Yoga from India is marginal. For instance, out of 21 Cochrane reviews, only one has come from India (from Apollo Hospitals); four each are from the United Kingdom, United States, and Canada.

Yoga has gained scientific credence, and has moved rapidly toward EBM by going through the rigors of scientific research. Yoga has the subject of experiments, and tests, and has been explored through sophisticated modern tools, including functional magnetic resonance imaging at the Universities of California, at San Diego, Los Angeles, and San Francisco; Harvard University; Stanford University; the University of Texas; Cleveland Clinic; Yale University; and the Mayo Clinic, among many others. In the process, Yoga has enhanced our current knowledge of neurophysiology and has spurred new research in areas such as meditation, cognition, and consciousness [35]. Ayurvedic medicine can take lessons from its Yoga component in evolving strategies and creating a road map to move it toward an integrative evidence base.

EVIDENCE FOR AYURVEDA

Every system of medicine that has endured the test of time has demonstrated evidence of efficacy and safety. Evidence of benefits can come from drug-based interventions, or because of a holistic approach involving diet and lifestyle. Ayurveda might not comport with the Cochrane EBM approach, but that does not mean that there is no evidence supporting it as a valid approach to health. Whether or not Ayurveda can be considered EBM is not the right question. The real question is whether the Ayurvedic fraternity has an open mind and the vibrancy required for the system to grow and evolve. Among other factors, this will involve the use of modern scientific methodology. There must also be a conducive environment for encouragement, nurturance, and respect for scholarship that facilitates the emergence of true scientists who can take any science to unending heights. Thousands of years of stagnancy in this sector raises several challenges to be met by this ancient science of life.

In India, Ayurveda is part of the culture. It is present in every family as home remedies. When it comes to doctors and patients, the picture is more complex. The value of Ayurveda as perceived by patients is very high, and so are the expectations. The majority of patient's have a favorable perception of Ayurveda [36]. However, the costs of Ayurvedic treatments are increasing because of commercialization in wellness industry and medical tourism [37]. A lack of insurance coverage is adding to the costs to be paid out of pocket and the treatments are becoming unaffordable [38]. There are hardly any systematic studies in India on patients' perceptions, behaviors, values, expectations, or acceptance, or on the economics and sociology of Ayurveda.

In the case of Ayurveda, the evidence might be drawn from two main sources. The first source of evidence is based on historical, classical, and contemporary clinical practice. The documentation of practice to support various claims is crucial; mere reference to classical texts is not sufficient evidence for practice. The second source of evidence can be based on scientific research to support various theories, medicines, and procedures used in Ayurvedic medicine. A critical situation analysis of the current status of clinical practice and scientific research on Ayurvedic medicine might be necessary at this stage.

Clinical Practice

Arguably, the clinical practice of classical Ayurveda is rare. Ayurvedic practitioners adopt allopathic practices in order to find more acceptance in urban settings [39]. Although a huge knowledge and wisdom resource is available in the form of classic Ayurvedic books, systematic data on actual use and evidence of reproducible outcomes is not available in the public domain. Standard treatment protocols for practitioners are not available. Systematic documentation and reliable data on pharmacoepidemiology and pharmacovigilance for clinical practice, safety, and adverse drug reactions is not available as open access, although a modest beginning has been made [40]. The status of professional [41] and continuing education [42], as well as attitudes of practitioners toward safety [43], are concerns.

In India's current regulatory scheme, no scientific or clinical data is required for the manufacture and sale of classical Ayurvedic medicines. Technically sound pharmacopoeia, good manufacturing practices, quality control, and pharmaceutical technologies for Ayurvedic medicine are still evolving [44,45]. Issues related to appropriate research methodologies, or treatment protocols for Ayurveda also have not been properly addressed. Many critics are demanding better coordination between stakeholders, a continuous dialog with the scientific community [46], a total overhaul of the curriculum and pedagogy, and cross-talks between different modern medicines and Ayurveda and Yoga [41]. A recent report on the status of Indian medicine and folk healing indicates a need to strengthen research on and the use of Ayurveda, Yoga, Unani, Siddha, and homeopathy systems in the national health care [47,48]. The need for innovation is stressed by thought leaders in this sector [49]. In short, the evidence base to support good clinical practice, guidelines, and documentation in Ayurvedic medicine remain scant, and grossly inadequate.

Scientific Evidence

Controlled clinical trials provide the highest level of evidence. Ayurveda lags far behind in scientific evidence, both in quantity and quality of RCTs, and in systematic reviews. For instance, of 7864 systematic reviews in The Cochrane Library, Ayurveda has just 1, while homeopathy and T&CM have 5 and 14, respectively. Substantial grants have been allocated to ambitious national projects involving reputed laboratories. However, the design, methodology, and quality of clinical trials on Ayurvedic medicines lack the expected rigor [50]. Of course, this does not mean that the RCT model is suitable to clinical research in Ayurveda. RCTs have already been subjected to criticism [51].

One of the basic principles behind RCTs is *clinical equipoise*, which agrees genuine uncertainty about treatment effects. This principle provides ethical basis to assign patients for comparisons in treatment arms—one with best known option as positive control and the other with a study drug with unknown effects. Obviously, in the case of T&CM drugs, where documentation and clinical experience is known in some form, the principle of clinical equipoise is not met. Therefore, critics argue that the RCT model might not be suitable to evaluate clinical benefits of T&CM drugs.

Therefore, while discussing about evidence for T&CM, the value of observational studies cannot be ignored. Certainly, there is a need to develop appropriate research methodology for complex, whole-system, whole-person-centered clinical trials as an alternative to RCTs. Already, scientists are advocating robust clinical study designs based on a personalized approach and metabolomics with only one patient [52]. Thus, the nonsuitability of RCTs to Ayurvedic research should not be used as an excuse for avoiding rigorous scientific research and clinical documentation.

Ayurvedic medicine continues to remain subcritical in scientific research publications, which is an important indicator of external evidence [53]. The current scientific evidence in support of Ayurvedic medicine remains extremely poor. The House of Lords and the European Union have put several restrictions on Ayurvedic medicines [54]. Many articles lamenting the quality of Ayurvedic medicines, particularly citing the presence of heavy metals and other safety-compromising substances, have been published [55,56].

Noteworthy attempts related to research and practice of Ayurveda include a national program on Ayurvedic biology [57,58], Ayugenomics [59], whole systems clinical research [60–62], good clinical practices guidelines, the establishment of a digital helpline [63], a clinical decision support system AyuSoft, and a systematic reporting standards on lines with Consolidated Standards of Reporting Trials (CONSORT) for Ayurveda [64,65]. Recent efforts to develop robust clinical protocols for comparing the effectiveness of complex Ayurvedic and conventional treatments are laudable [66]. Other notable efforts related to integrative therapy for leishmaniasis have been able to generate sufficient scientific evidence [67]. Admittedly, many of these efforts have not produced any remarkable products, processes, or protocols, and the desired impact on the scientific community is yet to be seen. The need to enhance the collaborative culture between the Ayurvedic and modern, scientific communities rightly has been stressed by many researchers who have experienced benefits of integrative approaches [68].

CONCEPTS OF EVIDENCE IN AYURVEDA

The original knowledge base of Ayurveda is evidence based. Ayurveda is not merely a collection of traditional experiences. Ayurveda expects logic and proof of causality in every therapeutic decision. The chance effect is not accepted in Ayurveda. Ayurveda describes evidence in context of *Pramana*, in four main categories. The term *Pramana* literally means "right perception" and "means of acquiring knowledge" [69]. The first category is known as *Pratyaksha*, or evidence generated from actual and direct observations. This is very similar to experimental evidence. This evidence is based on measurable parameters, which could be based on direct feelings, or understanding through the five senses. Measurements using any instrument are also included in this category. Essentially, what is measured using instruments ultimately has to be read and understood by our senses. In a way, instruments are extensions of our senses. The Western philosophy of scientific positivism only considers this category as evidence.

The second category of evidence is known as *Anumana*. This involves analyzing the observed and measured data, with the help of logic and causality. The third category is *Apta*, which comes from experiential wisdom, and the unbiased opinions of experts. According to Ayurveda, *Apta* are the authorities who are distinguished, enlightened, and absolutely free of bias and ignorance, and who have state-of-the-art knowledge pertaining to the past, present, and future. *Apta* is not necessarily applicable to any one individual, rather, it is a continuum of knowledge traditions. Ayurveda gives maximum evidence weightage to *Apta*.

The fourth category is *Yukti*, which is a tactical analysis. *Yukti* actually considers multiple variables, confounding factors, and systematic reviews of available knowledge. The sum and substance of these categories is used as evidence for any decision making. Obviously, this process may involve subjectivity and qualitative variables. However, any outcome is expected to be judged in the context of causation and association.

Ayurveda emphasizes a critical assessment of cause and effect through reproducible and unbiased evidence. It expects a consistent, reproducible association in every causality analysis. The reliability and validity of evidence is considered as proof of its theories. Any reliable association with proof of causation assessed in multiple trials by various observers is considered as theory.

Ayurveda has evolved through documentation of observations from nature: the results of experiments designed to test assumptions and documented clinical experiences. Ayurveda, in its true spirit, is open to questions and critique. The knowledge of Ayurveda is not restricted to the particular person, or a book. Any demonstrated conclusion of an argument is termed as *Siddhanta*, which requires sound proof through multiple, critical scrutiny before it is widely accepted. The *Siddhanta* term covers the following concepts related to evidence:

- Unbiased enquiry by impartial authorities
- Testable hypothesis
- Multiple and reproducible tests
- Consistent observations established by causality
- Overcoming counterfactual evidence
- Any statement requiring proof

Ayurvedic sage Charaka discusses the definition of management in the context of logic, and proof of efficacy. He terms any management "effects following certain logic and rules," rather than chance or haphazard phenomenon. He further condemns any practice that does not follow this approach as "quackery," and castigates the doctor who spreads diseases rather than providing health. The expectations in EBM match the qualities of a good doctor detailed in the *Samhita*, and there is no conceptual divide. Sushruta states that evidence endorsed by unbiased authorities, and which is perceived in direct observation, can be practiced as a truth, and treated as a sacred *mantra*.

Yoga describes a new paradigm for evidence in terms of self-realization; in Sanskrit self-realization is *Aatmanubhooti*. In addition to any external evidence, Yoga suggests "self-experiencing" as the highest level of evidence. It suggests eight steps to gradually move toward self-realization. In simple terms, before convincing others, the concerned individual should be convinced about the value of the knowledge; this conviction can be considered as evidence. For example, any doctor should prescribe only such drugs and procedures which he or she will be willing to accept as a safe and appropriate treatment. This requires a high level of professional competency and ethics. Therefore, self-experiencing is an acid test, which should be considered as the highest level of evidence.

Ayurveda has its own EBM. The key challenge is to strengthen its evidence base further to produce *Vaidya* scientists in the professional community who are Ayurvedic physicians (*Vaidya*) as well as well-versed with modern science. This is similar to physician–scientists who simultaneously undertake scientific research and clinical practice. Admittedly, the arguments for and against EBM and Ayurveda are based on Western biomedical standards. This includes systematic drug discovery, development with the help of RCTs, and statistical analysis to show that the prescribed drug prevails over a placebo. However, it is important to recognize that Ayurveda has a holistic approach based on thousands of years of experience. It is an integrated system that considers body, mind, and spirit. Ayurvedic therapeutics is not restricted to the mere use of drugs. While herbal drugs are used, lifestyle modification remains the bedrock of Ayurvedic therapeutics. Lifestyle modification prescribed by Ayurveda is not just diet and exercise, but encompasses the entire process of personality—including mental attitude and personal philosophy. This is how Ayurveda approaches diseases, and attempts to treat the root cause—by introducing change in the environment within a *milieu intérieur*, as described by Claude Bernard. This approach goes beyond eliminating or decreasing symptoms with drugs. In fact, RCTs for Ayurveda could be considered an oxymoron, and experts are questioning validity of these models to holistic systems like Ayurveda [70]. This has to be kept in mind when we talk about EBM for Ayurveda. There are no existing models to follow, and this is a great challenge for holistic systems.

Therefore, the authors believe that the Cochrane concept of EBM in its present form might not be suitable for T&CM systems. We strongly feel that "E" in EBM should not be restricted to scientific Evidence but should also include "Experience" and wisdom of generations. The epidemiological and outcome-based evidence should also be given due weightage. We feel that scientific *evidence*, and traditional *experience* should not be mutually exclusive. The

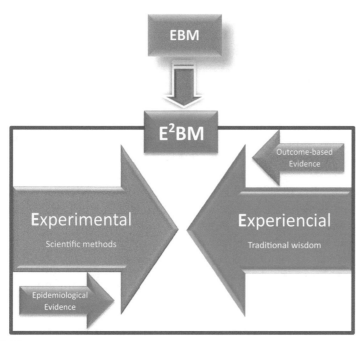

FIGURE 4.3

EBM needs to include experimental and epidemiological dimensions. Experiential wisdom is not enough unless it is backed by outcome-based evidence. Together, they can evolve a new approach known as evidence- and experience-based medicine (E²BM).

new EBM should also be experienced-based medicine. Appropriately, we might call it **E²BM**, wherein experience-based medicine and EBM are fully integrated (Figure 4.3).

THE NEW EBM

The concept of EBM has gone through many transitions ever since Gordon Guyatt, a Canadian professor of clinical epidemiology, and his colleagues coined the term over 20 years ago. *JAMA* has recently released the oral history of EBM [71]. While EBM has many advantages, it also poses some problems. Many experts feel that the evidence-based "quality mark" has been misappropriated, and distorted by vested interests—particularly by the pharmaceutical and medical devices industries [72]. We discussed this subject in Chapter 2 in sections related to medicalization. Industries seem to use financial means to not only set the research agenda but also influence definitions of disease conditions. They also often dictate normal values, and the nature of diagnostic tests. Sponsored clinical trials and using statistics to the commercial interests' advantage is a matter of great concern. The volume of evidence is becoming unmanageable. The inflexible rules, and technology-intensive protocols create health care management which is driven by the interests of partners who have competing interests, rather than by a focus on the good of the patient—which should be the primary concern. In the process, EBM has become more mechanical, protocol driven, and pathology diagnostics oriented. Ironically, in the process, EBM has become very expensive and unaffordable to many—especially to those who are not covered by insurance.

The new EBM should make the care of individual patient's top priority. The best course of action is likely to be different for different patients, who are in different environments, and have different disease conditions [73]. The new EBM should consciously and reflexively refrain from unethical practices and unnecessary, or avoidable, tests and medicines. It should engage with ethical and existential questions. Experts are suggesting that patients should be empowered to make correct and appropriate decisions regarding their own health care. Medical associations, guidelines, governments, and insurers should not unilaterally decide the fate of the patient. This is now possible due to the power of information science [74]. The new EBM should be individualized for the patient. It should be accepted in principle that even the "best evidence" may not apply to certain patients for a variety of reasons [75]. Every patient should be objectively evaluated by the clinician in an unbiased way, on case-by-case basis. Clinicians should use their own clinical acumen and experience and then judge their evaluation against the scientific evidence. Here also, the "first-do-no-harm" principle should be followed.

We feel that it is high time for the EBM movement to be extended to integrative health. A new idea needs to be conceptualized as evidence-based health (EBH) approach. The principles of EBH should be based on personalized and integrative approaches, bridging modern science and traditional knowledge. We must produce an integrative system of evidence-based care, which combines the strength of each system. For instance, Ayurveda possesses the proven benefits of body–mind melding Yoga and meditation, a rational diet, and nutrition and lifestyle doctrines, and it emphasizes restorative or rejuvenative medicine.

The future of medicine has in its grasp the capability to heal the increasingly sick planet. It needs both the Western biomedicine, as well as the Eastern Ayurveda and other Eastern traditional practices. A new system of integrative health is emerging where the barriers between traditions, disciplines, and systems are dissolving. It is hoped that this new integrative system will meet the daunting challenges of today and tomorrow!

Efforts to obtain scientific evidence in the T&CM sector require applying appropriate methods. Current evidence is based on our contemporary understanding and on the availability of scientific tools, both of which are continuously evolving. Therefore, just because we are not able to measure something using current scientific methods, it does not mean that it does not exist. It is generally accepted that "the absence of evidence should not be taken as evidence of absence" [76]. Current statistical practices, and double-blind approaches in clinical trials, where p values are considered symbols of scientific evidence to prove a drug's efficacy, are under debate. We need to develop novel statistical models based on Bayesian logic and systems analytics in order to evolve innovative scientific methodologies, create a new evidence base, and facilitate sensible decision making for Ayurveda.

At this stage, it is important to recall Sir Austin Bradford Hill's 1965 discourse where he listed experiential evidence as one among nine, key factors. We cannot make a final judgment call merely based on experimental evidence. Many decisions can be made based on epidemiologic evidence and outcome-based evidence. Sir Austin advised epidemiologists to avoid the overemphasis of statistical significance testing, because systematic error is often greater than random error. He also stressed the need to consider costs and benefits when making decisions. Sir Austin's presidential address of 1965 is even more relevant today [77].

HIGH EVIDENCE LOW ETHICS

We observe that while EBM aspires to reach a higher level of scientific evidence, practice of medicine has suffered from declining values and ethics. Several reasons for this decline can be identified. First, the fee-for-service culture, referral commission, cut practice (fee splitting/kickback practice), and other bad practices have resulted in the increased commercialization of the medical profession. Second, expensive medical education puts pressure on the practitioners to make an early recovery of these costs. Third, investment-intensive pathology and diagnostic equipment make breakeven difficult. Fourth, a desperate pharmaceutical industry seems to be promoting unethical practices (Figure 4.4).

According to a study by the watchdog group, Public Citizen, the pharmaceutical drug industry has surpassed the defense industry to become the biggest defrauder of the federal government. The Public Citizen report states that the drug industry paid nearly $20 billion in penalties during the past two decades for violations of the False Claims Act [78]. This unhealthy behavior, and the tendency to compromise ethics for profits, appears to be a global phenomenon. The big four companies—GlaxoSmithKline, Pfizer, Eli Lilly, and Schering-Plough—account for more than half of the violations. These leading violators are among the world's largest pharmaceutical companies.

The United States Food and Drug Administration has levied fines of $500 million for scientific fraud and dubious research outcomes on Indian companies like Ranbaxy. The whistle-blower, Dinesh Thakur, who single-handedly

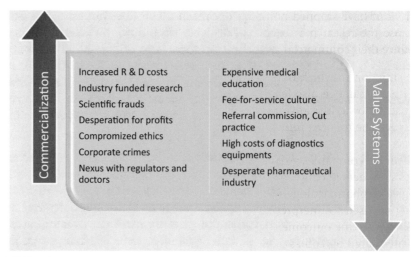

FIGURE 4.4
High in evidence low in ethics.

exposed this scam, was awarded $28 million in recognition of his role in exposing the fraud. A detailed case study of corporate crime in the pharmaceutical industry [79] by John Braithwaite of the University of California, Irvine, gives a very scary glimpse of unethical practices that seem to be deeply rooted in this industry—an industry which directly deals with the health of billions. Sadly, the role of scientists in this process is also being questioned. The analysis of the data in one recent systematic review suggests that the outcomes of industry-sponsored drug and device studies are often favorable to the sponsors' products [80]. Another systematic review concludes that industry-sponsored studies are biased in favor of the sponsors' products [81].

There exist not only biases in favor of sponsors' products but also instances where the scientific data has been fudged, twisted, and interpreted in such a way that a particular drug's actions are highlighted as favorable, while safety issues or side effects are downplayed. Several examples of such attempts are available. For instance, blockbuster drugs like new-generation antipsychotics have seen a fall in favor, because their predicted benefits were found literally to wear off [82].

In this context, a recent report titled Late Lessons from Early Warnings, released by the European Environmental Agency is very relevant [83]. Citing examples, the crux of this report is that, by and large, scientists have worked more for businesses and economics, and less for people and humanity. The best, or rather worst, example of such industry–scientist nexus is from the tobacco industry, which deliberately created confusion, and delayed public awareness of the dangers of tobacco use, and legislation regarding the advertising of tobacco products for decades. Many journals like *British Medical Journal* (*BMJ*), *Thorax*, and *Heart* have stopped publishing research funded by the tobacco industry, because the research is corrupted, and the companies back the research to advance their commercial aims.

The situation in which bottom-line, financial considerations outweigh public health and safety is also feared to be insinuating itself into the pharmaceutical industry. Now critics are extending the reasons and arguments against the way in which the tobacco industry comported itself in obfuscating the dangers of tobacco use, to research funded by the drug industry. The way in which pharmaceutical companies get the results they want has been extensively debated. Critics are deeply concerned about the use and misuse of statistical methodologies in pharmacology research [84]. It is of great concern that statistics is being misused to interpret results to support a desired outcome, rather than to show the true outcome. There seems to be strong evidence that companies obtain desired outcomes through backing research that will only support their interests, or skewing the interpretation of the data to suit their interests; a substantial number of the clinical trials published in major journals like *Annals of Internal Medicine, JAMA, The Lancet,* and *The New England Journal of Medicine* are funded by the pharmaceutical industry [85]. Critics ask the tough question: Are medical journals a marketing arm of pharmaceutical companies? [86]. It is suggested that journals should critique trials, and not just publish them. Experts are suggesting that, in the interest of patient safety, more transparent, and better ways to release results of trials must be implemented.

Today, prescription drugs are the third leading cause of death worldwide. It is a matter of great concern; investigation and introspection are required to understand why we have reached such a scenario. The present models of drug research and regulatory provisions, especially to evaluate safety and toxicity, are in need of serious overhaul. Vested interests which push unsafe drugs are certainly not in the interest of anyone. Earlier incidences from trials of cyclooxygenase-2 inhibitors for arthritis and selective serotonin reuptake inhibitors for depression are good examples. Critics are asking the question: Should journals stop publishing research funded by the drug industry [87]. In short, scientific evidence is important, but it should not compromise medical ethics at the cost of business.

We need a balance of scientific medicine, EBM, and traditional experience-based medicine. Against this background, we hope that the traditional knowledge and ethos can help to resolve the current impasse. This new balance will offer rich experience, knowledge and resources, new ideas, and leads and strategies. Drug discovery should be curiosity- and research-based approach where passionate individuals are committed to serve the humanity by developing affordable effective treatments. Present drug discovery is viewed as a *pipeline* process, which has been outsourced to mechanical productions where evidence is seen as a commodity, which can be manufactured.

REFERENCES

[1] Google Scholar search data for key words "Medicine", "Ayurveda" and "Traditional Chinese Medicine" [accessed on 11th May 2015] from Pune India.

[2] Hajarizadeh B, Grebely J, Dore GJ. Epidemiology and natural history of HCV infection. [Internet] Nat Rev Gastroenterol Hepatol 2013;10(9):553–62.

[3] Liu CJ, Kao JH. Global perspective on the natural history of chronic hepatitis B: role of hepatitis B virus genotypes A to J. Semin Liver Dis 2013;33(2):97–102.

[4] Mulrow CD. Rationale for systematic reviews. BMJ 1994;309(6954):597–9.

[5] Nuzzo R. Scientific method: statistical errors. Nature 2014;506(7487):150–2.

[6] Cochrane AL. Effectiveness and efficiency: random reflections on health. London: Nuffield Provincial Hospitals Trust; 1972.

[7] http://www.cochrane.org.

[8] L'Abbe KA, Detsky AS, O'Rourke K. Meta-analysis in clinical research. Ann Intern Med 1987;107(2):224–33.

[9] Evidence-Based Medicine Working Group. Evidence-based medicine. A new approach to teaching the practice of medicine. JAMA 1992;268(17):2420–5.

[10] Sackett DL, Rosenberg WMC, Gray JAM, Haynes RB, Richardson WS. Evidence based medicine: what it is and what it isn't. BMJ 1996;312(7023):71–2.

[11] Balshem H, Helfand M, Schünemann HJ, Oxman AD, Kunz R, Brozek J, et al. GRADE guidelines: 3. Rating the quality of evidence. J Clin Epidemiol 2011;64(4):401–6.

[12] Feinstein AR, Horwitz RI. Problems in the "evidence" of "evidence-based medicine.". Am J Med 1997;103(6):529–35.

[13] www.handbook.cochrane.org.

[14] Smith GCS, Pell JP. Parachute use to prevent death and major trauma related to gravitational challenge: systematic review of randomised controlled trials. BMJ 2003;327(7429):1459–61.

[15] Raha S. Foundational principles of classical Ayurveda research. J Ayurveda Integr Med 2013;4(4):198–205.

[16] http://www.compmed.umm.edu/cochrane_about.asp.

[17] Narahari SR, Aggithaya MG, Suraj KR. A protocol for systematic reviews of Ayurveda treatments. Int J Ayurveda Res 2010;1(4):254–67.

[18] Shang A, Huwiler-Müntener K, Nartey L, Jüni P, Dörig S, Sterne JA, et al. Are the clinical effects of homoeopathy placebo effects? Comparative study of placebo-controlled trials of homoeopathy and allopathy. Lancet 2005;366(9487):726–32.

[19] When to believe the unbelievable. Nature 1988;333(6176):787.

[20] Maddox J, Randi J, Stewart WW. "High-dilution" experiments a delusion. Nature 1988;334(6180):287–91.

[21] Calabrese EJ. Hormesis and homeopathy: introduction. Hum Exp Toxicol 2010;29(7):527–9.

[22] Calabrese EJ, Iavicoli I, Calabrese V. Hormesis: Its impact on medicine and health. Hum Exp Toxicol 2013;32(2):120–52.

[23] Bell IR. All evidence is equal, but some evidence is more equal than others: can logic prevail over emotion in the homeopathy debate? J Altern Complement Med 2005;11(5):763–9.

[24] Shaw DM. Homeopathy is faith healing without religion. Focus Altern Complement Ther 2014;19(1):27–9.

[25] Beecher HK. The powerful placebo. J Am Med Assoc 1955;159(17):1602–6.

[26] Hróbjartsson A, Gøtzsche PC. Placebo interventions for all clinical conditions. Cochrane Database Syst Rev 2010;(1):CD003974. http://dx.doi.org/10.1002/14651858.CD003974.pub3.

[27] Thangapazham RL, Gaddipati JP, Rajeshkumar NV, Sharma A, Singh AK, Ives JA, et al. Homeopathic medicines do not alter growth and gene expression in prostate and breast cancer cells in vitro. Integr Cancer Ther 2006;5(4):356–61.

[28] Clausen J, Albrecht H, Mathie RT. Veterinary clinical research database for homeopathy: placebo-controlled trials. Complement Ther Med 2013;21(2):115–20.

[29] Ernst E. Homeopathy: what does the "best" evidence tell us? Med J Aust 2010;192(8):458–60.

[30] Evans D. Placebo: the belief effect. London: HarperCollins; 2003.

[31] Singh SEE. Trick or treatment? Alternative medicine on trial. UK: Bentham Press; 2008.

[32] Pert CB. Molecules of emotion: why you feel the way you feel. Scribner Book Company; 1997.

[33] Ciaramella A, Paroli M, Poli P. An emerging dimension in psychosomatic research: the nocebo phenomenon in the management of chronic pain. ISRN Neurosci 2013:574526. http://dx.doi.org/10.1155/2013/574526.eCollection 2013.

[34] McCall MC, Ward A, Roberts NW, Heneghan C. Overview of systematic reviews: yoga as a therapeutic intervention for adults with acute and chronic health conditions. Evid Based Complement Altern Med 2013;2013:945895.

[35] Froeliger BE, Garland EL, Modlin LA, McClernon FJ. Neurocognitive correlates of the effects of yoga meditation practice on emotion and cognition: a pilot study. Front Integr Neurosci 2012;6:48. http://dx.doi.org/10.3389/fnint.2012.00048.

[36] Rastogi S. Effectiveness, safety, and standard of service delivery: a patient-based survey at a pancha karma therapy unit in a secondary care Ayurvedic hospital. J Ayurveda Integr Med 2011;2(4):197–204.

[37] Shetty P. Medical tourism booms in India, but at what cost? Lancet 2010;376(9742):671–2.

[38] Patwardhan B. Ayurveda: finding place in own house. J Ayurveda Integr Med 2012;3(3): 109–10.

[39] Nisula T. In the presence of biomedicine: Ayurveda, medical integration and health seeking in Mysore, South India. Anthropol Med 2006;13(3):207–24.

[40] Chaudhary A, Singh N, Kumar N. Pharmacovigilance: boon for the safety and efficacy of Ayurvedic formulations. J Ayurveda Integr Med 2010;1(4):251–6.

[41] Patwardhan K. Medical education in India: time to encourage cross-talk between different streams. J Ayurveda Integr Med 2013;4(1):52–5.

[42] Patwardhan B, Joglekar V, Pathak N, Vaidya A. Vaidya-scientists: catalysing Ayurveda renaissance. Curr Sci 2011;100(4):476–83.

[43] Milden SP, Stokols D. Physicians' attitudes and practices regarding complementary and alternative medicine. Behav Med 2004;30(2):73–82.

[44] Anantha ND. Approaches to pre-formulation R & D for phytopharmaceuticals emanating from herb based traditional Ayurvedic processes. J Ayurveda Integr Med 2013;4(1):4–8.

[45] Bhutani KK. Natural products: bench to bedside, an Indian perspective. Planta Med 2012;78(5). OP20.

[46] Valiathan MS. Towards Ayurvedic biology, decadal vision document. Bangalore: Indian Academy of Sciences, 2006.

[47] Chandra S. Status of indian medicine and folk healing. Part I. New Delhi: Department of AYUSH, Government of India; 2012.

[48] Chandra S. Status of indian medicine and folk healing. Part II. New Delhi: Department of AYUSH, Government of India; 2013.

[49] Singh RH. Perspectives in innovation in the AYUSH sector. J Ayurveda Integr Med 2011;2(2): 52–4.

[50] Beryl P, Vach W. Methodological considerations in evidence-based indian systems of medicine – a systematic review of controlled trials of Ayurveda and Siddha. J Altern Complement Med 2014;20(5):A137.

[51] Sanson-Fisher RW, Bonevski B, Green LW, D'Este C. Limitations of the randomized controlled trial in evaluating population-based health interventions. Am J Prev Med 2007;33(2):155–61.

[52] Van der Greef J, Hankemeier T, McBurney RN. Metabolomics-based systems biology and personalized medicine: moving towards n = 1 clinical trials? Pharmacogenomics 2006;7(7): 1087–94.

[53] Patwardhan B, Vaidya ABD. Ayurveda: scientific research and publications. Curr Sci 2009; 97(8):1117–21.

[54] Patwardhan B. European Union ban on ayurvedic medicines. J Ayurveda Integr Med 2011;2(2):47–8.

[55] Saper RB, Phillips RS, Sehgal A, Khouri N, Davis RB, Paquin J, et al. Lead, mercury, and arsenic in US- and Indian-manufactured Ayurvedic medicines sold via the Internet. JAMA 2008;300(8):915–23.

[56] Thatte UM, Rege NN, Phatak SD, Dahanukar SA. The flip side of Ayurveda. J Postgr Med 1993;39(4):179–82.

[57] Valiathan MS. Putting house in order. Curr Sci 2006;90(1):5–6.

[58] Lakhotia SC. Neurodegeneration disorders need holistic care and treatment–can Ayurveda meet the challenge? Ann Neurosci 2013;20(1):1–2.

[59] Patwardhan B. AyuGenomics – integration for customized medicine. Indian J Nat Prod 2003;19:16–23.

[60] Furst DE, Venkatraman MM, Krishna Swamy BG, McGann M, Booth-Laforce C, Ram Manohar P, et al. Well controlled, double-blind, placebo-controlled trials of classical Ayurvedic treatment are possible in rheumatoid arthritis. Ann Rheum Dis 2011;70(2):392–3.

[61] Chopra A, Saluja M, Tillu G, Sarmukkaddam S, Venugopalan A, Narsimulu G, et al. Ayurvedic medicine offers a good alternative to glucosamine and celecoxib in the treatment of symptomatic knee osteoarthritis: a randomized, double-blind, controlled equivalence drug trial. Rheumatology (Oxford) 2013;52(8):1408–17.

[62] Dieppe Paul, Marsden D. Managing arthritis: the need to think about whole systems. Rheumatology (Oxford) 2013;52(8):1345–6.

[63] Manohar PR, Eranezhath SS, Mahapatra A, Sujithra MR. DHARA: digital helpline for Ayurveda research articles. J Ayurveda Integr Med 2012;3(2):97–101.

[64] Narahari SR, Ryan TJ, Aggithaya MG, Bose KS, Prasanna KS. Evidence-based approaches for the Ayurvedic traditional herbal formulations: toward an Ayurvedic CONSORT model. J Altern Complement Med 2008;14(6):769–76.

[65] Tillu G. Workshop on a CONSORT statement for Ayurveda. J Ayurveda Integr Med 2010;1(2):158–9.

[66] Witt CM, Michalsen A, Roll S, Morandi A, Gupta S, Rosenberg M, et al. Comparative effectiveness of a complex Ayurvedic treatment and conventional standard care in osteoarthritis of the knee–study protocol for a randomized controlled trial. Trials 2013;14:149. http://dx.doi.org/10.1186/1745-6215-14-149.

[67] Narahari SR, Ryan TJ, Bose KS, Prasanna KS, Aggithaya GM. Integrating modern dermatology and Ayurveda in the treatment of vitiligo and lymphedema in India. Int J Dermatol 2011;50(3):310–34.

[68] Narahari SR. Collaboration culture in medicine. Indian J Dermatol 2013;58(2):124–6.

[69] http://spokensanskrit.de/.

[70] Brar JS, Chengappa KN. Does one shoe fit all? J Ayurveda Integr Med 2011;2(2):91–3.

[71] Smith R, Rennie D. Evidence-based medicine—an oral history. JAMA 2014;311(4):365–7.

[72] Greenhalgh T, Howick J, Maskrey N. Evidence based medicine: a movement in crisis?. BMJ 2014;348:g3725.

[73] Mongomery K. How doctors think: clinical judgment and the practice of medicine. Oxford University Press; 2006.

[74] McNutt R, Hadler NM. How clinical guidelines fail both doctors and patients. Scientific American Blog Network. Available from: http://blogs.scientificamerican.com/guest-blog/2013/11/22/how-clinical-guidelines-can-fail-both-doctors-and-patients/.

[75] Gigerenzer G, Gaissmaier W, Kurz-Milcke E, Schwartz LM, Woloshin S. Helping doctors and patients make sense of health statistics. Psychol Sci Public Interest 2008;8(2):53–96.

[76] Phil A. Absence of evidence is not evidence of absence: we need to report uncertain results and do it clearly. BMJ 2004;328(7438):476–7.

[77] Phillips CV, Goodman KJ. The missed lessons of Sir Austin Bradford Hill. Epidemiol Perspect Innov 2004;1(1):3.

[78] Goodman Amy GJ. A daily independent global news hour. Democracy now; Available from: http://www.democracynow.org/2010/12/17/pharmaceutical_drug_industry_tops_defense_industry.

[79] John B. Corporate crime in the pharmaceutical industry. Routledge & Kegan Paul; 2013. Available from: http://www.ivantic.net/Ostale_knjiige/Zdravlje/Braithwaite-John-Corporate-Crime-in-the-Pharmaceutical-Industry.pdf.

[80] Bero L. Industry sponsorship and research outcome: a Cochrane review. JAMA Intern Med 2013;173(7):580–1.

[81] Lundh A, Sismondo S, Lexchin J, Busuioc OA, Bero L. Industry sponsorship and research outcome. Cochrane Database Syst Rev 2012 Dec 12;12:MR000033. http://dx.doi.org/10.1002/14651858.MR000033.pub2.

[82] Lehrer J. The truth wears off. New Yorker 2010:1–10.

[83] Harremoes P, Gee D, Malcom M, Andy S, Jane K, Brian W, et al. The precautionary principle in the 20th century: late lessons from early warnings. Routledge; 2013.

[84] Marino MJ. The use and misuse of statistical methodologies in pharmacology research. Biochem Pharmacol 2014;87(1):78–92.

[85] Egger M, Bartlett C, Juni P. Are randomised controlled trials in the BMJ different? BMJ 2001;323(7323):1253–4.

[86] Smith R. Medical journals are an extension of the marketing arm of pharmaceutical companies. PLoS Med 2005;2(5):e138.

[87] Smith R, Gøtzsche PC, Groves T. Should journals stop publishing research funded by the drug industry? BMJ 2014;348:g171. http://dx.doi.org/10.1136/bmj.g171.

CHAPTER 5

Systems Biology and Holistic Concepts

The cure of the part should not be attempted without the cure of the whole.

Plato

THE PHILOSOPHICAL BASIS TO SYSTEMS APPROACH

In its preface, the latest edition of the most popular book, *Guyton and Hall Text Book of Medical Physiology* states: "Indeed, the human body is much more than the sum of its parts, and life relies upon this total function, not just on the function of individual body parts in isolation from the others." This clearly indicates how the thinking in mainstream biology is moving toward holistic approaches. The new recognition of a systems approach in biology has significant impact on health, wellness, diseases, therapeutics, and future health care. The systems approach is in consonance with the basic concepts of Ayurveda and Yoga.

The philosophy of systems is inspired by nonlinear approaches, network analysis, cybernetics, and nonequilibrium dynamics of open systems [1]. Actually, the systems biology concept is not new. Erwin Schrödinger's thermodynamic approach was based on systems thinking. Eminent physiologist, Claude Bernard, and the father of cybernetics, Norbert Wiener, among many others, also propounded or incorporated systems thinking in their work. Systems biology follows Hegel's dialectical principles of development from thesis, to antithesis, to synthesis. Systems biology, in particular, deals with the higher level analysis of complex, biological systems. Through the reductionist approach, scientists have acquired detailed knowledge of biochemistry, genetics, and molecular biology. But in the process, biology of organisms was reduced to components like tissues, cells, genes, and molecules. Fragmentation of whole systems into parts is actually the antithesis to the thesis of biology. While a detailed understanding of these fragments is absolutely necessary, unlike in physics or mathematics, the sum total of these parts does not make the whole biological organism. This realization has given impetus to systems thinking in biology.

Systems biology as a new approach, emerged in 1966 at the international symposium at the Case Institute of Technology in Cleveland, Ohio; it was hailed as a new way of holism. The successes of molecular biology, functional genomics, and other omics technologies have further strengthened systems approaches. It is possible to design and synthesize proteins, lipids, and carbohydrate complexes as constituent parts of a cell. Yet, scientists and technologists have not been able to create even a single living cell, de novo. Our cells, tissues, organs, and bodies are not mathematical equations. Any living system such as a cell is an integrated, interdependent, and complex ecosystem where the parts and the whole both play important roles in design, sustenance, and growth.

Aristotelian logic followed the path of reductionism, and gradually science and philosophy were separated. During the Renaissance, science and medicine continued on the reductionist path, and attempted analytical evaluation of natural phenomenon based on the assumption that complex problems are solvable by dividing them into smaller, simpler, and manageable units. In this process, big objects are "reduced" to smaller basic units. Undoubtedly, this approach was rewarding, and many path-breaking inventions, discoveries, and technologies emerged to make our life more comfortable. Today, the reductionist approach dominates the biomedical sciences, and governs the way toward diagnosis, treatment, and prevention of diseases (Figure 5.1). As a result, medicine relies

FIGURE 5.1
Reductionist approach in modern medicine.

more on explanatory models of scientific positivism based on sensory experience as an exclusive source.

However, many scientists are now challenging linear thinking based on reductionism. They argue, based on complexity theory, that analytical and predictive powers can be enhanced with an understanding of the whole system, rather than limiting them as parts to narrow disciplines. This is very important, especially in the process of clinical evaluation when patient, disease, environment, and many other confounding factors need to be considered together. As rightly put by Sweeney and Kervik, "it is better at times to be vaguely right, rather than precisely wrong" [2].

While reductionism was progressing rapidly in the West, a few philosophers were thinking differently. For instance, during the seventeenth century, René Descartes proposed a theory of dualism; this triggered discussions related to the mind and brain. He distinguished the brain as the seat of the intellect, and the mind as having consciousness and self-awareness. The theory of duality made scientists think about the brain and the mind together, as interdependent entities. Such efforts have given some impetus to systems thinking. Of late, several limitations of reductionism have been realized. The systems biology perspective appreciates that holistic problems can be better addressed through the use of computational and mathematical tools.

Any system is constituted by interconnected parts, which form a unified whole. Nature presents an apt example of complex ecosystems with interdependent biotic and abiotic components. Various components of the ecosystem have their own existence, but they are also interdependent in terms of survival and sustenance. Biological systems are often complex, with several, multifunctional elements interacting selectively in a nonlinear fashion.

Chromosomes, genes, proteins, and biochemicals constitute a cell. Many cells together can form tissues, different tissues can form anatomical systems, and many such systems make an organism. The organism is not just an assembly of cells, tissues, and organs—it has a specific structure, hierarchy, and networks. Newer disciplines like omics sciences are making modern biology a "data-rich" science, and systems biology, by making a composite picture, tries to make sense of the whole. Superspecialization and therapeutic precision in modern medicine is target specific, linear, and rigid, and does not consider the dynamic context of time and space. Such reductionist and fragmented thinking have added limitations to modern medicine; a systems biology approach may offer an alternative model.

In simple terms, a complex ecosystem like a forest cannot be explained merely by studying individual trees or insects. At the same time, a butterfly cannot be seen from an airplane without the use of high-resolution telescopes. The morphology of cells cannot be known without the use of a microscope and details of Saturn in the galaxy cannot be seen without the use of a telescope (Figure 5.2). In fact, application of the tool as a microscope or a telescope much depends on the direction in which one is trying to see. Both the views are important.

Chemicals present in cytoplasm are not known without the use of powerful, chromatographic analytical tools. Thus, in reality the reductionist and holistic approaches are complementary, and not mutually exclusive. Systems biology has tried to bridge the gap between these two valuable approaches. In short, systems biology studies microscopic details of cells and the diversity of molecules without losing the sight of the whole organism. Systems biology is a science that allows us to see parts in the whole and whole in the parts.

Reductionism is helpful and useful in understanding the components of any system. It provides a detailed understanding of a part down to the finest details.

FIGURE 5.2
Telescopic and microscopic approaches for studying ecosystems, both are looking at each other.

It is a way of dividing complex problems into solvable smaller, simpler, and thus more tractable units. In the collection of pieces of knowledge, a reductionist approach is required in almost every science. However, the subsequent divisions lead to a loss of important information about the whole. Reductionism is a natural process of gaining knowledge in modern science. Holism is the ability to look at the whole picture. Reductionism and holism are the two sides of a coin. Modern science adopted the reductionist approach, which is based on rigorous experimentation. The earlier notion that considers reductionism as the only way to know any phenomenon is being reconsidered in the light of systems biology.

Richard Dawkins, famous evolutionary biologist once said, "Reductionism is one of those things, like sin, that is only mentioned by people who are against it." The systems approach was helpful in overthrowing narrow thinking, and also changed the dogmas of modern medicine. Against the broader backdrop of systems understanding, the limitations of reductionism were more apparent. The debate on the utility and limitations of the reductionist approach has been discussed at various levels. For example, the *Stanford Encyclopedia of Philosophy* discusses "reductionism in biology." It discusses reductionism at three levels: ontological, epistemological, and methodological [3]. We feel that reductionism at methodological levels is necessary, but unless it considers epistemological level, there is a danger of losing sight of the whole picture. The systems approaches are now reconnecting science to philosophy!

BIOLOGICAL SYSTEMS

Every living thing is made up of cells, which are the smallest part, or building block of any organism. Every cell survives in a particular environment, and performs its independent function. In 1665, English philosopher Robert Hooke published his observations of experiments using the microscope. He observed the cell structure of cork, and called the individual compartments *cells*. The Latin meaning of this term cell is "small room." Later, in 1673, Antonie van Leeuwenhoek, the father of microbiology, with an improved microscope, observed living algal cells. These developments provided new insights in biology, and led to the birth of cell theory. Cell theory states that every organism is constituted of cells as its structural and functional units. Living cells reproduce and make similar cells.

The estimated number of total human cells ranges from 37.3 to 100 trillion [4]. Many similar cells can connect and organize to form tissues, which have unique structures and functions. There are four basic tissues: epithelial, connective, muscle, and nervous tissues. The aggregation of cells to form tissues has many advantages. Tissues display more properties than individual cells. A number of tissues may function together as organs, with specific structures and functions. The structures of organs have evolved to perform related functions. For example, the liver is a glandular organ, where all four basic tissues are involved. The liver secretes bile, metabolizes drugs, detoxifies unwanted substances, and controls the metabolism of protein, carbohydrate, and fat. Many organs work together in a coordinated fashion to function as a system. For example, the digestive system consists of the buccal cavity, tongue, esophagus, stomach, duodenum, small intestine, colon, and rectum—which are parts of the alimentary canal. In addition, organs like the liver, pancreas, and gall bladder, which support digestive processes, are also part of this system. Many systems together form an organism. Populations of similar organisms comprise a community and many communities together with animals, plants, and nonliving components make an ecosystem (Figure 5.3).

This is still not a complete story. As per other estimates, the human body is made up of about 10 trillion cells but it carries near 100 trillion microbial cells. Thus in terms of mere numbers microbes are 10 times more than human cells. These microbes are active partners in human body living in symbiotic system mostly in gut. This is known as the gut microbiome, which has become a new target of diagnosis and treatment of many diseases. Microbiome is very much a part of the human body and actively interacts with human cells, tissues, organs, and physiological systems.

Although conventional biological study stops at the organism level, medicine and social sciences consider the community as a unit for epidemiological attributes. The communities, and surrounding ecosystems, consisting of plant,

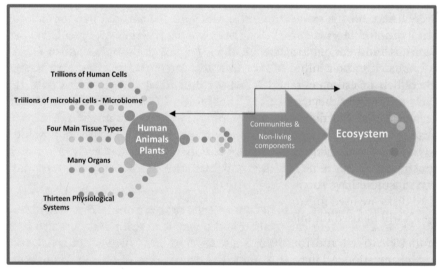

FIGURE 5.3
Whole organization of ecosystem from parts.

animal, and nonliving geographical components, constitute an ecosystem. There are many ecosystems making up the biosphere [5]. The complex network of interdependent, living, and nonliving forms existing in natural harmony—including a wide range of objects from microscopic cells, to complex ecosystems—is known as the web of life.

OMICS FOR STUDYING CELLS

During the past few decades, the work of several scientists has advanced the detailed understanding of cell structure, function, and the underlying biological processes. At the microlevel, the cell has amazing machinery. It contains many components, known as cell organelles. Each cell of the multicellular organism has a permeable membrane that surrounds a jellylike substance known as the cytoplasm, which contains various organelles such as the endoplasmic reticulum, Golgi bodies, lysosomes, mitochondria, ribosomes, and vacuoles. Plant cells have chloroplasts that are important in photosynthesis. The cell nucleus is the control center for the cellular functions. It contains deoxyribonucleic acid (DNA), a long, linear polymer arranged in a double helix structure. DNA is comprised of four chemical bases. They are arranged in pairs to form ladder-shaped, twisted DNA molecules. The two pairs of these four molecules are the purines, adenine (A) and thymine (T), and the pyrimidines, cytosine (C) and guanine (G). The threadlike coiled strands of DNA carrying genes are called chromosomes. The genes are units of inheritance that pass the traits to the offspring. Genes have "instructions," that is, codes, to make specific proteins. The process known as gene expression produces the respective proteins. Proteins are complex macromolecules. The genetic makeup of the cell is the genotype, and its expressed, resultant, observable traits are phenotypes. Each gene has hundreds of base pairs (AT, GC). The maximum number of base pairs is estimated to be 2 million [6]. The exact number of human genes is a subject of debate; estimates range from 20,000 to 23,000 [7,8]. Thus, the DNA of every cell carries a huge cache of coded information. An organism's full DNA sequence, that describes the order of genes in chromosomal sets, is known as the genome of the organism. Our understanding of information stored in DNA is now reshaping information science. Nature's way of storing information in such a tiny biological unit is really amazing! Scientists from Harvard's Wyss Institute have successfully stored 700 terabytes in 1 g of DNA [9]. Now, hard disks are used for storing information, but in the future, DNA-based disk drives will be used for information storage.

From DNA to cell, and from cells to organisms, there are various layers of biological organization. All these expanding levels are interrelated, and become more and more complex. The higher level has newer properties and controls than the lower levels. New properties emerge in expanding systems. For example, a tissue functions differently from the collective properties of its constituent cells. Aggregation of tissue leads to formation of organs, which have specific functions. Biological systems thus have emergent properties, which differentiates them from nonliving, engineered models. In such complex systems, physicochemical understanding of the lower levels may limit comprehension of the larger system. The linear, mathematical rules fail to describe the phenomenon of emergent properties. An understanding of expanding levels of biological organization requires an understanding of the holistic approach. Just looking at components in details is the reductionist view, which may describe the parts, but not necessarily the whole picture.

After the discovery of the microscope, with its ability to look at cells, as demonstrated by Robert Hooke in 1665, biological sciences took a major leap. This millennium has been dominated by several groundbreaking discoveries in cellular and molecular biology. Progress in analytical technology, coupled with computing and informatics, has resulted in generation and high-throughput data analysis. The human genome was decoded at the beginning of this millennium, which resulted in an explosion in genome data. Genomics emerged as a branch of genetics which studies the DNA sequences of various organisms. Ribonucleic acid (RNA) has important functions in coding, decoding, regulation, expression, and signaling. DNA sequences are translated into proteins. Since Robert Hooke first observed cells under the microscope, it took 300 years of rigorous research in cell and molecular biology to reach this level.

The term omics is used to denote powerful, or high-throughput technologies used to study various biological molecules on a large scale. Genomics is the study of whole DNA, and different genes. The DNA is copied into RNA for transcription. The set of all RNA molecules of an organism is the transcriptome, and its study is transcriptomics. Study of the structure and functions of proteins is proteomics—the study of the full set of proteins encoded by a genome. The various molecules produced in the metabolic processes are metabolites. The metabolites are the nutrition, structure, signaling, and regulatory functions of the organism. Metabolomics is a relatively new branch of science, which studies specific patterns of chemicals involved in the cellular processes of an organism. The large data sets emerging from omics sciences have enriched biology, and at the same time, have posed challenges of interpretation of huge amounts of information. Biology today is more than just a descriptive subject—it has become a data-rich science. Even a cursory look at the extent of the information generated in current biological sciences is mind-boggling.

The high-throughput studies of biological systems give rise to omics data of a gigantic scale. Six major omics of human data types constitute millions of components. As of May 2014, The National Center for Biotechnology Information gene database covers 47,758 human genes [10]. The transcriptomics data from protein-encoding, messenger RNA transcripts include 29,000 variants. Small, noncoding RNA molecules, also known as microRNA (miRNA), regulate transcription via gene silencing. The human genome might encode over 1000 miRNA. The proteomics database is composed of around 22,000–39,000 proteins, enzymes, receptors, and structural units of the cell. Phosphoproteomics data has 200,000 proteins, which are involved in cell signaling. The metabolomics database covers all small biomolecules in the cell, including lipids, nucleotides, dipeptides, and hormones. The number of all such metabolites is estimated to be over 40,000. In addition, epigenomic data suggests that nongenomic encoded DNA modifications may be as large as 20–30 million [11]. A typical gene chip, which is hardly the size of a coin, may carry up to 400,000 miniature reaction wells. Each reaction well might carry a DNA fragment with genetic information. The biological data is growing continuously, and developing a deeper understanding of parts poses more and more challenges of interpretation, relevance, and translation. This situation has given further impetus to the study of systems biology.

During the visualization and analysis of this data it is important to create a comprehensive biological insight without losing sense of intrinsic complexity. To interpret protein interaction, gene expression, and metabolic profile data, several visualization tools can be used [12]. High-throughput technologies require powerful computational support and new study models to manage the huge amount of information. In many cases, routine causality analysis might not be sufficient; the cause and effect relation might not be a linear phenomenon. Engineering and physical sciences also deal with complex circuits and networks. However, unlike machines and printed circuit boards, living organisms are not just an assembly of metabolites, proteins, nucleic acids, and genes.

OPEN SYSTEMS

Biological systems are open systems. They are not closed like engineered systems. Biological systems continuously interact with their surroundings, and they evolve. Biological systems have coherent behaviors, and the robustness to adapt in response to evolutionary changes through processes like homeostasis, and to tolerate these changes. Biological systems of a multifunctional nature interact selectively in nonlinear fashion to produce coherent behavior. The coherent behavior can be studied using network complexity. The phenomenon has to be interpreted in the context of multiple components, and the array of organizational and emergent properties. The direction and quantity of response may not be unidirectional. An input may affect multiple targets and loops; the response may be the upregulation or downregulation of components. Thus, more than just components, the nature and type of their interrelations are very important. The relations and components together form complex networks. Every biological entity—from the unicellular ameba to the mammoth elephant—are composed of such complex networks.

Every biological entity is in a continuous, dynamic state, and one of the important properties of a living system is its ability to maintain homeostasis. Every biological entity continuously responds to changes in the surrounding environment and actively tries to maintain its metabolic equilibrium. The homeostatic mechanisms involve the regulation of several complex, biological processes. For example, temperature regulation involves skin, sweat glands, blood capillaries, thermoreceptors, autonomic nervous system, hypothalamus, water transport, sweating, vasodilatation, muscles, and muscle contraction. This explains how a small stimulus can affect large responses in linear and complex systems.

THE EMERGENCE OF SYSTEMS BIOLOGY

The process of emergence of systems biology was initiated much before it was formally launched in 1966 at the Case Institute of Technology; prior to that, systems biology was viewed as a new form of holism [13]. Later, the term holism was dropped because it had different connotations.

Biological systems are subject to the process of evolution. For instance, the immune system as it exists today has evolved over millions of years. This capacity of adaptation, and the race for survival during the process of evolution, has given dynamic robustness to living organisms. Every organism tries to cope with the stress condition. Properties like coherent behavior, capacity of self-regulation, adaptation, and homeostasis make the living system flexible, and support their sustained existence. A physiological understanding, especially of prokaryotic animals, provides interesting insights into the modularity of components. Various components are arranged to perform specific functions. There are similar mechanisms which support each other as backup. The redundant mechanisms provide another advantage which make the systems more robust. Understanding the differences between engineered and living systems is crucial for effective management. In short, the reductionist approach of getting deeper and deeper understanding of parts, has led to rapid advances in the biological sciences. Moreover, the challenges of meaningfully managing the huge quantity of omics data have compelled scientists to rethink the whole picture. Systems biology has emerged as an attempt to strike a scientific balance between the two important and complementary approaches, namely, reductionist and holistic. Previously, these were considered to be alternative or exclusive approaches. The reductionist robustness of modern science and the holistic vision of the ancient sciences both find a place in the new systems biology approach.

The culture of holism in the East and concepts of reductionism in the Western world actually share a similar spirit, according to many experts and philosophers. In 1926, British military leader, and philosopher, Jan Smuts, proposed a concept of holism in his book, *Holism and Evolution*. He described holism as "the tendency in nature to form wholes that are greater than the sum of the parts through creative evolution." Interestingly, although Smuts was a believer in holism, he was also a racist who favored discrimination between black and white people [14].

In 1969, Austrian biologist, Ludwig von Bertalanffy described physical systems as closed systems, and biological systems as open systems. According to von Bertalanffy, a closed system does not allow certain types of transfers through the system, while an open system continuously interacts with its environment or surroundings, where laws of thermodynamics might not be applicable. He proposed a systems theory for self-regulating systems. This theory defines systems as a configuration of parts connected together through a web of complex relationships. He further proposed the methodologies and applications for studying open systems, and stressed the importance of illustrating interaction between all related domains, rather than focusing on the stand-alone components. The systems theory was applied to various sectors like engineering, psychology, and biology.

The complex system, and its regulation, has been a curiosity for scientists for many years. In 1948, American mathematician, Norbert Wiener, proposed the concept of cybernetics. He defined it as "the scientific study of control and communication between the animals and the machines." This concept tried to bridge the gap of micro- and macrolevels of big systems. Actually, systems theory, artificial intelligence, and cybernetics together seem to have led to complexity science during the mid-1960s. Gradually, the concepts of dynamic systems paved the way to new thinking on nonlinear systems, fractal geometry, and chaos theory [15].

The theory of complexity has progressed during the past five decades to encompass an understanding of self-organization, adaptive fuzzy logic, robotics, and network sciences. All these developments are the result of scientists' efforts to try to understand unexplainable phenomenon from biological and computational sciences. Scientists were trying to connect the dots on a wide canvas of science. Today, many of these disciplines are converging, and new concepts like big data, cloud computing, global network society, and multilevel complex systems are evolving.

Interestingly, because of technological breakthroughs the distance between humans and machines is narrowing. Machines are becoming more and more intelligent, and the human being is becoming more and more mechanical. This realization has triggered a new interest in holistic approaches.

Concepts of holism, cybernetics, and systems theory are intermingled with information, technology, molecular biology, and the omics sciences, in general. This comprehensive approach is now known as systems biology. This approach focuses on an understanding of the whole through connecting the components generated as a result of reductionist approaches. The inputs from systems theory for transdisciplinary investigation, in the context of time and space, are crucial. Systems biology has enabled a nonlinear, dynamic, and holistic study, and has stimulated many new vistas in the biosciences.

The systems theory actually was most relevant for applications in the field of medicine. However, it quickly meshed with biology, and emerged as a new discipline of synthetic biology. These efforts went in dual directions: first, to understand the complexity of natural biological structures and processes and second, to use available technology to create these biological structures artificially.

THE LIVING CELL AS SYSTEM

The living cell itself is a complex system which continues to be a challenge for biologists. Efforts to create artificial cells have been ongoing for a long time. Systems-biology-based understanding of the cell organelles at the molecular level is helpful in modeling cellular mechanisms and metabolic pathways. The first attempt to create an artificial cell was made in 1965 by a Canadian physician–scientist, Thomas Chang. He used ultrathin membranes of nylon and cross-linked protein which allowed ingress and egress of small molecules [16]. The result was not a cell. It was more like a specialized, permeable pouch containing biological materials like hemoglobin, enzymes, and proteins.

During the mid-1990s, biodegradable, encapsulated artificial red blood cells were developed and used for the treatment of a diabetic patient in a clinic [17]. Since then, genetically engineered cells are being studied for use in tissue regeneration.

In 2006, the United States National Science Foundation announced a grand challenge to build a mathematical model of the whole cell [18]. Craig Venter, one of the founders of Celera Genomics, which sequenced the human genome, made a successful attempt to use the synthetic genome to transfect a cell. In May 2010, Venter claimed to have created "synthetic life." He synthesized the DNA of the genome of one bacteria, and introduced it into another cell [19]. In 2011, scientists from Harvard reported the creation of an artificial cell membrane [20].

In 2013, Venter also indicated that "scientists would soon be able to use 3D printers to create synthetic life, possibly even recreating alien genomes." In 2014, a team of scientists from the United States, France, and India made the first artificial chromosome [21]. This discovery is celebrated as a landmark in synthetic biology, which will revolutionize the field of medical and industrial biotechnology.

These new-generation god-men also happen to be scientists. Their aspirations are sky high. However, in reality, what has been attempted so far cannot be called artificial life or an artificial organism. For an organism to qualify as artificial, all core elements should have been initially constructed from chemically simple, nonliving components. In all these efforts scientists have made some kind of a hybrid, where some part of the structure is natural and some part is synthetic. They have tried to transfect artificial components with natural components. Transfection is the process of introducing a genetic material into eukaryotic cell. Even the Craig Venter Institute admits that "it is not creating life from scratch, but an attempt to create new life from already existing life using synthetic DNA." So far, scientists may have been able to create some new life structures, but they have no clue as to how to create life itself, making living creatures rather than life structures.

Earlier efforts in genetic engineering have been manipulative. With the advent of synthetic biology, genetic engineering is becoming more creative. Obviously, this has huge ethical and moral significance [22]. For instance, normal, natural DNA has four nucleotides (A, T, G, and C) and two base pairs. This double helix structure of DNA carries genetic code based on 64 codons (a specific sequence of three adjacent nucleotides), each one of which encodes for 1 of the 20 amino acids used in the synthesis of proteins. Now scientists have created cells with six nucleotides—DNA with three base pairs—which can have 216 codons capable of handling up to 172 amino acids [23]. This is like creating more and newer words if, for example, you have 40 alphabets instead of 26. One can only imagine what chaos the English language would suffer in such a scenario; this is rightly feared, and people are raising serious concerns. There is no satisfactory answer to the question: "Are we creating monsters?" In any case, such efforts certainly look overambitious, scary, and unethical.

Artificial cells may be useful as targeted drug delivery systems; liposomes in therapeutics can help the ecosystem by creating better biodegradability, or through their use in new biofuels. However, making any genetic modification in unknown areas must be done with extreme caution. When humanity is already facing so many problems related to ecological and environmental situations, as well as facing lifestyle, behavioral, and mental health challenges, investing so much effort in gray areas like artificial life might call for a better justification. The knowledge, techniques, and technologies could be better used to explore new applications of systems biology in integrative health care. Certainly humanity does not want ninja mutants and species such as those in movies to exist in real life.

SYSTEMS MEDICINE AND HOLISTIC HEALTH

Systems biology concepts are relevant to health care and clinical practice. In Chapter 2 we discussed how the prevailing strategies to health care and medical care are more reactive than proactive. Medicine practice starts only when someone is sick or has a disease. Systems medicine has emerged as an interdisciplinary field, which considers the human body as part of an integrated, whole dynamic system involving biochemical, physiological, and environment interactions, which are responsible for health and homeostasis. Systems medicine considers body, mind, and spirit in a way quite similar to holistic health—but in light of their genomics, behavior, and the social environment. Knowing details of disease biology at the molecular level can strengthen the diagnoses, prognoses, and predictive nature of medicine. Alterations in physiological and pathological processes and modulations of metabolic networks can be understood better by studying profiles of proteins and metabolites. With the help of specific markers, which are as yet unknown, the development of diseases like diabetes, asthma, and cancer might be predicted.

Biological networks are dynamic, and constantly changing. Any experiment designed to study a system might just take a snapshot at a specific time. Splicing these snapshots together might allow us to see the dynamic and complex nature of biological networks. Systems behavior requires consideration of three important parameters: context, time, and space. The systems approach has provided a perspective useful in addressing network complexity challenges; it has made scientists rethink the reductionist approach and the linear articulation of "one gene, one risk, one disease."

A disease can be diagnosed with the help of clinical examination, pathological investigations, diagnostic biomarkers, and imaging techniques. Sometimes, only diagnostic factors are considered, and the clinical picture might not accurately reflect the patient's whole physical state. In such a case, doctors may treat the symptoms, instead of the patient. This approach is very mechanical. The focus is on regaining the acceptable levels of biomarkers, which becomes the only target. In this process, the underlying biological or pathological environment often takes a backseat. In reality, any abnormal finding needs to be assessed in the context of time and space. The systems approach suggests a totalistic view involving the behavior of the affected system, its linkage with other systems, and the patient as a whole. As a practical matter, consideration of the whole picture becomes challenging due to complexity. The analysis requires nonlinear, sensitive, and probability-based programs.

Systems biology is all set to change our present concepts and definitions. Health is not just a state, or stage of normality. Health is not necessarily obtained by reducing risks. Health is more about the capacity for robustness, and intention to adapt.

The reductionist approach is very powerful, but cannot be considered as the only solution. In the process of reductionism, any problem or object is divided into parts in order to study them in great detail. However, this may lead to a loss of crucial information about the whole. Moreover, it may ignore part-to-part dynamic interactions of system-wide behavior. By using reductionist methods, one can build an airplane or a computer by joining several well-defined parts in a well-defined sequence. This is a linear assembly line operation such as that which occurs in any automobile or appliances factory. Standard operating process and protocols help in doing the same process again and again with great precision. Such processes can be better handled by robots.

It is known that a reductionist approach does not work to build even a simple single cell. Today, cell biologists know most of the parts of a cell. They know how they are connected to each other. They are able to artificially prepare these parts in a laboratory. But still they have not been able to artificially create even a single cell. This is because, unlike physical systems, biological systems cannot be understood only by knowing all the details of the parts.

Contemporary physics and mathematics have limitations in their understanding of biology. Eminent scientists like Einstein and Schrödinger acknowledged these limitations. Today's physics can address simpler and mechanical systems. However, as rightly pointed out by senior physiologist F.E. Yates from the University of California, Los Angeles, complex systems possess self-organizing capabilities known as homeokinetics. Homeodynamics is the biological application of homeokinetics, which helps in knowing patterns of energy flow, and their transformations in metabolic networks of living systems [24]. The concept of homeostasis is central to modern medicine. However, systems scientists feel that the term stasis needs to be reconsidered. Homeostasis is a dynamic activity, and not a static entity. Therefore, Biologists from Cardiff University Lloyd et al. proposed the term homeodynamics, which seems more appropriate [25]. Homeodynamics is related to regulatory mechanisms, feedback control, structural stability, redundancy, and range of adaptations of the systems in response to stress. Homeodynamics reflects the Ayurveda theory of *Dosha* much better than the concept of homeostatis.

According to current theory, an illness is the result of an imbalanced homeostatic mechanism. Treatment is aimed at regaining the balance by correcting deviations. This corrective treatment approach is applicable to a range of disorders from hypothyroidism, to hypokalemia, to hyperglycemia and diabetes. Correcting deviations to regain homeostasis is a typical reductionist perspective. Here, the emphasis is mainly on correcting the deviated parameter without a systems view. If one has elevated blood glucose—lower it; if one has high blood pressure—lower it; if one has high fever—reduce it; if one has low hemoglobin—increase it.

Fixing the symptom is required, but not sufficient. Such a reductive approach has limitations, and is even harmful. For example, in some conditions, calcium supplementation can cause adverse effects—even in a hypocalcemic state [26]. Sometimes, lowering elevated blood pressure can be harmful [27]. A selective focus on maintaining normal ranges also undermines the importance of dynamic stability and interactions between parts and systems. Biological systems are not merely collections of static components; rather, they are in dynamic, stable states, such as oscillatory or chaotic behavior. For example, circadian rhythms relate to oscillatory behavior [28], and complex heart rate variability relates to chaotic behavior [29]. Thus, homeostasis, or more appropriately, homeodynamic states should be considered during therapeutics; failing to do so may lead to ineffective, or even detrimental, outcomes. Limits to reductionism in medicine have been well discussed by Drs Russell Phillips and Andrew Ahn of the Harvard Medical School. According to them, present clinical medicine focuses on an understanding of parts and symptoms, while systems medicine considers the whole person in the context of time and space [30].

Against this background, it will be interesting to discuss a few examples from health and disease. By and large, the principles of systems biology can best be applied to multifactorial, chronic diseases such as metabolic syndrome, diabetes, coronary artery disease, arthritis, asthma, dementia, and Alzheimer disease, in which there is not a single causative factor, a single biomarker to diagnose, or a single target to treat. Importantly, in modern medicine these are considered as difficult-to-treat diseases.

Current treatments of such conditions are mainly targeted at disease symptoms, and are attempts to regain the normal status of the patient. Most of the time, the treatment is additive. Various drugs are added, based on various symptoms. For example, a patient with diabetes, hypertension, and dyslipidemia might be prescribed a combination consisting of oral antihyperglycemic drugs, antihypertensive drugs, and statins. Of late, a few companies have launched such commonly used combinations as polypills. However, this cannot be considered a holistic, system, or even a multitargeted approach. A polypill might provide convenience to the patient, but it is not a therapeutic innovation; rather, it is merely a mechanical combination.

The complex interactions between various etiopathological factors, and the dynamics of their interactions, cannot be understood well using only a reductionist approach. However, holistic or systems approaches can help to understand complex interrelationships between multiple factors, and so are better suited for difficult-to-treat chronic diseases. Most of these chronic diseases are polygenic, with lifestyle and psychosomatic etiologies. Most interestingly, traditional, complementary, and alternative medicine is the preferred choice of patients with these conditions.

On the other hand, many acute conditions like simple headache, fever, appendicitis, and infectious diseases like tetanus, diphtheria, diarrhea, and urinary tract infection are driven primarily by specific or single pathology making it easier for medical interventions. Thus, in short, reductionism can work for specific symptoms and problems with known causes, where a quick and effective treatment is possible, whereas a systems approach is suitable for the management of chronic and complex diseases.

THE EXAMPLE OF DIABETES

Treatment of a complex disease such as diabetes is challenging because it is a multidimensional disorder. Many factors, including insulin resistance, inflammation, genetics, and lifestyle play a crucial role. Many markers like insulin and glucagon balance, peroxisome proliferator-activated receptor-γ, leptin, and cortisol, are involved, in addition to other parameters such as stress, diet, and obesity, in the pathogenesis of diabetes. The metabolomics of diabetes has become extremely complex.

Type I diabetes is supposedly caused by the inability of body to produce insulin. Type II diabetes (T2D) is considered to be the result of insulin resistance, when beta cells cannot produce enough insulin to counter insulin resistance. This situation leads to an increase in blood sugar, and can cause several complications which are mainly due to inflammatory changes that harm blood vessels, eyes, kidneys, and heart. Genetic predisposition and obesity have been considered as possible factors. However, a new hypothesis suggesting the behavior-mediated, evolutionary origin of insulin resistance has been proposed. The pathophysiology of T2D is hypothesized to originate from the brain and from behavior, rather than from diet and energy imbalance [31]. Recent research on genome-wide association studies indicate that insulin resistance can be triggered due to environmental factors, and genetically programmed beta cell dysfunction can result in diabetes [32].

Metabolic syndrome is known to affect over 24% of adult Americans. It is a premorbid condition which carries the risk of T2D and cardiovascular disease (CVD). The effective control of risk factors such as less physical activity and abdominal obesity can reduce the risk for T2D and CVD and prevent metabolic syndrome. A metabolic syndrome triad of obesity, diabetes, and coronary artery disease needs to be considered from a systems biology perspective. Today, it is known that only glycemic control is not sufficient to prevent all the complications of diabetes. A meta-analysis of 13 randomized controlled trials has shown that glucose-lowering treatment has no significant effect on all-cause mortality or death from cardiovascular causes in patients with T2D [33]. Thus, T2D might not only be about insulin and glucose homeostasis. In fact, a few critics are challenging the scientific rationale of using insulin in T2D, where the main problem is insulin resistance. Moreover, a majority of glucose uptake takes place in the brain, where the glucose transport mechanism is noninsulin dependent. Theoretically, due to the action of insulin-dependent glucose transport, administering more insulin in such a condition can push glucose into skeletal, muscle, liver, and adipose tissue. Obviously, an excess of glucose will lead to increased fat deposition, mainly in adipocytes, leading to obesity [34]. Actually, adipocyte- or fat-cell-specific insulin resistance might be good for health, while hepatic and brain insulin resistance might be deleterious. Different triggers for insulin resistance might affect different tissues in different ways. Insulin sensitivity and insulin secretion respond to a large number of signaling molecules including sex hormones, endorphins, myokines, and many others. A better understanding of the clinical progression of insulin resistance can help better clinical management of diabetes [35].

Thus, prescribing hypoglycemic or secretogogue drugs or insulin just to maintain blood sugar levels within limits might not be universally useful. Management of diabetes requires a comprehensive treatment. Here, the systems approach may be appropriate—it recognizes the presence and interplay of these complex factors in disease management. This approach is indeed very complicated. Yet, because of a deeper knowledge of underlying pathophysiological processes, and the availability of computational tools, it now seems to be possible. The medical community needs to be open to this holistic or systems understanding, and should develop the ability to appreciate the possible application of multitargeted, holistic, and integrative approaches. In attempting this approach it is necessary to keep in mind three crucial variables: time, space, and context.

Present diagnosis of diabetes relies on a measurement of fasting and postprandial glucose levels. However, these measurements are obtained at a single point in time. In such case, diagnosis is done after the beginning of the underlying abnormality.

To get a more accurate diagnosis, it is necessary to know glucose–insulin interdependent variability over a larger time frame. A whole body simulated model of the glucose–insulin–glucagon regulatory system has been developed to understand the pharmacokinetic and pharmacodynamic effects [36]. Still, time-sensitive interplay is not obtained through prevailing methods. It is now known that healthy individuals may show a pattern of periodic oscillations of insulin secretions, which can range from 6 to 10 min. It is also known that people with T2D have abnormal insulin oscillations. Impaired insulin oscillations can also be observed in first-degree relatives who may be metabolically normal [37]. This clearly suggests that time-sensitive, continuous evaluations might be better in detecting beta cell dysfunction. As we know, blood glucose levels are controlled by the interplay of insulin and glucagons, along with growth hormone and epinephrine. The glucose regulatory pathways are closely interconnected, and any dysfunction can affect the glucose/insulin dynamics. Therefore, time-sensitive measures might provide valuable information for diagnosis and treatment of T2D. Consideration of time-sensitive measures is not being done in today's evidence-based, modern medicine.

The most common method of measuring blood glucose levels is using a glucometer with strips. If a strip reading from the right finger is equivalent to that from the left finger, it leads to the assumption that the distribution of glucose in the body is uniform. However, plasma glucose is known to show spatial differences [38], which normally are not considered in clinical practice. Similarly, insulin injections in the thigh are assumed to be as effective as those in the abdomen. However, insulin absorption and distribution is known to differ at different sites. Currently, the recognition of such spatial variations in treatment or diagnosis is lacking in clinical practice.

As the analytical techniques are becoming more sensitive, it is possible to ascertain the best sites for insulin injections. A better understanding of bodily glucose distribution can help to predict diabetes risk. We might need to know how and why certain foods modulate stimulation of beta islet cells.

Today's medicine is more focused on symptoms and diseases rather than on the individual. In diabetes, the focus is on hyperglycemia, which is a symptom. Treatments are aimed at lowering the glucose levels. Interestingly, the systems approach to medicine might shift attention from the elevated glucose levels to the contextual milieu responsible for this. It could be genetics, behavior, lifestyle, dietary habits, sleeping patterns, immunity, and many such factors which all need to be integrated into the treatment. The individual is always at the central position in systems medicine.

This discussion may seem utopian to many modern medical practitioners. How can a diabetes patient not be given glucose-lowering drugs? The systems biology knowledge indicates that complex diseases might have different etiological processes, and might have different treatment options. Different processes do exist for specific diseases, and each process requires a different treatment. This perception also supports the personalized approaches in medicine. It is true that some patients with T2D might respond well to insulin therapy, while a few others might not; some might not require hypoglycemic drugs, but might benefit from lifestyle modification, diet planning, and meditation.

According to researchers from Harvard Medical School, Ahn et al., "the decision regarding appropriate treatments is possible through understanding of complex factors peculiar to each and every patient" [39]. This approach has a striking similarity to Ayurvedic diagnosis and treatment. Literally hundreds of variables are considered including the *Prakriti* of the patient and of the disease. An illustration of what happens when a patient visits a modern medical clinic and an Ayurveda clinic speaks volumes (Figure 5.4).

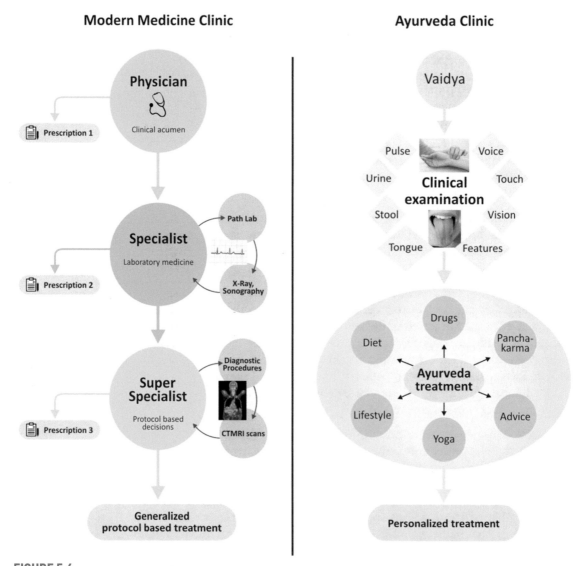

FIGURE 5.4

The figure sketches diagnostic and treatment activities in typical clinics of modern medicine in Ayurveda.

THE FUTURE OF SYSTEMS MEDICINE

Over the past two decades, systems biology has gained acceptance as an important branch of science. Much work is underway to apply systems principles to health promotion, disease prevention, and medical treatment. The National Institutes of Health supports over 10 systems biology centers in the United States. Systems-based approaches integrating genomic, molecular, and physiological data complement conventional genetic and biochemical approaches in understanding the complexity of conditions like metabolic syndrome [40]. The existing systems biology methodologies require refinement, development, and the integration of new technologies such as next-generation sequencing, DNA methylome, miRNA, and metabolomics.

The involvement of multiple cell types, tissues, and organs in complex diseases and syndromes necessitates the exploration of cross-tissue interactions. A recent study has shown that communications and signaling across multiple tissues—such as heart muscle, sympathetic nervous system, bone marrow, and spleen—are involved in the enhanced inflammatory response which induces atherosclerosis acceleration after myocardial infarction [41]. The recognition of such a relationship is possible only when organism-wide, systems-level data is integrated. Few studies related to cross-tissue networks have been reported, but this area certainly needs better, and more efficient methodologies. As we have discussed, most of the current methodologies capture static information—a snapshot of disease status.

Scientists at the Institute of Systems Biology (ISB) in Seattle are involved in developing new tests based on protein patterns in blood. This is expected to reveal the health status of every major organ in the body. With this knowledge, and systems understanding, early predictions of disease causation and progression are possible. Leroy Hood, founder of the ISB, has proposed a model known as P4 Medicine. The four 'Ps' include Predictive, Preventive, Personalized, and Participatory medicine. According to Hood, the systems approach can help in demystifying diseases and democratizing health care [42]. An ambitious project known as the 100K Wellness Project plans to study 100,000 individuals for 20 to 30 years. This project attempts to capture a variety of health-related data types when individuals are healthy, in a disease state, or in the transition between disease and wellness. Scientists will have an opportunity to closely study transitional states of common diseases such as cardiovascular diseases, cancer, and neurodegenerative disorders.

The 100K Wellness Project will carry out a series of investigations including whole genome sequencing to identify key genomic variants; detailed clinical chemistries; quantified self-measurements, including heart rate, respiration rate, quality of sleep, weight, blood pressure, and calories expended; gut microbiome; as well as organ-specific blood proteins from the brain, heart, and liver. This detailed profiling will help in the prognosis of very early stage, wellness-to-disease transitions, or vice versa. This will certainly have great importance in disease prevention, health, and wellness promotion.

The 100K project is an excellent example of how the principles of systems biology can be holistically applied to medicine. Of course, this might not be as simple as it looks. There will be difficulties in translating rapidly growing data into clinically usable knowledge for daily wellness, or at the bedside. Efforts to bridge the gap between systems biology and medicine call for collaborative research between scientists from the theoretical, experimental, computational, and clinical domains [43].

Dynamic models to capture the ever-changing nature of disease progression are needed. Only a comprehensive understanding of the whole body system in relation to its environment can lead to effective diagnostic, preventative, and therapeutic strategies—especially for disabling, deadly, and difficult-to-treat diseases [44].

The efforts of scientists in such ambitious projects can provide better insight into the future of systems biology and medicine. However, there is a danger of getting buried under the gigantic data sets where complexity will again supersede. This is what happened in the past when scientists were dreaming about genomic and molecular medicine. The hopes and aspirations created by the hype of omics technologies have not been turned into reality, especially in the field of drug discovery and molecular and personalized medicine. The smarter way is going back to traditional wisdom to get new insights, understandings, and leads so that we are able to disrupt the vicious cycle of technology–data–technology.

We feel that the integration of ancient experiential sciences like Ayurveda can help to expedite ambitious projects like 100K, P4 Medicine, Horizon 2020, P3 Medicine, person-centered medicine and whole person healing approaches. Against this background, it is interesting to review systems approaches in Ayurveda. Although some details are also available in the primer, information relevant to systems approach is discussed here for ready reference.

THE SYSTEMS APPROACH OF AYURVEDA

Traditionally, the relationships between the parts and the whole have been recognized in many ancient cultures as the macrocosm and microcosm relationship. The Vedic concept of *Pinda* (microcosm) and *Brahmanda* (macrocosm) suggest a harmonious continuum of complex interdependent relationships between various living and nonliving forms in the interconnected world and the universe. The microcosm generally represents any living form: the smallest of microscopic cells, to the developed organism, or evolved plant and animal species; macrocosm represents the nonliving materials like the smallest atoms, dust particles, rocks, mountains, rivers, oceans on the Earth, and stars or galaxies in the universe. The *Mahabhoota* theory explains how the primordial elements like earth, water, fire, energy, air, and ether are common between the *Pinda* and *Brahmanda*. If one imagines smallest of the smallest unit in brain known as neurons as *Pinda* and the whole universe as *Brahmanda*, the comparison might not be fictional. They astonishingly look similar (Figure 5.5). Scientists think that the number of neurons in the human brain is the same as the number of stars in the galaxy. Thus the entire universe is interconnected. This is one of the fundamental principle of Indian philosophy and Ayurveda.

Ayurvedic concepts of *Pinda*/microcosm and *Brahmanda*/macrocosm entities considers human being as a *whole*, involving body, mind, and spirit, and not merely a composite of cells, tissues, or organs. While doing so, it may not offer a detailed understanding of genes, molecules, cells, and biological processes. On the other hand, modern science and biology considers living or nonliving entities as parts in isolation and can offer a deeper understanding at the atomic, molecular, and cellular levels. However, in the process of getting deeper into parts, it loses sight of the whole. Therefore, the modern biological approach is considered reductionist, and Ayurveda, holistic. In the interconnected world it is becoming increasingly clear that assumptions based on limited understanding of parts might not be relevant or valid to the whole system. The whole need not be equivalent to the sum of the parts. This realization has given rise to systems thinking among modern scientists. Actually, the modern systems approach is an extended form of the traditional holistic approach; in reality, both are important.

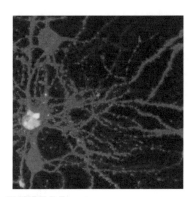 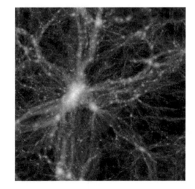

FIGURE 5.5
The structures and organization of macarocosm and microcosm look astonishingly similar. Microscopic structure of mouse neuron (left) and computerized model of the universe (right).

Against this background, it is important to take a closer look at the holistic or systems nature of Ayurveda and Yoga. A complex one-to-one, one-to-many, and many-to-many relationship diagram in the form of graphical notation describes the schema of the Ayurveda knowledge base. This logic chart, titled Systems Ayurveda, gives a comprehensive view of concepts and entities, including cause and effect relationships. The immense data consisting of over 200 variables, describing the logical flow and concepts of "Systems Ayurveda," have been presented using the systems biology graphical notation (SBGN) approach for processes, entity relationships, and activity flow [45] (Figure 5.6). This chart broadly describes logical flow, forward loops, backward loops, and entity relationships, and their application to health and disease.

The foundation and logic of Ayurveda is mainly based on ancient Indian schools of philosophy. According to this philosophy, all matter is composed of five primordial elements. The manifested parts of matter are microcosm

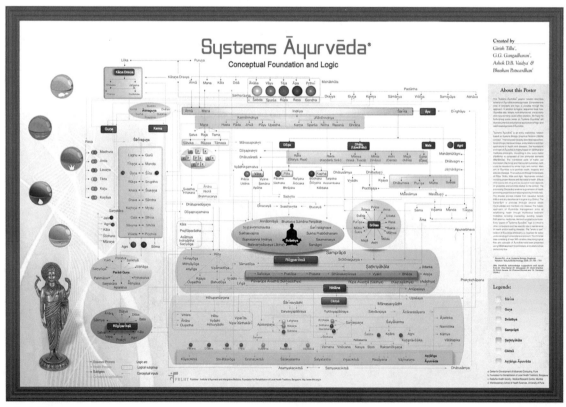

FIGURE 5.6

Conceptual foundation and logic of Ayurveda.

(the living cell) and macrocosm (universe); both could be assessed using similar logic and methods. The main aim of Ayurveda is to promote health and longevity and alleviate diseases. This is achieved through homeostasis, and the balancing of *Dosha*, *Dhatu*, *Mala*, and *Agni*. Appropriate conduct, including proper lifestyle and diet, leads to health. The Ayurveda approach enables augmentation of health-promoting properties and reducing the toxicity of medicines. The causative factors and disturbances in organs take several steps, which progress toward disease manifestation. The holistic approach of Ayurveda aims to establish health through multifarious treatment modalities, including counseling, diet advice, *Panchakarma*, herbal medicines, and surgical procedures.

Any substance (*Dravya*) has some attributes (*Guna*) and actions (*Karma*). Ayurveda explains every action in terms of *Dravya*, *Guna*, and *Karma*. The description of the physiology, pathogenesis, and pharmacology of Ayurveda is based on this concept. For example, the drug *Piper longum* has a hot (*ushna*) property, which alleviates *Kapha* and aggravates *Pitta* (Figure 5.7).

The logic and ontology of Ayurveda is similar to systems markup language, or SBGN. It specifies various types of relationships and logical group(s) of components, along with their entity relationships. Dr Iris Bell and colleagues from the University of Arizona have proposed an interesting hypothesis that aptly captures the systems approach of Ayurveda. According to Bell et al., nonlinear dynamic complex approach and whole systems approach of medicine like Ayurveda do not necessarily treat a symptom directly but try to modulate the imbalance to regain health through dietary and lifestyle modifications in accordance with *Prakriti* [46].

In this context, it will be interesting to see how Ayurvedic clinical evaluation is done utilizing the whole systems approach—where the patient is at the center and in the context of the internal and external environment in relation to the disease (Figure 5.8).

Historically, many philosophers, scientists, and physicians have recognized the importance of the whole system. Plato and Aristotle, the pioneers of reductive logic, suggested that in life systems, the relationship of the whole and the parts—microcosm and macrocosm—cannot be undermined. After discovering more and more details about the parts, modern science once again realizes the need for the whole picture. This is even more relevant and crucial in modern medical health care. Systems biology is an effort to relink the part and whole understanding, without compromising the value of details and without losing the reality of the whole.

A quote from Charaka Samhita restates Plato's admonition: *"No knowledge is complete unless it is studied as a whole."*

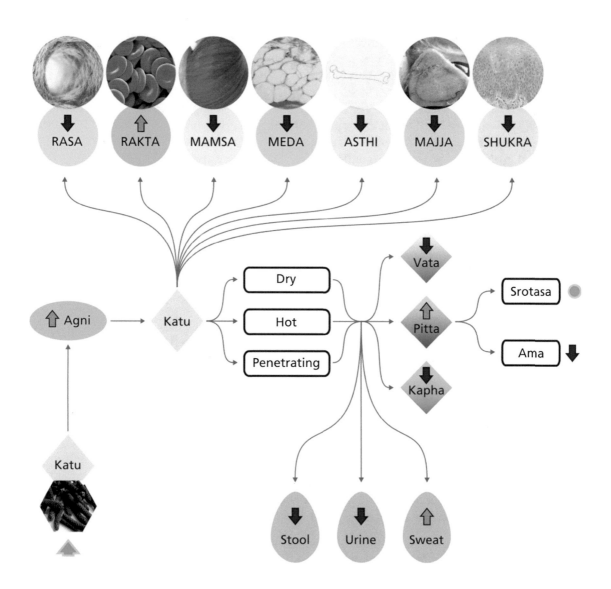

Upregulation
Downregulation

FIGURE 5.7

A drug *Piper longum* with *Katu Rasa* creates multiple systemic effects. It facilitates the process of bioregulation, depletes tissues, induces vasodilatation, causes sweating, and improves digestion.

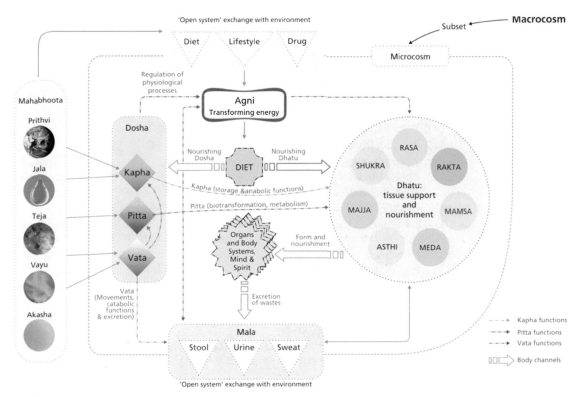

FIGURE 5.8

The figure explains systems approach of Ayurveda. The biological entities are open systems that interact continuously with the external environment. The bioregulatory mechanisms are represented as functions of three *Dosha* which govern nutrition, biotransformation, and excretion mechanisms.

REFERENCES

[1] Saks V, Monge C, Guzun R. Philosophical basis and some historical aspects of systems biology: from Hegel to noble – applications for bioenergetic research. Int J Mol Sci 2009;10(3):1161–92.

[2] Sweeney K, Kernick D. Clinical evaluation: constructing a new model for post-normal medicine. J Eval Clin Pract 2002;8(2):131–8.

[3] http://plato.stanford.edu/entries/reduction-biology.

[4] Bianconi E, Piovesan A, Facchin F, Beraudi A, Casadei R, Frabetti F, et al. An estimation of the number of cells in the human body. Ann Hum Biol 2013;40(6):463–71.

[5] Lobo I. http://www.nature.com/scitable/topicpage/biological-complexity-and-integrative-levels-of-organization-468.

[6] http://ghr.nlm.nih.gov/handbook/basics/gene.

[7] Pertea M, Salzberg SL. Between a chicken and a grape: estimating the number of human genes. Genome Biol 2010;11(5):206. http://dx.doi.org/10.1186/gb-2010-11-5-206.

[8] http://arxiv.org/abs/1312.7111.

[9] http://www.extremetech.com/extreme/134672-harvard-cracks-dna-storage-crams-700-tera-bytes-of-data-into-a-single-gram.

[10] http://www.ncbi.nlm.nih.gov/gene/statistics.

[11] Berg EL. Systems biology in drug discovery and development. Drug Discov Today 2014;19(2):113–25.

[12] Gehlenborg N, O'Donoghue SI, Baliga NS, Goesmann A, Hibbs MA, Kitano H, et al. Visualization of omics data for systems biology. Nat Methods 2010;7(3 Suppl):S56–68.

[13] Rosen RA. Means toward a new holism. Science 1968;161(3836):34–5.

[14] Smuts JC. Holism and evolution. N & S Press; 1926.

[15] Kia B, Murali K, Motlagh MRJ, Sinha S, Ditto WL. Synthetic computation: chaos computing, logical stochastic resonance, and adaptive computing. In: International conference on theory and application in nonlinear dynamics. Springer International Publishing; 2014.

[16] Chang TM. Semipermeable microencapsulation. Science 1964;146(3643):524–5.

[17] Soon-Shiong P, Heintz RE, Merideth N, Yao QX, Yao Z, Zheng T, et al. Insulin independence in a type 1 diabetic patient after encapsulated islet transplantation. Lancet 1994;343(8903):950–1.

[18] Omenn GS. Grand challenges and great opportunities in science, technology, and public policy. Science 2006;314(5806):1696–704.

[19] Gibson DG, Glass JI, Lartigue C, Noskov VN, Chuang R-Y, Algire MA, et al. Creation of a bacterial cell controlled by a chemically synthesized genome. Science 2010;329(5987):52–6.

[20] Budin I, Devaraj NK. Membrane assembly driven by a biomimetic coupling reaction. J Am Chem Soc 2012;134(2):751–3.

[21] Annaluru N, Muller H, Mitchell LA, Ramalingam S, Stracquadanio G, Richardson SM, et al. Total synthesis of a functional designer eukaryotic chromosome. Science (80-) [Internet] 2014;344(6179):55–8.

[22] Douglas T, Powell R, Savulescu J. Is the creation of artificial life morally significant? Stud Hist Philos Sci 2013;44:688–96.

[23] Malyshev DA, Dhami K, Lavergne T, Chen T, Dai N, Foster JM, et al. A semi-synthetic organism with an expanded genetic alphabet. Nature 2014;509(7500):385–8.

[24] Yates FE. Homeokinetics/homeodynamics: a physical heuristic for life and complexity. Ecol Psychol 2008;20:148–79.

[25] Lloyd D, Aon MA, Cortassa S. Why homeodynamics, not homeostasis? Scientific World Journal 2001;1:133–45.

[26] Zaloga GP, Sager A, Black KW, Prielipp R. Low dose calcium administration increases mortality during septic peritonitis in rats. Circ Shock 1992;37(3):226–9.

[27] Oliveira-Filho J, Silva SCS, Trabuco CC, Pedreira BB, Sousa EU, Bacellar A. Detrimental effect of blood pressure reduction in the first 24 hours of acute stroke onset. Neurology 2003;61(8):1047–51.

[28] Scheer FA, Czeisler CA. Melatonin, sleep, and circadian rhythms. Sleep Med Rev 2005;9(1):5–9.

[29] Wu GQ, Arzeno NM, Shen LL, Tang DK, Zheng DA, Zhao NQ, et al. Chaotic signatures of heart rate variability and its power spectrum in health, aging and heart failure. PLoS One 2009;4(2):e4323. http://dx.doi.org/10.1371/journal.pone.0004323.

[30] Ahn AC, Tewari M, Poon CS, Phillips RS. The clinical applications of a systems approach. PLoS Med 2006;3(7):e209.

[31] Watve M. Diabetes in a textbook. In: Doves, diplomats, and diabetes. New York: Springer; 2013.

[32] Jain P, Vig S, Datta M, Jindel D, Mathur AK, Mathur SK, et al. Systems biology approach reveals genome to phenome correlation in type 2 diabetes. PLoS One 2013;8(1):e53522.

[33] Boussageon R, Bejan-Angoulvant T, Saadatian-Elahi M, Lafont S, Bergeonneau C, Kassaï B, et al. Effect of intensive glucose lowering treatment on all cause mortality, cardiovascular death, and microvascular events in type 2 diabetes: meta-analysis of randomised controlled trials. BMJ 2011;343:d4169.

[34] Salvi S. Story of diabetes: how the world got misled into the wrong understanding of this deadly disease (Hypothesis presentation at Interdisciplinary School of Health Sciences, Savitribai Phule University of Pune). August 12, 2014.

[35] Watve MG, Yajnik CS. Evolutionary origins of insulin resistance: a behavioral switch hypothesis. BMC Evol Biol 2007;7:61.

[36] Schaller S, Willmann S, Lippert J, Schaupp L, Pieber TR, Schuppert A, et al. A generic integrated physiologically based whole-body model of the glucose-insulin-glucagon regulatory system. CPT Pharmacometrics Syst Pharmacol 2013;2:e65.

[37] O'Rahilly S, Turner RC, Matthews DR. Impaired pulsatile secretion of insulin in relatives of patients with non-insulin-dependent diabetes. N Engl J Med 1988;318(19):1225–30.

[38] Stahl M, Brandslund I. Measurement of glucose content in plasma from capillary blood in diagnosis of diabetes mellitus. Scand J Clin Lab Invest 2003;63(6):431–40.

[39] Ahn AC, Tewari M, Poon CS, Phillips RS. The limits of reductionism in medicine: could systems biology offer an alternative? PLoS Med 2006;3(6):709–13.

[40] Lusis AJ, Attie AD, Reue K. Metabolic syndrome: from epidemiology to systems biology. Nat Rev Genet 2008;9(11):819–30.

[41] Dutta P, Courties G, Wei Y, Leuschner F, Gorbatov R, Robbins CS, et al. Myocardial infarction accelerates atherosclerosis. Nature 2012;487(7407):325–9.

[42] Hood L, Price N. Demystifying disease, democratizing health care. Sci Transl Med 2014;6(225):225ed5.

[43] Clermont G, Auffray C, Moreau Y, Rocke DM, Dalevi D, Dubhashi D, et al. Bridging the gap between systems biology and medicine. Genome Med 2009;1(9):88. http://dx.doi.org/10.1186/gm88.

[44] Meng Q, Mäkinen V-P, Luk H, Yang X. Systems biology approaches and applications in obesity, diabetes, and cardiovascular diseases. Curr Cardiovasc Risk Rep 2013;7(1):73–83.

[45] Le Novère N, Hucka M, Mi H, Moodie S, Schreiber F, Sorokin A, et al. The systems biology graphical notation. Nat Biotechnol 2009;27(8):735–41.

[46] Bell IR, Koithan M, Pincus D. Methodological implications of nonlinear dynamical systems models for whole systems of complementary and alternative medicine. Forsch Komplementärmedizin 2012;19(1):15–21.

CHAPTER 6

Lifestyle and Behavior

Doctors give drugs of which they know little, into bodies, of which they know
less, for diseases, of which they know nothing at all.

Voltaire

THE EPIDEMIC OF LIFESTYLE DISEASES

Throughout history, the world has seen severe epidemics of deadly infectious
diseases like small pox and the plague. With improved hygiene, sanitation,
and better nutrition, coupled with effective antibiotics and vaccines, many
such communicable diseases were controlled and eradicated. Now in the
twenty-first century, the world is witnessing major epidemics of noncommunicable diseases (NCDs), which are mainly the result of changes in lifestyle and
human behavior. These diseases include high blood pressure, heart disease,
obesity, diabetes, chronic lung diseases, cancer, Alzheimer disease, and others.
Many NCDs, also known as lifestyle diseases, are known to appear as a cluster
of diseases resulting in more complex conditions like metabolic syndrome.

A World Health Organization (WHO) report which gives an overview of
NCD reveals that more and more countries are adopting strategies to control and monitor known risk factors like unhealthy diet, physical inactivity,
tobacco, and the harmful use of alcohol. According to WHO Director General,
Dr Margaret Chan, 85% of premature deaths in developing countries are
because of NCDs. The challenges presented by these diseases are enormous.
Every year, 38 million people die from NCDs, of which about 28 million are
from developing countries. Nearly 16 million of these die before the age of
70. Since the beginning of the new millennium, the number of deaths due to
NCD has increased worldwide. In 2013, during the World Health Assembly,
194 member countries accepted the WHO Global Action Plan for the Prevention and Control of NCDs. This plan hopes to attain at least a 25% reduction
in premature mortality from NCDs by the year 2025.

The challenge to control the present epidemic of NCDs or lifestyle diseases
seems to be even more difficult than the battle against communicable diseases.
Mainly because of the fact that NCDs cannot be controlled merely by discovering powerful drugs; the root causes of these diseases must be controlled.

141

Integrative Approaches for Health. http://dx.doi.org/10.1016/B978-0-12-801282-6.00006-1

NCDs cannot be controlled unless healthy lifestyles, and behavioral modifications are adopted by people. Therefore, preventive strategies through governmental policies, and participatory approaches involving communities are required. This may involve promoting healthy eating habits, physical exercise, and avoiding mental stress. Behavioral modifications may involve refraining from use of harmful substances like tobacco, drugs, and alcohol, and strengthening the mind to be able to cope better with adversity. In addition, reducing environmental pollution and enhancing general safety is also necessary.

LIFESTYLE MODIFICATION

Many disorders, diseases, and syndromes have their origin in modern sedentary lifestyle. Due to technological advances and automation, the extent of physical activities and exercise has significantly reduced. Lifestyle diseases can be better managed with healthy lifestyle. Mere medicine will not suffice. Therapeutic interventions through diet, physical exercise, and lifestyle modifications have emerged as important prevention and treatment strategies. A lot of research is underway in various institutions worldwide.

A cluster of diseases involving high blood pressure, high blood sugar level, excess body fat around the waist, and abnormal cholesterol levels are often found together. This condition is now recognized as Syndrome X, or metabolic syndrome, which increases the risk of heart disease, stroke, and diabetes. Metabolic syndrome is considered as a typical outcome of the modern sedentary lifestyle, inadequate physical exercise, and the consumption of unhealthy junk foods. Aggressive life style modifications—improving diet, nutrition, and physical activity—have been shown to play a preventive role. However, what type of diet and exercise is to be adopted needs careful consideration.

Many times vitamins, minerals, and antioxidant supplements are used as preventive or curative agents, however, no substantial evidence to support their use is available. In many studies it has been reported that instead of such dietary supplements and nutraceuticals, consumption of natural foods, fruits, and vegetables has been found to be much better. Similarly, exercise does not mean merely joining a health club, building muscles, burning calories through physical activity, or practicing aerobic and anaerobic exercises. Exercise science is now more advanced, and uses various combinations of physical and mental exercises depending on specific conditions, and individual needs. Yoga practice involving stretching, breathing, and meditation has been shown to be very beneficial. In short, life style modifications involving appropriate diet, nutrition, and exercise is valuable in the prevention and control of metabolic syndrome, and many other associated conditions, including hypertension, ischemic heart disease, diabetes, arthritis, asthma, cancer, and polycystic ovarian syndrome. Nutritional supplementation consisting of vitamins, minerals, and antioxidants can be useful, but do not suffice. More discussions on health supplements are available in Chapter 8. A short review of research on lifestyle and behavior modification provides more insight on this subject.

BODY, MIND, CONSCIOUSNESS

The relationship between the mind and brain is still a central philosophical question. Neuroscientists might say that the brain causes mind. But where consciousness arises is still not known in modern science. Many scientists believe that the brain and the mind are different types of entities. The brain is a physical entity, and the mind is mental. The mind–brain relationship is also explained based on a theory of consciousness, which is not fully rejected by science. Many scientists believe that consciousness is created by electrochemical activity within the brain. How the functions of the brain can produce consciousness is not clearly known. The conscious mind can realize human existence.

The brain is not merely a computer. If that were so, near-death experiences, out-of-body experiences, past-life memories, transcendent states, and spirituality would never have existed. However, many of these phenomena have been well documented scientifically. The brain-as-computer approach has been challenged by neuroscientists who attempt to understand the scientific basis for seemingly spiritual mental states [1].

Admittedly, modern science understands the brain better than the mind. Scientists can track brain activities with the help of sophisticated equipment like functional magnetic resonance imaging (MRI) and single photon emission computerized tomography. Advances in psycho-neuro-endocrinology have helped to understand how the brain can control many vital functions of the body. Scientists can measure the levels of neurotransmitters, and relate them to various biological activities affecting body functions. It is easier to measure brain activities, so following the principle of management, it is easier to manage the brain, as opposed to managing the mind.

Not surprisingly, the current focus of discussions about health and disease is primarily on brain and body. This is probably because they are relatively better known to science. Although the importance of mental health is known, our current understanding about it is preliminary. In biomedicine, psychiatric diseases are treated mostly with the help of drugs acting on the brain and nervous system; this is because of a better understanding of neurobiology and endocrinology.

Many chronic, difficult-to-treat diseases involve mental status (psycho), and the physical body (somatic), together—resulting in pathophysiologic changes. Such diseases cannot be treated by addressing the physical body, or mental health, in isolation. Such diseases need holistic consideration involving body and mind. It is normal practice to use the term *holistic* for many traditional health practices like Ayurveda, Yoga, Sowa rigpa, Traditional Chinese Medicine (TCM), Tai chi, Kampo, and others. This is because these disciplines consider body–mind–spirit as a whole and not in isolation. They might use natural materials such as somatic medicines for *body*; for psychiatric conditions of *mind*, meditation can be good medicine; and for *spirit*, spiritual practices can serve as medicine. In reality, the spirit needs no medicine, still spiritual practices help to maintain the balance of mind and body.

Leon Eisenberg, a leader in psychiatry, and professor at Johns Hopkins and Harvard Medical School, pioneered a study on the effects of stimulants on attention deficit disorder in children. He admits that "during the late twentieth century, psychiatry practice moved from a state of *brainlessness,* to one of *mindlessness.*" Eisenberg indicated that earlier, the focus of psychiatry was on Freudian approaches in psychology, and recognized the role of the *mind*. The current focus is on the physical *brain* and gives more consideration to the role of neurotransmitters [2]. Psychiatric drugs were discovered through advances in neuropharmacology and brain chemistry, where the focus on mind was greatly compromised. In the process, in clinical settings, psychology was replaced with psychiatry.

Essentially, modern psychiatry uses depressants and stimulants as therapeutic strategies. Mental illness medications were first introduced in the mid-twentieth century, with antipsychotic drugs like chlorpromazine. Since then, many drugs have been developed and used primarily because of their effects on different neurotransmitters. These include new generation blockbuster drugs like Prozac, a selective serotonin reuptake inhibitor (SSRI); typical mood stabilizers, such as lithium carbonate; anticonvulsants like carbamazepine; popular anxiolytics like diazepam; barbiturates; and stimulants like caffeine. Western psychiatry tends to end here.

While these drugs have substantially helped in some of the mental disorders, their side effects and risks of abuse are quite worrying. According to the report from the Citizens Commission on Human Rights, global annual sales of antidepressants, stimulants, anxiolytic, and antipsychotic drugs are estimated to reach over $76 billion. In reports published by this commission, many critics and experts have raised serious questions related to conflicts of interests, and the nexus between industry and professional bodies.

For instance, the American Psychiatry Association gets one-fifth of its funding from the drug industry. According to Robert Whitaker, senior medical writer, and author of a book *Anatomy of an Epidemic,* in 2009 alone, just one company, Eli Lilly, paid $551,000 to the National Alliance on Mental Illness, $465,000 to the National Mental Health Association, $130,000 to an ADHD patient-advocacy group, and $69,250 to the American Foundation for Suicide Prevention. This example is an eye opener, and explains the intensity of this problem.

Marcia Angel, former Editor-in-Chief of the *New England Journal of Medicine* has rightly opined that psychoactive drugs are not the only option for the treatment of mental illness or emotional distress. Nonpharmacological interventions like psychotherapy, exercise, relaxation techniques, yoga, meditation, and behavioral and lifestyle medicine were found to be comparably efficacious to drugs for depression. Nonpharmacology interventions are safe and affordable with longer-lasting beneficial effects. However, for obvious reasons, the industry is not interested in promoting such easy, and affordable alternatives. Therefore, instead of supporting nonpharmacology measures, a systematic propaganda campaign asserts that modern drugs are potent and inevitable. According to Marcia Angel, more rigorous research is needed to explore various alternatives to current psychoactive drugs. Proper scientific and factual information about the strengths and limitations of psychoactive drugs need to be a part of medical education [3].

The market for antidepressants is quite large. Debilitating depressive disorders are common in all age groups, from children to senior citizens. Depression can affect individuals, families, societies, and functioning of a professional individual. A systematic review of 10 studies, involving 1235 participants with different severities of disorders does not establish the superiority of antidepressant drugs over psychological therapy [4]. Another systematic review has shown that in patients with chronic low-back pain, the efficacy of antidepressants was almost the same as that of a placebo [5]. In such situation, instead of antidepressants, the use of mind–body medicine involving yoga and mindfulness might be a better, affordable, and safer option.

EMERGING EVIDENCE FROM RESEARCH

During the last decade, many clinical trials on lifestyle and behavioral approaches have been published. Powerful tools like meta-analysis, and systematic reviews are being used to synthesize evidence from available research data. A systematic review has concluded that lifestyle modification intervention was effective in the management of metabolic syndrome—with a reduction in blood pressure, fasting blood glucose, triglycerides, and waist circumference [6]. The lifestyle modification interventions involved dietary modifications through the Mediterranean, or dietary approaches to stop hypertension like healthy diet and suitable forms of physical activity.

Type 2 diabetes is a typical disease of lifestyle. A review of eight trials involving 2241 participants who were put on a diet and exercise regime showed a reduced risk of diabetes, as compared with standard treatment group of 2509 participants. This study also reported favorable effects of diet and exercise on weight, body mass index, waist circumference, waist-to-hip ratio, blood lipids, and blood pressure. These interventions decreased the incidence of T2D in high-risk groups [7].

Polycystic ovary syndrome (PCOS) is a common condition affecting women. Overweight women can have reduced ovulation frequency, irregular menstrual cycles, and reduced fertility. Increased levels of testosterone is known to cause acne, and excess hair growth on the body and face. PCOS is known to be associated with hyperinsulinemia, insulin resistance, and abnormal cholesterol levels. PCOS affects quality of life, and can cause psychological conditions such as depression and anxiety. A review of six studies on 164 participants with PCOS has indicated that adopting a healthy lifestyle can reduce testosterone levels, improve insulin resistance, and reduce body weight, and abdominal fat. However, healthy lifestyle did not have a significant effect on cholesterol or glucose levels [8].

Rheumatoid arthritis and asthma are polygenic, immune-pathological, psychosomatic diseases. Diet and lifestyle are considered as one of the causative factors. Suitable modifications might help in the control of these chronic, and difficult-to-treat diseases. However, a systematic review of 14 trials with a total of 837 patients has not shown any conclusive evidence in support of dietary manipulation in rheumatoid arthritis [9].

A study of 38 patients with chronic asthma has shown that a calorie-controlled diet can be beneficial as an adjuvant to drug therapy, with no serious adverse effects. However, the impact of a calorie-controlled diet in the general asthmatic population has not been established [10].

Patients with coronary heart disease can substantially benefit from lifestyle intervention programs. The evidence summarized in a meta-analysis of 23 trials and 11,085 patients confirms that lifestyle modification programs are much better compared to routine clinical care given to coronary heart disease patients [11]. It is known that plasma homocysteine levels are associated with cardiac disease risk. A study reveals that lifestyle interventions, and vitamin B intake might lower homocysteine, and reduce the risk of Cardiovascular Diseases (CVDs) [12].

Lifestyle interventions involving healthy diet, and exercise can be much more effective if they are personalized to individual needs. Personalized lifestyle interventions and advice specific to *Prakriti*-based classification, with the consideration of time and seasons can be of great value as predictive, and preventive medicine. These concepts are also discussed in Chapters 6 and 11.

Recent reports indicate that about 37% of adults in the United States have neuropsychiatric symptoms, which are difficult to treat with standard treatments. Researchers have examined the usefulness of mind–body therapies in neuropsychiatric symptoms. This large study on 23,393 adults compared the use of mind–body therapy like yoga, meditation, deep-breathing exercises, biofeedback, energy healing, guided imagery, hypnosis, relaxation therapy, qi-gong, and tai chi. The study population was suffering from symptoms like headaches, daytime sleepiness, anxiety, depression, insomnia, memory, and attention deficit. This study concluded that adults with more than one neuro-psychiatric symptom tend to use mind–body therapies more frequently [13].

Psychological stress in cancer patients has been identified as a major problem in oncology. Mind–body therapies, including yoga, mindfulness, qigong, tai chi have shown potential to reduce stress, and improve the quality of life of cancer patients and survivors [14]. Researchers from the Mayo Clinic in Rochester have studied the effectiveness of a mind–body therapy involving stress management, and resiliency training (SMART) program on irritable bowel syndrome patients; stress management and SMART showed potential to improve anxiety, life satisfaction, and gratitude [15]. A joint clinical study was conducted by the University of California and University of Washington on 435 patients with low back pain. The study reported that patients who practiced yoga or meditation, and body awareness had better recovery from pain [16]. In an interesting meta-analysis of 1193 papers, involving 58 trials examining the effectiveness of therapies for hot flushes, it was found that yoga, and mindfulness-based behavioral modifications significantly reduced hot flushes, and improved cognitive symptoms, more than exercise alone [17].

A meta-analysis of 34 studies from 39 clinical trials involving a total of 2219 participants has shown evidence that mind–body therapies can increase the immune response to vaccination. Mind–body therapies also resulted in a reduction in inflammation markers, and improved virus-specific immune responses [18].

Thus, for psychiatric, psychosomatic, and lifestyle diseases, the role of *mind* needs to be given predominance over body and *brain*. The psychological or mental disease is not only in the brain—it might be the case that the mind is involved. The emergence and wide acceptance of mind–body therapy, mindfulness, and behavioral medicine is a clear indication of the growing awareness of this reality. Modern psychiatry must address this issue seriously, and judiciously balance pharmacological interventions involving drugs, and nonpharmacological interventions involving yoga, meditation, mind–body, and mindfulness practices.

YOGA AND MEDITATION

Vedic and Buddhist traditions have given much emphasis to the mind. The role of yoga and meditation to enhance self-awareness and consciousness has been well-recognized in the promotion of mental and physical health. It is possible to train our minds to become more flexible, and adaptable without losing focus on the goal. Yoga suggests the means of controlling the mind through a detached watchfulness of our thoughts. The health tips from yoga for achieving a sound mind suggest refraining from greed, grief, fear, anger, jealousy, attachment, and malice. Ayurveda also reiterates the same concept, and states that behavioral errors may lead to diseases.

Yoga offers various ways to achieve health, increase wellness, and prevent disease at physical, mental, and spiritual levels. Yoga and meditation techniques are becoming increasingly popular worldwide. Yoga is not a quick fix for health. Yoga can benefit those who are ready to put forth effort [19]. The spiritual and biological effects—together with physical and mental benefits—of yogic practices are also important. There is fair evidence supporting the belief that biomedicine and yoga may complement each other. Yoga practices have been shown to be useful in the prevention of several psychological problems, chronic diseases, and musculoskeletal disorders. Yoga has helped scientists to explore new paradigms for obtaining insights into physiological states and mind–body interactions [20]. The study of yoga and meditation, with the help of positron emission tomography, has helped to identify neural networks in brain regions which are active in different states of consciousness. Scientists have shown that systematic breathing exercises can alter cerebral hemisphere activity, neuroendocrine, and autonomic functions [21].

The mind–body relationship is becoming better understood through research on psycho-neuro-endocrinology aspects. The next step appears to be its extension to the deeper understanding of the concept of spirituality. Although, the term *spirituality* has yet to find a place in the official definition of health, a fairly good number scientific papers suggest a supportive role of meditation and spirituality—especially for cancer patients. A preliminary study published in *JAMA* has indicated benefits of yoga postures in the treatment of carpal tunnel syndrome [22]. Chanting yoga mantras has been reported to induce psychological and physiological effects [23]. In a randomized controlled study on lymphoma patients, a yoga program was found to be feasible for patients with cancer, and significantly improved sleep-related outcomes [24]. Another prospective, randomized trial has suggested that yoga can be complementary to the conventional treatment of pulmonary tuberculosis [25].

While pharmaceutical drugs are thought of as medicines for disease of the body, meditation can be thought of as a tonic for mind. Yoga places much importance on meditation. Meditation is about various practices and techniques designed to promote peace, tranquility, compassion, love, patience, generosity, and forgiveness. Through meditation, the attempt is made toward relaxation, and to strengthen willpower, and internal energy. Meditation can help improve concentration, and mind activity. Meditation can act as a medicine by enhancing the power of the natural healing force [26].

Meditation is often used to relax the mind, and heal many stress-related diseases like hypertension, anxiety, and depression. Meditation can be done in various ways. According to yoga, one can do meditation through prayers, worshiping, and total devotion; it can be done by extreme dedication, involvement, and commitment to work, or a worthy cause. Meditation actually is an emotional state, which may involve chanting a mantra, and closing the eyes to attempt critical introspection, and sense the inner voice. In short, there are several variants, meanings, perceptions, and prescriptions for meditation. Meditation can be practiced for specific, desired benefits, and also as an attempt to attain ultimate salvation, or *nirvana*. In general, meditation is an individual or group attempt to train the mind, and to induce a state of consciousness [27].

Many times meditation relates to spirituality, but spirituality has a much broader meaning, which will be discussed later in this chapter. Meditation and spirituality have become a global market. Practically every religion has some form involving chanting, rituals, or practices, which can be considered meditation. Many variants of meditation are promoted by yoga celebrities and spiritual gurus—creating cults, or giving it a mystical aura. Many physicians and scientists have devised easier methodologies for specific purposes, and have given these methods proprietary names such as transcendental meditation, and mindfulness meditation. We will discuss a representative case of mindfulness meditation in the following section.

MINDFULNESS

The concept of mindfulness has its roots in yoga and meditation practices from Buddhist and Zen traditions. Mindfulness is about maintaining a continuous awareness of our thoughts, feelings, sensations, and external environment. Mindfulness is a wholesome mental state capable of developing deep insight into the nature of reality. More research on neuroscience and consciousness is unraveling the science of mindfulness [28].

Mindfulness is an art that attempts to achieve awareness to individual thoughts and feelings, and to spur the subject to get involved in a process of believing or judging them as right or wrong. A new field known as mindfulness-based interventions is emerging [29]. Yoga, Zen and Buddhist traditions are sometimes linked to particular societies, religions, and cultures so there might be certain inhibitions for their universal acceptance. Therefore, using basic principles from these important traditions, a secular practice of mindfulness has been promoted by Dr Jon Kabat-Zinn, Professor of Medicine at the University of Massachusetts Medical School. In 1979, Dr Jon established the Center for Mindfulness in Medicine, and launched a program known as Mindfulness-Based Stress Reduction (MBSR). Since then, many studies have been conducted on the physical, and mental health benefits of mindfulness. The scientific evidence of benefits from many studies resulted in the growing acceptance of the MBSR programs in various places like hospitals, veterans' centers, schools, and prisons. The regular practice of mindfulness meditation for a few weeks can lead to many physical, psychological, and social benefits.

The effect of mindfulness training on resilience mechanisms in active-duty marines preparing for deployment showed greater reactivity and enhanced recovery of heart rate, and breathing rate after stressful training; lower plasma neuropeptide Y concentration; and decreased blood-oxygen-level-dependent signal. This study indicates that mechanisms related to stress recovery can be modified in healthy individuals prior to stress exposure. Thus, mindfulness training might have important implications in the prevention of stress, and in effective treatments for better mental health [30]. Participants in mindfulness stress reduction programs showed improvements in worry severity, and improvements in memory. An aging population may have anxiety and mood disorders as common problems, where stress reduction intervention can provide great public health value [31]. In general, there is convincing evidence that mindfulness is safe and therapeutic for psychosis [32].

A meta-analysis of 47 trials with 3515 participants indicated that mindfulness meditation programs had moderate evidence of improved anxiety, depression, and pain, and some benefit to mental health-related quality of life. This study concluded that small to moderate reductions of psychological stress can be achieved through meditation programs. Researchers suggested that the clinicians should talk to patients about the positive role of meditation in addressing psychological stress [33]. Mindfulness meditation is shown to reduce cognitive rigidity because it overcomes the tendency to be "blinded" by experience [34]. Studies have also shown that mindfulness was related to the somatic marker circuit of the brain [35].

Another meditation practice similar to mindfulness meditation practice, known as *Vipassana*, is popular in India, and a few other countries. Based on Buddhist tradition, *Vipassana* is an introspective practice in which the practitioner tries to develop insight into the true nature of reality. *Vipassana* practice includes contemplation, introspection, and the observation of bodily sensations. This technique of analytic meditation necessitates silence with no external communication on the part of the practitioner; this triggers a deep internal dialogue, and introspection about the meaning of changes in self-concept, ego defense, life, death, and decomposition. A few studies have shown beneficial neurobiological, and clinical changes following *Vipassana* meditation [36].

Tai chi chuan (TCC) is a Chinese martial art practice, which is also used for health benefits. A systematic review of seven studies involving 391 participants has shown that fibromyalgia symptoms can be improved by tai chi practice [37]. In another interesting study, compared with controls, TCC practitioners showed significantly thicker cortex in the precentral gyrusin of the right hemisphere, and in the superior temporal gyrus of the left hemisphere [38].

Thus, many scientific studies are indicating the usefulness of meditation—be it through mindfulness, *Vipassana*, tai chi, Zen, or yoga.

YOGA AND EXERCISE

Yoga breathing techniques, and physical postures are very sophisticated exercises, designed for the whole body. Specific postures are meant to target specific tissues, organs, and systems. There are 84 classical *Asana*; thousands of their variants are in practice. Again, many proprietary variants of yoga are being promoted in different parts of the world; just to mention a few—iyengar yoga, siddha samadhi yoga, kundalini yoga, and power yoga. On June 21, 2014— which happens to be the longest day of the year—a record number of 11,000 people practiced yoga in Times Square in New York City. The yoga wave can be witnessed all over the world.

In Indian tradition, the sun is given much importance as the source of vital energy. Salutation to the sun is considered a healthy practice. This practice has been cleverly used by sages to create one of the most simple, yet effective exercises for all age groups. This is known as the "Sun Salutation" exercise, or *Surya Namaskara*; it packages key postures from complicated yogic processes (Figure 6.1). It is actually a set of *Yogasana* with coordinated breathing patterns. Every day, just 12 or more repetitions according to the stamina of the individual, are recommended as daily exercise. Many studies have concluded that *Surya Namaskara* can be an ideal exercise to keep oneself in optimum level of fitness [39]. The regular practice of sun salutations as daily exercises is probably the easiest, most convenient, and economical exercise; it can be practiced singly, or in groups, and can help in disease prevention, and health promotion.

Yogasana-based exercise methods like *Surya Namaskara* are designed to effect the internal organs. Research on *Surya Namaskara* has revealed its efficacy in controlling bronchial asthma [40]; and its potential to improve pulmonary functions, muscle strength, and endurance, and cardiovascular parameters [41]. In another study, the practice of *Surya Namaskara* has shown to result in better synchronization of muscular movements with breathing, sympathetic arousal, and muscular exertion [42]. An integrative practice involving *Surya Namaskara*, along with *Pranayama*, *Yogasana*, and meditation, has shown significant improvement in social adaptation, and intelligence quotient—including that of mentally retarded children [43]. The speed of *Surya Namaskara* performance, and associated breathing patterns have specific effects. The effects

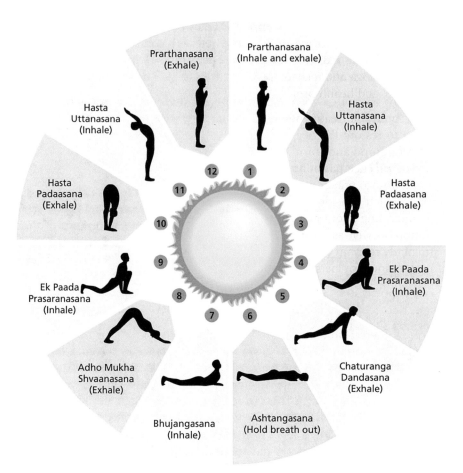

FIGURE 6.1
Sun salutations.

of performing *Surya Namaskara* at a fast pace are similar to physical, aerobic exercises, whereas the slow performance of *Surya Namaskara* is similar to those of yoga training [41]. The ancient Sanskrit verses (*Akal mrutyu haranam sarvavyadhi vinashanam*) tell us that the regular practice of *Surya Namaskara* has the potential to prevent untimely, premature death, and can cure most diseases. *Surya Namaskara* seems to be the most effective, most easily done, and most inexpensive practice for the protection of health.

Recent studies have demonstrated that even short-term training in techniques like yoga, and exercise programs can influence the sympathetic nervous system, and immune system. Healthy volunteers practicing these techniques have shown profound increases in the release of epinephrine, which leads to increased production of anti-inflammatory mediators [44]. Yoga interventions such as performing gentle postures, breathing exercises, and meditation have definitive advantages over other interventions, such as nature walks, and listening to relaxing music. A comparative study with these two interventions showed that yoga and related practices might provide longer-term physiological benefits, and health protection [45].

Typical Ayurveda treatments include drug, diet, and lifestyle modifications; all help to restore the health of the body, and treat somatic diseases. Yoga focuses on mind and spirit. This helps in the strengthening of mental status. Ayurveda practice has adopted many concepts from yoga, and offered them in easier-to-do, practical procedures. This makes Ayurveda a holistic health science, which considers mind–body–spirit. Ayurveda offers detailed advice, and counseling about avoiding causative factors, diet restrictions, sleep, and digestion, as well as a list of dos and don'ts. Two branches of Ayurveda are particularly relevant to lifestyle and behavioral medicine. Primarily, they provide doctrines for personalized diet, and healthy, lifestyle modifications— which chime with daily routine, seasonal changes, and the purification of body systems. These disciplines, known as *Swasthavritta* and *Panchakarma*, are described here as case studies of the preventive and therapeutic potentials of Ayurveda interventions.

Swasthavritta

Ayurveda advocates keys to health promotion through advice on diet and lifestyle. This branch is known as *Swasthavritta*—it provides dictums to follow for attaining health. Any deviation from the dictums of *Swasthavritta* may lead to an imbalance in *Dosha*, resulting in ill-health. *Swasthavritta* incorporates several dictums pertaining to age, nutritional status, metabolic attributes, and individual tolerance and sensitivity—all in the context of environmental and seasonal variations. There are three main components of *Swasthavritta*: daily routine, seasonal variation, and behavior. According to Ayurveda, the three *Dosha* undergo chronobiological changes during the day and night cycle. *Dosha* are also affected by the pattern of daily routine, foods habits, and behaviors. Ayurveda suggests matching daily food and behavior in such a way that the *Dosha* are maintained in a balanced state. For example, one should consume food when *Pitta Dosha* is at the highest peak, exercise when *Kapha* is increased, and rest when *Vata* is aggravated. If daily activities are not synchronized with *Dosha* levels, the resultant imbalance can cause disease. Hence, the history of a patient's diet and lifestyle is important in the diagnosis. The treatment includes detailed advice about when, and what to eat—according to the attributes of diet which result in keeping *Dosha* balanced. An asthmatic patient might be advised to drink medicated water before bedtime, in order to avoid the aggravation of *Kapha* during the night. Sometimes, a useful, therapeutic substance can be harmful if consumed at the wrong time.

The daily routine suggested by Ayurveda aims at protecting body, mind, and senses. Daily activities lead to accumulation of wastes in the internal organs, and to the degeneration of body tissues. There is overuse of some organs and body systems. Ayurveda advocates simple, but effective daily measures for replenishing, and repairing tissues. Many daily routine measures can have protective effects on body and mind. Interestingly, these effects seem to be linked with chronobiology.

Morning wake-up time: A person desiring health should get up early. The general suggestion is that one should get up at least 3 h before sunrise. It is also important to note that Ayurveda advises early bedtime, and the avoidance of late-night work. Many proverbs in various Indian languages describe early awakening as one of the healthy practices. Another dictum concerns digestion. One should ensure that food is digested (no heaviness in abdomen), sleep is complete, and the person is feeling energetic. Awakening at proper times is also helpful for bowel movements. Generally, the period before sunrise is *Vata* dominant, hence it is the best time for elimination of waste. Regular and timely elimination of urine and feces is important to keep systems clean. Proper day–night cycle automatically helps to keep the person healthy. Ayurveda also suggests the relation of sleep with digestion, and the immune system—leading to optimum physical and mental performance.

Sleep: Sleep is one of the important activities. It is known that adequate sleep prevents neurobehavioral deficits, improves memory, and helps to maintain a thoughtful mind [46]. Sound sleep provides nutrition to tissues, and rejuvenates the body. It endows strength, good complexion, growth, and longevity.

Food: The person should consume food only after the digestion of the previous meal, and when experiencing a proper sensation of hunger. The quantity of food should be limited (according to digestive capacity). Food items should vary according to a changing environment (details of food are covered in Chapter 7—Food and Diet).

Oral hygiene: Teeth, gums, and the oral cavity should be cleaned with brushes or powders having bitter, astringent, or pungent tastes. The herbs with these tastes decrease *Kapha*, and removes wastes (oral secretions and dental coating). The specific herbs indicated for dental cleaning remove tongue coating, open pores of salivary glands, and maintain oral health. Gargling with decoctions strengthen jaw, teeth, and gums; improve the quality of voice, reduce tastelessness, and heal ulcers of the gums. Regular gargling with medicated oils strengthens the jaw and teeth, and heals oral ulcers.

Sensory organs: Specific eye-cleansing lotions are advised to maintain the health of eyes. Oil drops in ear prevent torticollis, lock jaw, and deafness. Medicated oils, and clarified butter drops in the nostrils are helpful in improving respiratory health; to prevent aging; brain degeneration; and to strengthen the head and neck. More information about the nasal administration of medical oil is discussed in Chapter 10—Longevity, Regeneration, and *Rasayana*.

Skin rub: Ayurveda advocates that after exercise, the body should be massaged with dry herbal powders. The skin rub with such powders helps fat reduction, and provides stability and strength. Interestingly, a skin rub with a dry towel is popular in Japan for preventing respiratory infections; it is a traditional therapy for health promotion known as *Kanpu Masatsu*. Research on skin rubbing shows striking results in immunological and metabolic variables. After a skin rubdown, pulse rate, body temperature, PO(2), SO(2), pH were increased and rouleaux formation by red blood cells was reduced. Adrenaline and noradrenaline levels increased, indicating sympathetic nerve dominance, with increase in granulocytes. It is suggested that a skin rub with dry towel can have a beneficial effect on the autonomic nervous system, improvement in the immune system, and give benefits such as those obtained with systemic, aerobic exercises [47].

Bath: In addition to skin and hair cleaning, the bath improves metabolism, blood circulation, energy, appetite, sexual vigor, enthusiasm, strength, and longevity. Research on the effect of a sauna bath on human physiological parameters suggested that skin blood flow, and the concentration of oxygenated hemoglobin is improved by a sauna bath. It also significantly reduces muscle fatigue [48]. Taking into account the interpretation of current research trends on massage, skin rubs, and bathing, we can say that sparing a few minutes for exercise, combined with massage, rubbing medicinal powders, and a hot water bath can prevent several metabolic, and infectious diseases. The research trends also hint that innovative discoveries about daily activities are possible if we interpret Ayurveda in terms of modern biology.

Oil massage: Oil massage relieves fatigue, stops aging, and rejuvenates tissues. It also improves the capacity of the body to adapt to stress, and strenuous work, and makes the body able to sustain exercise, trauma, and hard work. Usually, in Indian tradition, any physical exercise is followed by an oil massage. Regular massage reduces drowsiness, fatigue, lethargy, body odor, and lax skin. Massaging feet with oil prevents leg pain, sprains, sciatica, muscle spasms, and cracks. The description of the effect of foot massage includes important effects on eyesight. Ayurveda suggests massaging feet with cow ghee for eye strain, and burning eyes. Head massage is effective in preventing headache, baldness, and early graying of hair. It improves the health of sense organs, and produces sound sleep. Thus, oil massage is one of the important treatments for insomnia, and degenerative and neurological diseases. Regular massage is one of the important secrets of Ayurveda for longevity. Massage is one of the important parts of Indian health traditions.

However, oil application is contraindicated in a state of indigestion, inflammation of the ear, and ear drum perforation. Ayurveda suggests that the ideal timing for massage is in the early morning on an empty stomach (after passing stools), when the body is ready to accept, and digest the oil applied on the skin.

Emerging research on massage is adding various pieces of information. In a pilot study on massage, the participants showed significant reductions in stress with just one session of massage treatment. Reductions in heart rate and blood pressure were also observed [49]. In another study, massage therapy resulted in the reduction of vascular endothelial adhesion molecules 1, and intracellular adhesion molecules 1, in hypertensive women, suggesting its role in the prevention of cardiovascular diseases [50].

Integrative applications of massage have been shown to improve patient satisfaction during hospitalization. In a randomized trial postoperative cardiac surgery patients receiving massage therapy reported significantly decreased pain, anxiety, and tension. These patients showed high levels of satisfaction with massage. According to Dr Brent Bauer from Mayo Clinic in Rochester, massage therapy may be an important component of the healing experience for patients after cardiovascular surgery [51].

A recent meta-analysis reports that massage combined with antihypertensive drugs might be more effective than antihypertensive drugs alone [52]. Massage with aromatic oils applied to middle-aged women with hypertension also showed remarkable benefits in terms of significant reduction in blood pressure, sleep quality, and overall quality of life [53]. Massage therapy was found to be a safe, effective, and cost-effective intervention for controlling blood pressure in prehypertension women [54]. Researchers have tried to assess the duration of massage effects. The beneficial effects of massage on blood pressure may be sustained up to 72 h [55].

Massage has shown remarkable effects in diabetes patients. The diabetic patients, who were treated with gentle massage, showed changes in metabolic markers. They showed improvements in adiponectin, and ratios of adiponectin-to-leptin, and adiponectin-to-HbA1c levels [56].

In India, baby massage is a traditional practice. Massaging in infants results in a higher daily weight gain, and increased potential to fight viruses, and malignant cells [57].

Exercise: Ayurveda emphasizes exercise as a routine. Daily exercise maintains healthy weight and creates a feeling of lightness and strength. It facilitates proper digestion, removes excess fat, and bestows a distinct physique. Everyone except the aged, children, those whose food is not digested, and those having *Vata* and *Pitta* diseases should do exercise. One should exercise up to "half of the one's strength". The body should be massaged after exercise.

Traditional Indian exercises are quite different to typical aerobic exercise. The aim of exercise is not just to increase the heart rate, blood supply, to dilate vessels, and tone muscles. These exercises are easy to learn and practice. *Surya Namaskara* is the best example, and has been discussed in earlier sections. These exercises do not require any special gadgets, equipment, or facilities. The exercises are subtle and include specific cycles of stretching, relaxation, breathing, and meditation. They target the entire body—especially abdominal and thoracic organs, and the mind and brain.

Ayurveda advises specific changes in diet and lifestyle according to various seasons. This is known as *Ritucharya*. Specific interventions like *Panchakarma* are also advocated in particular seasons. The guidelines for daily routine, seasonal changes, and treatment regime are specific to *Dosha* changes, and aim to achieve a dynamic balance of *Dosha*. The accompanying charts of *Dinacharya* and *Ritucharya* are useful in understanding the specificity and richness of Ayurvedic advice for health promotion.

LIFESTYLE AND BEHAVIORAL MEDICINE

The concepts of *Swasthavritta* overlap with the modern concepts of behavioral medicine, and lifestyle interventions. During the past four decades, behavioral medicine has emerged as an interdisciplinary field. It deals with behavioral, sociocultural, psychosocial, and biomedical aspects—all relevant to health and illness. The systematic use of this knowledge has begun in diagnosis, prevention, and the treatment of disease, and in health protection, promotion, and rehabilitation [58]. The Academy of Behavioral Medicine Research was founded in 1978, by the eminent psychologist, Dr Neal Miller. Around the same time, the Society of Behavioral Medicine was established in the United States to promote interdisciplinary studies on the interactions of behavior with biology, and the environment to improve health and well-being. Today, the International Society of Behavioral Medicine has a membership from 29 countries. The field of lifestyle medicine has grown over the past two decades, mainly because of the efforts of scientists like Dr James Rippe, Professor of Biomedical Sciences at the University of Central Florida, and Dr David Katz, the founding director of the Prevention Research Center at Yale University. The pioneering, scientific work of scientists such as Dr Elizabeth Blackburn and Dr Dean Ornish has also played a key role in drawing the attention of the scientific community to this emerging discipline. In the United Kingdom, the Society of Behavioral Medicine has been set up for behavioral interventions, and research.

Many scientific studies and reviews indicate the benefits from behavioral medicine. A meta-analysis has shown that the psychological preparation of patients undergoing surgery is very important. This can be done by giving procedural information, and behavioral instructions. This simple measure has reduced the use of analgesics, and the length of hospital stay [59]. Behavioral therapy has been shown to reduce chronic pain better, as compared to alternative, active treatments [60]. Behavioral interventions also help in disease prevention. An interesting meta-analysis comparing behavioral interventions such as intensive physical activity, and weight loss, with conventional drug therapy, has shown that the incidence of diabetes was significantly reduced, as compared with the conventional drug metformin [6].

Although behavioral medicine has an important role in the prevention and treatment of many diseases, sadly, it does not get due priority in research, policy, and practice. The nature of medical practice may be responsible for this. The medical school curriculum focuses on pharmacology, use of drugs, and surgery to manage diseases. The role of nonpharmacology measures such as behavioral and lifestyle interventions hardly finds a place in the curriculum. Moreover, the marketing tactics of pharmaceutical industries are very aggressive and attractive.

Although, behavioral and lifestyle interventions are simple and cost-effective, industry and government funding to research such nonpharmacology approaches remains very poor. For instance, in 2003, in the United Kingdom, the pharmaceutical industry spent about £3550 million on the research for drug discovery. This is more than two times the amount spent by the Medical Research Council, and major charities put together. The pharmaceutical industry strongly influences health care delivery strategies. There is reasonable evidence that industry funding can lead to biased results in favor of its products [61]. The WHO has admitted that global health priorities of preventing and managing chronic disease will not be achieved by using drugs and prescription, alone.

Behavioral and lifestyle interventions to improve health may not bring the immediate results to alleviate symptoms, as compared to powerful drugs. But these drugs also bring many side effects, which behavioral interventions do not. Lifestyle intervention benefits might be modest, but they are real, and at much reduced costs, increased empowerment, and autonomy for individual health. Experts are attempting to change hesitant attitudes toward behavioral medicine in science and health policy [62].

According to the American College of Lifestyle Medicine, the use of lifestyle interventions in the treatment and management of disease include diet, nutrition, exercise, stress management, smoking cessation, and many other nonpharmacologic modalities. Increasing scientific evidence indicates that lifestyle intervention is becoming an essential component in the treatment of chronic, difficult-to-treat, psychosomatic disorders, and diseases. Lifestyle and behavioral modifications can be as effective as drugs and medication. In a controlled study, 3234 nondiabetic persons were assigned to either a placebo, metformin, or a lifestyle-modification program. This study concluded that in the prevention and control of diabetes, the lifestyle intervention was more effective than metformin [63]. If used with proper care, lifestyle intervention is safe and devoid of serious side-effects. Lifestyle medicine is the preferred choice for the prevention and treatment of obesity, insulin resistance syndrome, type 2 diabetes, coronary heart disease, hypertension, osteoporosis, cancer, and other chronic diseases.

SALUTOGENESIS

A professor of medical sociology from Israel, Dr Aaron Antonovsky, coined the term salutogenesis as a new approach. The salutogenic model considers relationships between health, stress, and ways of coping. This branch of science studies how and why different individuals respond differently to stress. Dr Antonovsky worked to find out how some people successfully manage stress, and why some people show negative health outcomes due to stress. It was observed that many people survive, adapt, and overcome even the most difficult life-stress experiences. These observations, particularly on women who were in concentration camps, led to Antonovsky's salutogenic model [64].

Much work has been brought to bear in order to understand the phenomenon of salutogenesis, which is considered as the power of "inner strength," and a phenomenon of "coping with." Salutogenesis incorporates concepts of resilience, sense of coherence, hardiness, purpose in life, and self-transcendence—all connected to inner strength. Salutogenesis considers the ability to transcend, and the value of a firm stand, and moral support given by family, friends, society, nature, and spiritual dimensions. Inner strength is also about shouldering responsibility to endure and confidently deal with adversity [65].

Interestingly, the basic philosophies of salutogenesis, lifestyle medicine, and behavioral medicine have drawn substantially from traditional knowledge systems. The concept of salutogenesis draws substantially from yoga. The yogic practices of postures, meditation, and behavioral modifications help to recognize and strengthen *inner strength*, to cope up, and adapt—countering stress, and other adversaries. Lifestyle and behavioral measures of Ayurveda, and salutogenesis prevention strategies share a common perspective [66].

These new efforts have certainly brought to bear scientific understanding and clarity, and have packaged ancient concepts so that they become acceptable to contemporary cultures. For many psychiatric, psychosomatic, and chronic conditions, behavior medicine, and lifestyle medicine might have an advantage over psychotropic, or other symptom-specific drugs.

Simple interventions through behavioral medicine have been shown to reduce risk factors such as hyperlipidemia, and hypertension. Cardiac rehabilitation programs put a major emphasis on nonpharmacological interventions like yoga, meditation, exercise, and behavioral and lifestyle modifications. Depression is known to be associated with increased mortality, and nonfatal, cardiac events in patients with coronary heart disease. Depression is also associated with heart failure, and coronary artery bypass graft surgery. The standard therapy in such cases includes second-generation antidepressants such as SSRI. However, recent studies have indicated that for depression, antidepressants have marginal benefits as compared with a placebo. Instead, drugs like SSRI, behavioral therapy, and exercise together might play a better role in cardiac disease patients with depression [67].

Yoga continues to remain attractive to researchers due to the fact that it is a simple and safe intervention. Many teachings of yoga are now practiced as *behavioral medicine*. Yoga interventions can be easily integrated into any medical system. Nonpharmacological approaches from yoga, meditation, mindfulness, mind–body medicine, lifestyle, and behavioral medicine can play a substantial role in the treatment of chronic, psychosomatic, and psychiatric disease. Yoga offers a path for disease prevention, and health protection. More than just health, it advocates a comprehensive approach toward life. This approach changes outlook and behavior, and results in relieving stress, achieving psychological balance, and clarity in thought processes.

Panchakarma

A novel treatment approach for the elimination of vitiated *Dosha*, leading to physiological purification, and body detox is known as *Shodhana*. Various ways to attempt the purification of different body systems by removing toxic substances can be attempted with the help of specialized procedures known as *Shodhana*.

Panchakarma procedures bring a sophisticated confluence of pharmacological and physiological interventions. *Panchakarma* consists of five main procedures for maintenance of health, and management of disease pathology. *Panchakarma* procedures are primarily based on physiological processes in the body. Various procedures are included in *Panchakarma*: therapeutically controlled emesis, or vomiting known as *Vamana*; therapeutically controlled purgation, known as *Virechana*; nasal administration of drugs, known as *Nasya*; medicated enema, known as *Basti*; and bloodletting using medicinal leeches, known as *Raktamokshana*. These approaches are used for specific conditions. For example, *Vamana* is a choice of treatment for vitiation of *Kapha* leading to respiratory, digestive, and metabolic diseases; *Virachana* is administered for diseases of *Pitta*, leading to skin diseases, and acid peptic diseases; *Basti* is indicated for diseases of *Vata*, like rheumatic, musculoskeletal, and neurological diseases; *Nasya* is indicated in diseases of the head, sinus, and sensory organs; *Raktamokshana* is used for diseases of the blood.

Medicated fats, oils, clear butter, or ghee, are administered before *Panchakarma* for reliving vitiated *Dosha*. Fomentation is used to facilitate the expulsion of toxic wastes from *Dhatu*. Selected herbs are used to prepare hundreds of medicinal oils indicated for specific conditions. Ayurveda describes 13 different types of fomentations, such as local, general, with or without steam, nourishing, and anti-inflammatory. Medicated oils, or ghee can be administered as a steady drip on the center of the forehead. This is known as *Shirodhara*. *Panchakarma* procedures are also indicated for healthy people as seasonal therapies. *Vamana* is indicated after the winter, *Virechana* is advised before summer, and *Basti* is suggested in the rainy season to counter adverse effects on *Kapha*, *Pitta*, and *Vata Dosha*. *Vamana* and *Virechana* can be used as detox processes for purification— useful in the prevention and treatment of various diseases like asthma, irritable bowel syndrome, and skin diseases.

Each disease stage requires a specific treatment approach. For example, treatment of a malnourished patient requires not only nutrition, but also improved appetite, digestion, absorption, and metabolism. This can be done by the restoration of digestive fire, which can be achieved by specific herbal medicines, and *Panchakarma* processes.

Panchakarma are very special therapeutic procedures. They require the skills of an experienced physician, who can use appropriate drugs for the season. Experienced therapists to assist the physician, and the patient's cooperation are also necessary. Improperly performed *Panchakarma* can cause adverse effects. The beneficial effects of *Panchakarma* in chronic, and difficult-to-treat diseases are often experienced. Many hospitals, and treatment centers—especially in South India—are known for the classical treatment of *Panchakarma*.

Ayurveda therapeutics includes preconditioning of the body by processes which revitalize systems, and remove toxins. These interventions can be simple, cleansing procedures, such as medicated oil baths, massage, fasting, and special diet.

Ayurveda has prescribed *Raktamokshana* as a systematic procedure of bloodletting through venipuncture, medicinal leeches, and drainage of cysts and abscesses. The use of medicinal leeches (*Hirudo medicinalis*) is one of the important instruments in Ayurveda. Lord Dhanvantari, who is the creator of Ayurveda, is shown holding leeches in one of his hands. This is a symbolic description of the importance of use of leeches in surgical procedures. Leeches are indicated for skin diseases, pain management, infections, wound management, nonhealing ulcers, erysipelas, herpes, abscesses, and tumors. Ayurveda physicians use leeches for symptom management of wounds, pain, inflammation, and paresthesia.

Bloodletting may still be considered as a crude, ancient treatment; however, the use of leech therapy has received the attention of researchers in recent years. Many components having anti-inflammatory, thrombolytic, anticoagulant, and blood and lymph circulation-enhancing properties have been found in leech saliva [68]. The first clinical report on using leeches for reduction of the venous congestion of skin flaps was published in *British Journal of Plastic Surgery* in 1960 [69]. Venous engorgement, and reduced microcirculation are the most common problems in plastic and reconstructive surgery. A systematic survey of 62 plastic surgery units in the United Kingdom, and the Republic of Ireland, reported the trend of leech use in postoperative wound care [70]. Thus, leech therapy has had a renaissance in reconstructive microsurgery. The animal has become a favorite "instrument" of plastic surgeons.

Leeches have been used in various indications other than plastic surgery. In a trial on osteoarthritis patients, leech therapy showed significant, and sustaining effects [71]. In patients with varicose veins, leech therapy showed the decongestion, reversal of edema, reduction in hyperpigmentation, and healing of varicose ulcers. The surgeons also studied blood gasses of these patients, and found that pO_2 of blood sucked by the leech was similar to that of venous blood. It is suggested that leeches suck venous blood, and facilitate the healing of venous ulcers [72]. The precise mechanisms of leech application are under scrutiny. However, a case report suggests increased perfusion, and hyperemia when studied with bone scintigraphy. It hints at possible vasodilatation effects on superficial veins, after leech therapy [73]. Researchers have studied recombinant hirudin as an anticoagulant therapy for prophylaxis of thromboembolic complications [74]. The topical application of hirudin was reported to heal bruises and hematoma [75].

Use of medicinal leeches was discussed by the General and Plastic Surgery Devices Panel (GPS Panel) of the United States' FDA, in 2005. The panel unanimously recommended medicinal leeches as nonexempt, Class II medical devices, with special controls [75,76]. After FDA approval, Ricarimpex SAS, a French firm, which has specialized in leech farming since 1835, got permission to market leeches as medical devices in United States [77].

Panchakarma as Physiological Interventions

It is interesting to know how *Panchakarma* techniques work as physiological intervention. They are aimed at balancing *Dosha*, and maintaining physiological functions at optimum levels. *Panchakarma* procedures also facilitate adaptogenic, and homeostasis mechanisms. *Panchakarma* treatments can be loosely translated as body cleansing, detox procedures. They are actually sophisticated, physiological interventions to promote health, and prevent and treat diseases. They can modulate gut microbiota, and have stimulating effects on the enteric nervous system. According to a senior physician–scientist, Dr Ashok Vaidya, many *Panchakarma* procedures might have its effect through vagus nerve stimulation. The vagus nerve is the tenth cranial nerve, and is involved in autonomic, parasympathetic control of the heart and digestive tract. It also branches, and becomes the pharyngeal nerve, recurrent laryngeal nerve, and esophageal and pulmonary plexus. The vagus nerve hypothesis might explain the beneficial effects of *Vamana* on pulmonary functions. The following few examples will give an idea of the rationale for, and sophistication of, *Panchakarma* procedures.

Vamana: Modulating Vagus Reflux

Vamana is therapeutically controlled emesis with the help of specific medicinal preparations. Modern physiological understanding of simple emesis—vomiting or the urge to throw up—is that it is related to natural reflexes mediated through the nervous system. The stimulation of a chemoreceptor trigger zone in the brain, known as the area postrema, can lead to vomiting. This involves a trigger from motor, sympathetic, and parasympathetic nervous system. This zone has many receptors of neurotransmitters, such as dopamine D2, serotonin 5HT3, opioid, acetylcholine, and substance P. The area postrema is outside the blood–brain barrier, and can be stimulated by chemical substances, which can cross this barrier and reach the blood. Many chemicals have the capacity to stimulate or suppress this zone, resulting in the induction or suppression of vomiting. Vomiting can be the result of various triggers such as psychological stress, acute infection, motion sickness, radiation, and chemotherapy. Vomiting is understood mainly as a disease symptom. It is also a physiological mechanism to remove toxins, and unwanted substances. Ayurveda has innovatively used this physiological reflex for therapeutic purpose.

Virechana: Modulation of Gut Environment

Virechana is controllable, induced purgation. It is not mere purgation, but consists of sequential, stepwise bowel cleansing, and is a rejuvenating procedure with much larger physiological impact. *Virechana* is suggested as a treatment for many diseases related to GI tract, skin, liver, and blood. As it controls metabolic processes, it is an important therapy for management of *Pitta*. The patient is advised to consume medicated oils, or clarified butter (*ghee*) in escalating doses for three to 7 days in order to condition the system. When the patient reports oily skin, soft stools, and nausea, a purgative herbal formulation is given to induce loose motions. The entire process is under the observation of an experienced *Vaidya*. The purgative formulation has an antidote to control its effect, and to avoid dehydration and complications. After the session, the patient is advised to keep a strict diet, and advised to consume a liquid diet for 1–2 days, and a light semisolid diet after 3 days. The diet is subsequently changed so that the Agni, or digestive fire, of the patient can be slowly restored. The patient can reassume a regular diet within a week. Although the detailed physiological mechanisms of *Virechana* are not known, it might act by altering the gut microbiota.

Basti: Beyond Body Cleansing

Basti is drug administration through anal route, but it is not a mere enema. *Basti* is not like the mechanical cleaning of drainage pipe. It is mainly indicated in diseases of *Vata*. Ayurveda practitioners hold *Basti* in very high esteem in terms of its health benefits. *Basti* is indicated in diseases of the nervous system such as dementia, hemiplegia, and cerebral palsy; degenerative diseases such as osteoarthritis; autoimmune diseases such as rheumatoid arthritis; musculo-skeletal diseases like sarcopenia, and muscular dystrophy; and metabolic diseases such as malabsorption syndrome.

There are many types of *Basti* targeted for specific tissue nourishment, *Dosha* alleviation, rejuvenation, fat reduction, wound healing, aphrodisiac, *Dosha* provocation, and as treatment of specific diseases. Generally, a retention enema of medicated oil and enema with herbal decoctions in alternate sequence is given for elimination of toxic wastes. The ingredients of the enema are pre-scribed according to the patient, and disease status. *Basti* for health may con-tain a nourishing enema with herbal decoctions, oils, honey, rock salt, meat soup, and such materials.

Ayurveda clinicians often use *Basti* to reduce pain and stiffness in musculo-skeletal conditions. Recently research to study the role of *Basti* on markers of inflammation suggest a reduction in pro-inflammatory cytokines [78]. Recent research indicates that *Basti* can modulate immune responses, reduce pro-inflammatory cytokines, increase immunoglobulin levels, and alter functional properties of T cells.

OIL MASSAGE

Body massage with specialized oils can tone muscles, increase blood circula-tion, and stimulate neuromuscular function, in addition to providing relax-ation, and a feeling of well-being. A small, descriptive, observational, clinical study by the Osher Research Center of Harvard Medical School found posi-tive effects of *Panchakarma*-based intervention in behavioral, and psychosocial measures like adherence to positive lifestyle practices, self-efficacy, and the reduction of anxiety. This study suggests its efficacy in stress-related diseases [79]. A double-blind, randomized clinical trial involving the treatment of hemiplegia patients with nourishing massage and fomentation, suggests improvement in muscle tone and strength, tendon reflexes, range of movement, and functional abilities [80]. Cerebral palsy patients, when treated with massage, fomentation, with a bolus of drugs prepared with boiled rice, and nourishing enema, showed remarkable improvement in motor functions, and memory retention [81].

Panchakarma treatment may be useful in arresting degenerative changes of the brain. Indian researchers from the National Institute of Mental Health and Neurosciences treated patients of progressive cerebellar ataxia with a typical, holistic Ayurveda treatment composed of *Shirobasti, Shirodhara*, oil massage and steam bath, and oral medication. They observed improvement in balance within 14 days [82]. The effects of *Shirodhara* on psychological and physiological response were observed in healthy volunteers. *Shirodhara* demonstrated reduction in blood pressure, and heart rate, and an increase in alpha rhythm in EEG. The results suggest a wakeful relaxation response, and point to the potential of *Shirodhara* as a management for anxiety, neurosis, hypertension, and stress mechanisms [83].

The Maharishi Ayurveda Health Center of Norway, studied effects of *Panchakarma*, and diet and lifestyle modifications in patients diagnosed with fibromyalgia. In this study, 31 patients, were treated for 2 years. The treatment reduced impairment of ability to work, pain, tiredness, morning tiredness, stiffness, anxiety, and depression [84].

Thus, *Panchakarma* procedures can be used as physiological interventions for health promotion, and in the treatment of several diseases. More intense research is required to understand the underlying mechanisms of these valuable detox procedures.

SPIRITUALITY—BEYOND BEHAVIOR

Lifestyle and behavior in humans is influenced by body–brain–mind, and spirit. New disciplines such as behavior and lifestyle medicine are welcome developments in our current, medicalized society. Lifestyle can influence behavior, and vice versa. Both of them, inter alia, are influenced by the brain and mind, which both produce somatic, and cognitive modulations. Lifestyle and behavioral changes can cause diseases, while their modification in the right direction can prevent, and treat, these diseases too. However, efforts in the area of lifestyle and behavior medicine should not follow the same track that pharmacological drugs have followed. Instead of the focus being on commercialization, the focus should be on empowering people to take their health into their own hands. This can be achieved by devising proprietary, patented modifications in the form of packages which combine traditional knowledge such as Ayurveda and yoga, or cultural practices such as those in Okinawan communities. People should be told about the power of their own mind, brain, and body. The natural potential of healing should be rediscovered. According to eminent Indian physician and cardiologist, Dr B.M. Hegde, occult methods were successfully used for healing in many civilizations. They probably worked by allowing our body to heal naturally [85].

This doesn't mean the mumbo jumbo should be brought back; it only means that if spiritual practices like prayers can help improve the mental status of people, the least we should do is not ridicule these practices. The commercialization of religions in the name of the god should be condemned. Hippocrates' statement "nature is the best healer" should not be forgotten.

The use of spirituality in daily life can be one way to empower people. Although the existence of spirit is a debatable question among scientists, during the past few decades, the role of spirituality in health has been recognized by philosophers. Many renowned neuroscientists, including Sir John Eccles and Candace Pert, have formally acknowledged the importance of spirituality in human life—especially for health, tranquility, bliss, and peace. The meaning and concept of spirituality differs in different regions, religions, and cultures. Historically, spirituality relates to religious ideals, and can influence a process of personal transformation. Many god-men, and gurus pose as spiritual leaders. They preach philosophy, and theology. However, since the last century, spirituality has been considered in relation to psychological development. No more is spirituality restricted to religion, but we see now that it relates to any meaningful activity, or blissful experience.

The famous philosopher, Ralph Waldo Emerson, pioneered the idea of spirituality as a separate field. The ethical outlook of having love and compassion for others is considered spiritual. Caring and sharing is spirituality in essence. The purity of thoughts, selfless attitude, respect to others who are around us, and balancing need and greed, are some of the key features of spirituality. Spirituality also induces a process of self-analysis, introspection, and search for meaning and purpose of life. Spiritual practices include meditation, mindfulness, prayer, reading sacred texts, and chanting mantras.

Spirituality is normally explored during end-of-life conditions like terminal cancer, when the reality of death becomes evident. However, the role of spirituality is also relevant in health promotion, and disease prevention. Spirituality may positively influence immunity, and patients' approach toward disease, hence it should be considered as an important dimension of health. A general tendency to think only of spirituality at the late stages of disease, and in advanced age, limits its advantages during the early, productive years of life.

Integrated medicine approach performed in a clean, and green environment is shown to be potentially useful for the emotional, and spiritual well-being of cancer patients [86]. Another study in end-of-life conditions has suggested that patients who are supported by religious communities are less likely to opt for aggressive medical interventions. They also depend less on hospice care, as compared with those who are treated in usual care, without community support [87]. Daily yogic meditation intervention is shown to reduce stress, inhibit nuclear factor (NF)-κB-related transcription of proinflammatory cytokines, and decrease interferon regulatory factor1-related transcription of innate antiviral response genes [88].

To conclude, the global disease burden is significantly shifting from the dominance of infectious, communicable diseases, to the chronic, lifestyle diseases. Obviously, the earlier strategies of drug-based therapeutics also need to change. Infectious diseases were controllable and treatable with help of vaccines, antibiotics, and chemical agents. However, treating lifestyle diseases requires something more than drugs. This realization has resulted in the development of many new initiatives, and disciplinary branches such as behavioral medicine, lifestyle medicine, and mind–body medicine, in which nonpharmacological approaches are receiving more attention. Today, predominant diseases like diabetes, asthma, obesity, cardiovascular diseases, cancer and many others, are the result of changes in human lifestyle and behavior. Obviously, their prevention, control, and treatment cannot be expected to be achieved by pharmaceutical drugs, unless the root causes are addressed by suitable modifications in lifestyle and behavior. We cannot go on consuming junk food, do less physical work, pollute environments, and in doing so hope to cure obesity, diabetes, and cancer with drugs, alone.

Imagination leads to discoveries and inventions. Today's ideas become a reality tomorrow. Ideas progress with available information, interdisciplinary environment, resources, dreaming, and desires. It might be dangerous if technology is not used with wisdom. Indian philosophy—led by Ayurveda and yoga—suggests "more responsibility for those who have more capacities." The concept of *Sadvritta* suggests compassion with all living beings; sacrifice; controlling body–mind, and speech; and respecting the interests of others as minimum expectations.

The twentieth century mindset of treating disease with drugs is not valid in the twenty-first century, unless it is backed by a healthy lifestyle. Ayurveda calls this *Sadvritta*, meaning "good behavior." *Sadvritta* involves honesty, ethical mindset, respect to others, ability to stay away from greed, the search for truth, and purpose of life. Such pure, moral, and spiritual behavior can keep stressors at bay. We feel the experiential wisdom available from Ayurveda and yoga, integrated with a psycho-neuroendocrinology understanding of biomedicine can provide better avenues to address the serious epidemic of diseases related to lifestyle, and behavior.

Unless a positive change is brought about in lifestyle and behavior, prevention and treatment of many diseases will not be possible. Alone, pharmaceutical drugs and surgeries cannot give health. Health is not a commodity that can be bought in the market. Health naturally occurs within everyone. Health has to be earned, or rather, hard-earned. Doctors are the facilitators; behavior and lifestyle are the real medicine for gaining health. There is a need to systematically translate the lifestyle medicine approach into various levels of health care—from individual to public health. This could be a possible strategy to addressing the present problem of the *medicalization* of the society.

> Compassion with all living beings, sacrifice, controlling activities of body, speech, and mind, and helping others—these are the minimum expectations of good conduct required to maintain health.
>
> **Vagbhata**

REFERENCES

[1] Moreira A, Alexander FSS. Exploring frontiers of the mind-brain relationship. Springer; 2011.

[2] Eisenberg L. Mindlessness and brainlessness in psychiatry. Br J Psychiatry 1986;148:497–508.

[3] Angell M. The illusions of psychiatry. N Y Rev Books 2011:1–13.

[4] Cox GR, Callahan P, Churchill R, Hunot V, Merry SN, Parker AG, et al. Psychological therapies versus antidepressant medication, alone and in combination for depression in children and adolescents. [Internet] Cochrane database Syst Rev 2012;11:CD008324.

[5] Urquhart DM, Hoving JL, Assendelft WW, Roland M, van Tulder MW. Antidepressants for non-specific low back pain. [Internet] Cochrane Database Syst Rev 2008;(1):CD001703.

[6] Yamaoka K, Tango T. Effects of lifestyle modification on metabolic syndrome: a systematic review and meta-analysis. BMC Med 2012;10:1–10.

[7] Orozco LJ, Buchleitner AM, Gimenez-Perez G, Figuls MR, Richter B, Mauricio D. Exercise or exercise and diet for preventing type 2 diabetes mellitus. Cochrane Database Syst Rev 2008;(3):CD003054.

[8] Moran LJ, Hutchison SK, Norman RJ, Teede HJ. Lifestyle changes in women with polycystic ovary syndrome. Cochrane Database Syst Rev 2011;(7):CD007506.

[9] Hagen KB, Byfuglien MG, Falzon L, Olsen SU, Smedslund G. Dietary interventions for rheumatoid arthritis. Cochrane Database Syst Rev 2009;(1):CD006400.

[10] Cheng J, Pan T, Ye GH, Liu Q. Calorie controlled diet for chronic asthma. Cochrane Database Syst Rev 2005;(5):CD004674.

[11] Janssen V, De Gucht V, Dusseldorp E, Maes S. Lifestyle modification programmes for patients with coronary heart disease: a systematic review and meta-analysis of randomized controlled trials. Eur J Prev Cardiol 2013;20(4):620–40.

[12] DeRose DJ, Charles-Marcel ZL, Jamison JM, Muscat JE, Braman MA, McLane GD, et al. Vegan diet-based lifestyle program rapidly lowers homocysteine levels. Prev Med Balt 2000;30(3): 225–33.

[13] Purohit MP, Wells RE, Zafonte R, Davis RB, Yeh GY, Phillips RS. Neuropsychiatric symptoms and the use of mind-body therapies. J Clin Psychiatry 2013;74(6):e520–6. http://dx.doi.org/10.4088/JCP.12m08246.

[14] Elkins G, Johnson A, Fisher WSJ. Efficacy of mind-body therapy on stress reduction in cancer care. In: William C. S. Cho, editor. Evidence-based non-pharmacological therapies for palliative cancer care. Netherlands: Springer; 2013. p. 153–173.

[15] Sharma V, Saito Y, Amit S. Mind-body medicine and irritable bowel syndrome: a randomized control trial using stress reduction and resiliency training. J Altern Complement Med 2014;20(5):A94.

[16] Mehling WE, Price CJ, Daubenmier J, Mike A, Bartmess E, Stewart A. Body awareness and the practice of yoga or meditation in 435 primary care patients with past or current low back pain. J Altern Complement Med 2014;20(Suppl. 5):A63–4.

[17] Woods NF, Mitchell ES, Schnall JG, Cray L, Ismail R, Taylor-Swanson L, et al. Effects of mind-body therapies on symptom clusters during the menopausal transition. Climacteric 2014;17(1):10–22.

[18] Morgan N, Irwin MR, Chung M, Wang C. The effects of mind-body therapies on the immune system: meta-analysis. PLoS One 2014;9(7):e100903.

[19] Morris K. Meditating on yogic science. Lancet 1998;351(9108):1038.

[20] Shannahoff-Khalsa DS. An introduction to Kundalini yoga meditation techniques that are specific for the treatment of psychiatric disorders. J Altern Complement Med 2004;10(1):91–101.

[21] Shannahoff-Khalsa DS, Kennedy B, Yates FE, Ziegler MG. Low-frequency ultradian insulin rhythms are coupled to cardiovascular, autonomic, and neuroendocrine rhythms. Am J Physiol 1997;272(3 Pt 2):R962–8.

[22] Garfinkel MS, Singhal A, Katz WA, Allan DA, Reshetar R, Schumacher HR. Yoga-based intervention for carpal tunnel syndrome: a randomized trial. JAMA 1998;280(18):1601–3.

[23] Bernardi L, Sleight P, Bandinelli G, Cencetti S, Fattorini L, Wdowczyc-Szulc J, et al. Effect of rosary prayer and yoga mantras on autonomic cardiovascular rhythms: comparative study. BMJ 2001;323(7327):1446–9.

[24] Cohen L, Warneke C, Fouladi RT, Rodriguez MA, Chaoul-Reich A. Psychological adjustment and sleep quality in a randomized trial of the effects of a Tibetan yoga intervention in patients with lymphoma. Cancer 2004;100(10):2253–60.

[25] Visweswaraiah NK, Telles S. Randomized trial of yoga as a complementary therapy for pulmonary tuberculosis. Respirology 2004;9(1):96–101.

[26] Khalsa DS, Stauth C. Meditation as medicine: activate the power of your natural healing force. NewYork: Simon and Schuster; 2002.

[27] Lutz A, Slagter HA, Dunne JD, Davidson RJ. Attention regulation and monitoring in meditation. Trends Cognitive Sci 2008;12(4):163–9.

[28] Paulson S, Davidson R, Jha A, Kabat-Zinn J. Becoming conscious: the science of mindfulness. Ann NY Acad Sci 2013;1303:87–104.

[29] Cullen M. Mindfulness-based interventions: an emerging phenomenon. (N Y) Mindfulness 2011;2:186–93. http://dx.doi.org/10.1007/s12671-011-0058-1.

[30] Johnson DC, Thom NJ, Stanley EA, Haase L, Simmons AN, Shih PA, et al. Modifying resilience mechanisms in at-risk individuals: a controlled study of mindfulness training in Marines preparing for deployment. Am J Psychiatry 2014;171(8):844–53.

[31] Lenze EJ, Hickman S, Hershey T, Wendleton L, Ly K, Dixon D, et al. Mindfulness-based stress reduction for older adults with worry symptoms and co-occurring cognitive dysfunction. Int J Geriatric Psychiatry 2014;(10):991–1000. http://dx.doi.org/10.1002/gps.4086.

[32] Chadwick P. Mindfulness for psychosis. Br J Psychiatry 2014;204:333–4.

[33] Goyal M, Singh S, Sibinga EMS, Gould NF, Rowland-Seymour A, Sharma R, et al. Meditation programs for psychological stress and well-being: a systematic review and meta-analysis. JAMA Intern Med 2014;174(3):357–68.

[34] Greenberg J, Reiner K, Meiran N. "Mind the trap": mindfulness practice reduces cognitive rigidity. PLoS One 2010;(5).

[35] Murakami H, Nakao T, Matsunaga M, Kasuya Y, Shinoda J, Yamada J, et al. The structure of mindful brain. PLoS One 2012;7(5):e36206. http://dx.doi.org/10.1371/journal.pone.0036206.

[36] Chiesa A. Vipassana meditation: systematic review of current evidence. J Altern Complement Med 2010;16(1):37–46.

[37] Raman G, Mudedla S, Wang C. How effective is tai chi mind-body therapy for fibromyalgia: a systematic review and meta-analysis. J Altern Complement Med 2014;20(5):A66.

[38] Wei GX, Xu T, Fan FM, Dong HM, Jiang LL, Li HJ, et al. Can Taichi reshape the brain? A brain morphometry study. PLoS One 2013;8(4):e61038. http://dx.doi.org/10.1371/journal.pone.0061038.

[39] Bhutkar MV, Bhutkar PM, Taware GB, Surdi AD. How effective is sun salutation in improving muscle strength, general body endurance and body composition? Asian J Sports Med 2011;2(4):259–66.

[40] Nagarathna R, Nagendra HR. Yoga for bronchial asthma: a controlled study. [Clinical research ed.] BMJ 1985;291(6502):1077–9.

[41] Bhavanani A, Madanmohan UK, Ravindra P. A comparative study of slow and fast suryanamaskar on physiological function. Int J Yoga 2011;4(2):71–6.

[42] Bhavanani AB, Ramanathan M, Balaji R, Pushpa D. Immediate effects of Suryanamaskar on reaction time and heart rate in female volunteers. Indian J Physiol Pharmacol 2013;57(2):199–204.

[43] Uma K, Nagendra HR, Nagarathna R, Vaidehi S, Seethalakshmi R. The integrated approach of yoga: a therapeutic tool for mentally retarded children: a one-year controlled study. J Ment Defic Res 1989;33(Pt 5):415–21.

[44] Kox M, van Eijk LT, Zwaag J, van den Wildenberg J, Sweep FCGJ, van der Hoeven JG, et al. Voluntary activation of the sympathetic nervous system and attenuation of the innate immune response in humans. Proc Natl Acad Sci USA 2014;111(20):7379–84. http://dx.doi.org/10.1073/pnas.1322174111.

[45] Qu S, Olafsrud SM, Meza-Zepeda LA, Saatcioglu F. Rapid gene expression changes in peripheral blood lymphocytes upon practice of a comprehensive yoga program. PLoS One 2013;8(4):e61910. http://dx.doi.org/10.1371/journal.pone.0061910.

[46] Shekleton JA, Flynn-Evans EE, Miller B, Epstein LJ, Kirsch D, Brogna LA, et al. Neurobehavioral performance impairment in insomnia: relationships with self-reported sleep and daytime functioning. Sleep 2014;37(1):107–16.

[47] Watanabe M, Takano O, Tomiyama C, Matsumoto H, Kobayashi T, Urahigashi N, et al. Skin rubdown with a dry towel, 'kanpu-masatsu' is an aerobic exercise affecting body temperature, energy production, and the immune and autonomic nervous systems. Biomed Res 2012;33(4):243–8.

[48] Lee S, Ishibashi S, Shimomura Y, Katsuura T. Physiological functions of the effects of the different bathing method on recovery from local muscle fatigue. J Physiol Anthropol 2012;31:26. http://dx.doi.org/10.1186/1880-6805-31-26.

[49] Basler AJ. Pilot study investigating the effects of Ayurvedic Abhyanga massage on subjective stress experience. J Altern Complement Med 2011;17(5):435–40.

[50] Supa'at I, Zakaria Z, Maskon O, Aminuddin A, Nordin NA. Effects of Swedish massage therapy on blood pressure, heart rate, and inflammatory markers in hypertensive women. Evid Based Complement Altern Med 2013;2013:171852. http://dx.doi.org/10.1155/2013/171852.

[51] Bauer BA, Cutshall SM, Wentworth LJ, Engen D, Messner PK, Wood CM, et al. Effect of massage therapy on pain, anxiety, and tension after cardiac surgery: a randomized study. Complement Ther Clin Pract 2010;16(2):70–5.

[52] Xiong XJ, Li SJ, Zhang YQ. Massage therapy for essential hypertension: a systematic review. J Hum Hypertens 2014; http://dx.doi.org/10.1038/jhh.2014.52.

[53] Ju MS, Lee S, Bae I, Hur MH, Seong K, Lee MS. Effects of aroma massage on home blood pressure, ambulatory blood pressure, and sleep quality in middle-aged women with hypertension. Evid Based Complement Altern Med 2013;2013:403251.

[54] Moeini M, Givi M, Ghasempour ZSM. The effect of massage therapy on blood pressure of women with pre-hypertension. Iran J Nurs Midwifery Res 2011;16(1):61–70.

[55] Givi M. Durability of effect of massage therapy on blood pressure. Int J Prev Med 2013;4(5): 511–6.

[56] Wändell PE, Ärnlöv J, Nixon Andreasson A, Andersson K, Törnkvist L, Carlsson AC. Effects of tactile massage on metabolic biomarkers in patients with type 2 diabetes. Diabetes Metab 2013;39(5):411–7.

[57] Ang JY, Lua JL, Mathur A, Thomas R, Asmar BI, Savasan S, et al. A randomized placebo-controlled trial of massage therapy on the immune system of preterm infants. Pediatrics 2012;130:e1549–58. http://dx.doi.org/10.1542/peds.2012-0196.

[58] Keefe FJ. Behavioral medicine: a voyage to the future. Ann Behav Med 2011;41(2):141–51.

[59] Johnston M, Vögele C. Benefits of psychological preparation for surgery: a meta-analysis. Ann Behav Med 1993;15:245–56.

[60] Morley S, Eccleston C, Williams A. Systematic review and meta-analysis of randomized controlled trials of cognitive behaviour therapy and behaviour therapy for chronic pain in adults, excluding headache. Pain 1999;80(1-2):1–13.

[61] Lexchin J, Bero LA, Djulbegovic B, Clark O. Pharmaceutical industry sponsorship and research outcome and quality: systematic review. BMJ 2003;326(7400):1167–70.

[62] Marteau T, Dieppe P, Foy R, Kinmonth AL, Schneiderman N. Behavioural medicine: changing our behaviour. BMJ 2006;332(7539):437–8.

[63] Knowler WC, Barrett-Connor E, Fowler SE, Hamman RF, Lachin JM, Walker EA, et al. Reduction in the incidence of type 2 diabetes with lifestyle intervention or metformin. N Engl J Med 2002;346(6):393–403.

[64] Antonovsky A. Unraveling the mystery of health – how people manage stress and stay well. San Francisco: Jossey-Bass Publishers; 1987.

[65] Lundman B, Aléx L, Jonsén E, Norberg A, Nygren B, Santamäki Fischer R, et al. Inner strength—A theoretical analysis of salutogenic concepts. Int J Nurs Stud 2010;47(2):251–60.

[66] Morandi A, Tosto C, Roberti di Sarsina P, Libera DD. Salutogenesis and Ayurveda: indications for public health management. EPMA J 2011;2(4):459–65.

[67] Blumenthal JA. New frontiers in cardiovascular behavioral medicine: comparative effectiveness of exercise and medication in treating depression. Cleve Clin J Med 2011;78 (Suppl. 1):S35–43.

[68] Koeppen D, Aurich MRT. Medicinal leech therapy in pain syndromes: a narrative review. Wien Med Wochenschr 2014;164(5–6):95–102.

[69] Derganc M, Z F. Venous congestion of flaps treated by application of leeches. Br J Plast Surg 1960;13:187–92.

[70] Whitaker IS, Izadi D, Oliver DW, Monteath G, Butler PE. Hirudo medicinalis and the plastic surgeon. Br J Plast Surg 2004;57(4):348–53.

[71] Stange R, Moser C, Hopfenmueller W, Mansmann U, Buehring M, Uehleke B. Randomised controlled trial with medical leeches for osteoarthritis of the knee. Complement Ther Med 2012;20(1-2):1–7.

[72] Bapat RD, Acharya BS, Juvekar S, Dahanukar SA. Leech therapy for complicated varicose veins. Indian J Med Res 1998;107:281–4.

[73] Ozyurt S, Koca G, Demirel K, Baskın A, Korkmaz M. Findings of bone scintigraphy after leech theraphy. Mol Imaging Radionucl Ther 2014;23(1):25–7.

[74] Eriksson BI, Kalebo P, Ekman S, Lindbratt S, Kerry R, Close P. Direct thrombin inhibition with rec-hirudin CGP 39393 as prophylaxis of thromboembolic complications after total hip replacement. Thromb Haemost 1994;72(2):227–31.

[75] Stamenova PK, Marchetti T, Simeonov I. Efficacy and safety of topical hirudin (Hirudex): a double-blind, placebo-controlled study. Eur Rev Med Pharmacol Sci 2001;5(2):37–42.

[76] http://www.fda.gov/AdvisoryCommittees/CommitteesMeetingMaterials/MedicalDevices/MedicalDevicesAdvisoryCommittee/GeneralandPlasticSurgeryDevicesPanel/ucm124755.htm.

[77] http://leeches-medicinalis.com.

[78] Thatte UM, Chiplunkar SV, Bhalerao SS, Kulkarni AM, Ghungaralkar R, Panchal FH, et al. Immunological and metabolic responses to a therapeutic course of basti in obesity. Indian J Med Res. (accepted for publication).

[79] Conboy L, Edshteyn I, Garivaltis H. Ayurveda and Panchakarma: measuring the effects of a holistic health intervention. ScientificWorldJournal 2009;9:272–80.

[80] Aggithaya MG, Narahari SR, Vijaya S, Sushma KV, Anil Kumar NP, Prajeesh P. Navarakizhi and Pinda Sweda as muscle-nourishing ayurveda procedures in hemiplegia: double-blind randomized comparative pilot clinical trial. J Altern Complement Med 2014;20(1):57–64.

[81] Shailaja U, Rao PN, Debnath P, Adhikari A. Exploratory study on the ayurvedic therapeutic management of cerebral palsy in children at a tertiary care hospital of karnataka, India. J Tradit Complement Med 2014;4(1):49–55.

[82] Sriranjini SJ, Pal PK, Devidas KV, Ganapathy S. Improvement of balance in progressive degenerative cerebellar ataxias after Ayurvedic therapy: a preliminary report. Neurol India 2009;57(2):166–71.

[83] Dhuri KD, Bodhe PV, Vaidya AB. Shirodhara: A psycho-physiological profile in healthy volunteers. J Ayurveda Integr Med 2013;4(1):40–4.

[84] Rasmussen LB, Mikkelsen K, Haugen M, Pripp AH, Fields JZ, Førre ØT. Treatment of fibromyalgia at the Maharishi Ayurveda Health Centre in Norway II - a 24-month follow-up pilot study. Clin Rheumatol 2012;31(5):821–7.

[85] Hegde BM. Science meets spirituality? J Indian Acad Clin Med 2014;15(3–4):170–1.

[86] Nakau M, Imanishi J, Imanishi J, Watanabe S, Imanishi A, Baba T, et al. Spiritual care of cancer patients by integrated medicine in urban green space: a pilot study. Explore NY 2013;9(2):87–90.

[87] Balboni TA, Balboni M, Enzinger AC, Gallivan K, Paulk ME, Wright A, et al. Provision of spiritual support to patients with advanced cancer by religious communities and associations with medical care at the end of life. JAMA Intern Med 2013;173(12):1109–17.

[88] Black DS, Cole SW, Irwin MR, Breen E, St Cyr NM, Nazarian N, et al. Yogic meditation reverses NF-κB and IRF-related transcriptome dynamics in leukocytes of family dementia caregivers in a randomized controlled trial. Psychoneuroendocrinology 2013;38(3):348–55.

CHAPTER 7

Food and Diet

Let food be thy medicine, and medicine be thy food.

Hippocrates

HISTORICAL ACCOUNT

Every living form requires food for survival and growth. Interestingly, all species have interdependence in terms of food—whether it is obtained from vegetable or animal sources. In this chapter we will deal with historical perspectives, current status, and future trends related to food, nutrition, and diet for human beings.

Food is an essential part of life needed to produce energy, metabolism, sustenance, and growth. Food contains essential nutrients known as proximate principles including carbohydrates, fats, proteins, vitamins, and minerals. The *Encyclopedia Britannica* defines food as "any substance consumed to provide nutrition for the body." However, in many traditions and holistic systems, the purpose of food is not only for nourishment of the body, but also to nourish mind and spirit. Probably, the expression "food for thought" is the result of such thinking. A detailed discussion on the philosophy of food is given by Harvard neurobiologist, Dr David Kaplan [1].

The history of food goes back to the history of mankind. Primitive populations were reliant on food mainly from water bodies, and forests. People secured food mainly through hunting and agriculture. Probably, sometime around 7000 BC, neolithic man acquired fire and cooking techniques. Dr Lynne Olver, famous food historian, started an interesting Web site The Food Timeline, which gives a very comprehensive account of the evolution of food, nutrition, diet, and culinary sciences. It describes use of various food items on a timescale that covers the use of salt, rice, and millet from ancient times (around 17,000 BC) until modern times with the example of cultured beef in the laboratory [2]. In earlier times, with the abundance of water, ice, salt, and forests primitive food consisted of different types of fish, squid, oysters, eggs, meat, insects, mushrooms, vegetables, and fruits. Rice seems to be the most ancient food grain, followed by millet—a generic name for cereals other than those of the genera of wheat, barley, rye, and oats [3].

173

Integrative Approaches for Health. http://dx.doi.org/10.1016/B978-0-12-801282-6.00007-3

Scientists estimate that nearly 10,000 years ago, rice cultivation was practiced in India and China. In around 1609, in the United States, rice cultivation began in Virginia, followed by South Carolina, and other parts of the country. Wheat emerged some time around 8000 BC, followed by many other foods like cereals, pulses, nuts, honey, ginger, chili peppers, maize, and dates. It took almost two millennia for milk, yogurt, and potatoes to emerge on the scene. The oldest known record of animals being kept in groups, and milked is found in cave paintings of 5000 BC in the Libyan Sahara. The Sumerians, around 3500 BC, and the Egyptians a few centuries later, used milk, and curdled milk products. At around 3000 BC chicken domestication, butter, palm oil, olive oil, onions, garlic, spices, and soybeans probably emerged. The beginning of processed food was also during this period, starting with popcorns. At around 2000 BC, many more meats, like ham and duck, and fruits and vegetables, like peaches, apples, and radishes, came into use. Recipes for pasta and noodles were originated. Around 1000 BC, chocolate, vanilla, sugar, mangoes, oats, tomatoes, celery, pickles made way. Probably from 4000 BC food processing and recipe science started developing with yeast fermented bread, pita bread, and ice cream was introduced some time in 3000 BC.

Ancient Egyptian and Mesopotamian recipes and customs dominated until the first century BC when ancient Romans introduced Italian cuisine, soups, puddings, omelets, fried chicken, and cheese cake. This was followed by Anglo-Saxon foods from the fifth to ninth centuries AD, and medieval foods from the tenth to fifteenth centuries. The sandwich was introduced by John Montagu in 1762, followed by a wide range of European, English, and Scottish foods such as pastries, muffins, and preserves.

An era of great American foods consisting of a large variety of irresistible and addictive junk foods began in the late nineteenth century. In 1853, the potato chip was developed as a new food product in New York. Pemberton developed and launched Coca-Cola in Atlanta, Georgia in 1886. The hamburger was first created by Louis Lassen in 1900. Junk food chains including McDonalds, Kentucky Fried Chicken, Burger King, Pizza Hut and others spread like wildfire in the United States. During the past two decades, the junk food industry has come on to the radar of public activists and scientists. The many health hazards from the consumption of junk food were noticeable, which led to increased awareness, and reduced consumption, especially in the United States. As a result, the junk food industry came under pressure to infiltrate new markets in developing countries. Today in India, chilled Coke and chips are available in remote areas where safe drinking water, and lifesaving drugs are scarce.

The lifestyle and dietary changes of the twentieth century had a deep impact on the social fabric in the United States. During this period, the United States witnessed substantial increase in automation, and an automobile-based way of life. A new culture of fast food consumption, and consumerism emerged.

Mass entertainment, starting with radio and television, followed by social networking, and the IT revolution had additional impact [4].

As a result, the family structure was disturbed, food habits were changed and social interactions now take place in a virtual mode. Physical activity and personal interactions have been greatly reduced. Home theaters have brought cinema into the living room, and the habit of eating out has supplanted the home kitchen. This double burden of food and lifestyle change has silently affected the American population, resulting in an epidemic of obesity—a greater than 75% rise in obesity prevalence has been reported between the years 1890 and 2000 [5]. The United States' National Health and Nutrition Examination Survey showed that more than one-third of children, and more than two-thirds of adults in the United States are overweight, or obese [6]. All agencies, including the National Institutes of Health (NIH), Centers for Disease Control and Prevention (CDC), and United States Department of Agriculture (USDA) emphasize the importance of improved nutrition, reducing calorie consumption, and increasing physical activity. Junk foods, red meat, sodas, and sugary drinks are going out of fashion. Healthy, organic, and nutritious food is gaining popularity. The rich world has just completed a full cycle. Nutrition is back to center stage.

FOOD AND NUTRITION

Food has two main functions. The first is to provide the nutrition to build every cell and tissue for body growth. The second function of food is to provide the necessary energy, in terms of calories, to keep the body functions running. Food consists of five proximate principles, including proteins, carbohydrates, fats, salts, and vitamins. Water is also an essential component of food, but has no caloric value. Proteins, carbohydrates, and fats are known as macronutrients, because they are required in large quantities. Vitamins and minerals are considered as the micronutrients, because they are needed in small quantities.

Carbohydrates are the main source of energy for our body. They constitute grain-based products, sugars or simple carbohydrates, starches or complex carbohydrates, and dietary fiber. The main sources of carbohydrates are cereals. Proteins are the building blocks vital for the structures and functions of the body. They are required for muscles mass, cell membranes, and for every tissue. All enzymes are proteins. The sources of proteins are legumes, pulses, milk, meat, poultry, and fish. Fats serve as the major storage of energy in the body. Excess fat is stored in the body in adipose tissues. The main sources of fats are butter, oils, nuts, and fish.

Vitamins regulate many functions in the body. They are essential for various metabolic processes, and building body tissues. They help in digesting carbohydrates, proteins, and fats. They prevent several deficiency diseases, promote healing, and good health. All fruits and vegetables, milk and milk products, nonvegetarian food, and cereals are good sources of vitamins. Minerals are also essential because they are required to build muscles, tissues, and bones; they are important components of hormones, and enzyme systems. There are two types of minerals: macro-minerals, which are required in large quantities, include calcium, phosphorus, magnesium, sodium, potassium, sulfur, and chloride; trace or micro-minerals, which are required in smaller quantities, include zinc, iodine, copper, manganese, fluoride, chromium, and selenium. The sources of minerals are all fruits, vegetables, milk and milk products, cereals, and nuts, and non-vegetarian food.

SCIENCE OF NUTRITION AND DIET

The process of food assimilation, and its use for sustenance, growth, and tissue regeneration is called nutrition. Nutrients are substances essential to life, which are available in food. Food is part of the ecosystem, and is required by all living, interdependent species. Thus, changes in the environment, and ecosystem have a direct impact on food, and hence, on nutrition. Hippocrates, in year 400 BC, said "Let thy food be thy medicine, and thy medicine be thy food." In those days foods were often used as cosmetics, or as medicine. Far-Eastern, and biblical writings make references to food and health. The famous philosopher and artist, Leonardo da Vinci, compared the process of metabolism in the body to the burning of a candle. This has a striking similarity to the concept of *Agni* in Ayurveda, which is elaborated in the primer, and discussed in later sections. In 1747, James Lind, a physician in the British navy, performed the first clinical trial as a scientific experiment. Lind gave six groups of sailors different supplements, along with the same diet. The six groups received cider, elixir of vitriol, vinegar, seawater, barley water, and oranges with lemon. He noticed that those receiving oranges and lemons were cured of the disease, scurvy. Vitamin C was discovered in the 1930s, and the nutritional value of lime was understood. Antoine Lavoisier, who is considered the father of nutrition, discovered the process of metabolism in 1770. He demonstrated that body heat is generated from the oxidation of food. In the early 1800s, it was discovered that food contains four elements, carbon, nitrogen, hydrogen, and oxygen. In 1840, Justus Liebig of Germany, showed that sugars are made up of carbohydrates, fats makeup fatty acids, and proteins are made up of amino acids.

In 1897, Christiaan Eijkman from Holland, observed that some of the natives in Java suffered from heart problems, and paralysis due to the disease, beriberi. He observed that when chickens were fed the native diet of white rice, they developed the symptoms of beriberi, but when they were fed unprocessed brown rice with the outer bran intact, there were no symptoms of the disease. When Eijkman fed brown rice to beriberi patients, they were cured. Later, it was discovered that the outer rice bran contains thiamine, or vitamin B1, and its deficiency leads to beriberi. This was the beginning of functional foods. Frederick Hopkins, an English biochemist discovered tryptophan, an important amino acid, which can't be synthesized by our body, and has to be supplied through proper diet. In 1912, he published the findings of his animal experiments. He demonstrated that the growth of rats is impeded when they were fed artificial mixtures of isolated casein, fat, carbohydrate, and salts. He demonstrated that the rats who were given additional fresh milk were able to maintain normal growth [7].

Sir Frederick concluded that, "no animal can live upon a mixture of pure protein, fat, and carbohydrate. Even necessary inorganic material like salt supplementation is not enough for any animal to grow and sustain." There are certain other ingredients, and minor factors that are vital for growth and sustenance. Later, some of these were categorized as vitamins. The requirement of these ingredients may be very small, but their presence in diet is vital to produce enzymes, hormones, and other substances, which are essential for growth, development, and survival. These are now known as micronutrients, and consist of several vitamins, minerals, and other growth factors. The consequences of micronutrient deficiency are severe. Eijkman and Hopkins were awarded the Nobel Prize for their discovery of vitamins.

In 1912, Dr Elmer McCollum, from the University of Wisconsin, discovered the first fat soluble vitamin, vitamin A. He demonstrated that rats fed on butter were healthier than those fed on lard. He showed that it was because of the vitamin A present in butter. That same year Dr Casimir Funk, a Polish biochemist coined the term vitamins—vital amines in the diet. Later, in 1930s, Dr William Rose, a professor of physiology from the University of Illinois discovered the essential amino acids as the building blocks of protein. In the 1940s, American chemist Russell Marker synthesized progesterone from diosgenin in wild yams. Subsequently, work on essential nutrients, and the role of vitamins and minerals as cofactors of enzymes and hormones, was advanced.

German physiologist, Carl von Voit is the father of modern dietetics. He demonstrated that protein turnover can be measured in the form of excreted urea. Another German physiologist, Max Rubner, proposed the isodynamic law of calories. This law stated that based on their caloric values, different foodstuffs can replace one another. This notion originated the much debated phase in modern nutrition "a calorie is a calorie." These two scientists helped in shaping modern dietetics. Rubner also stated the inverse relation between metabolic rate and longevity, based on his observations: "Larger animals have slow metabolic rate, hence they live longer than smaller animals." This theory was also debated in due course, however, and formed the relation between metabolic rate and longevity. Nobel laureate, Dr Linus Pauling, suggested that mega doses of vitamin C given intravenously can improve the quality of life of cancer patients. This was the beginning of nutraceuticals, and dietary supplements. Necessary regulations begun to be enforced from this time by FDA.

As discussed in Chapter 3, nutrition is one of the most important determinants of health. Both under-nutrition, and overnutrition may result in ill-health. This has led to concept of the balanced diet, which provides macro- and micronutrients in required quantities. Diet also provides energy in terms of calories which are required for running our systems. Foods rich in calories, but poor in nutrition are often termed junk foods. A good diet needs a balance of nutrients and calories. Of course, there is nothing like a universal balanced diet. The components and proportions may change depending on geo-cultural environments. Therefore, we need scientific evidence to know, and design healthy diets from available foods so that optimal nutritional needs are met. This is addressed in a discipline known as dietetics, and the job is done by dietitians.

The science and art of applying food and nutrition knowledge to design diets is known as dietetics. Dietitians are expert professionals who are able to advice the right diet for people in order to promote, and prevent disease. Food, nutrition, diet, and different cuisines have evolved over thousands of years. They have gone through individual's and communities' experimentation in relation to type of foods, processes, nutritive value, health benefits, taste, appearance, and many more considerations. This interdisciplinary science is known as gastronomy, which involves researching, discovering, understanding, tasting, experiencing, and documenting food preparations. It also studies the sensory qualities of human nutrition, and interfaces with broad sociocultural practices. A new branch, known as molecular gastronomy, deals with applying the knowledge of biology and chemistry to cooking, and the culinary arts.

DIGESTION AND GUT HEALTH

A review of the physiological mechanisms controlling energy balance is interesting, and an understanding of these mechanisms is propelling modern dietetics to become a holistic and interdisciplinary discipline. Modern nutrition and dietetics—that was previously confined to the approach, "a calorie is a calorie"—is adopting a comprehensive view based on the advancing understanding of physiology. Advancing physiology is adding to our understanding of the complex processes of hunger, food intake, digestion, and elimination of wastes.

The desire for food is hunger. This feeling is a basic instinct of every animal. Appetite is the word used for a feeling of craving the consumption of something. The delay between the swallowing of food, and its digestion may lead to overeating; hence the mechanism for the short-term control to prevent overeating is the sensation of satiety. Satiety is the feeling of satisfaction that limits further intake. These signals are triggered by centers in the hypothalamus, and regulate food intake. Hunger and satiety maintenance requires a balance of energy, and depend on several psychological, social, and environmental factors, such as nutrients and metabolic processes, gastric and intestinal contractions, blood levels of nutrients, and temperature.

The sensation of hunger is a complex phenomenon regulated by mechanical and hormonal stimuli. Ghrelin, a hormone produced by the empty stomach, sends signals to the brain that stimulates the hunger center. Orexin, another hormone secreted by the hypothalamus, also stimulates the hunger center. The amount of food intake is controlled by other hormones, and nervous controls. The distension of the walls of the stomach and duodenum causes vagus nerve stimulation that sends inhibitory signals to the feeding center. Cholecystokinin released mainly in response to the presence of fat and proteins, contributes to satiety, and prevents overeating. Neuropeptide Y, and insulin suppress appetite. Apart from these short-term controls of feeding, leptin is another hormone important for the long-term regulation of food intake, and body fat. Adipose tissue is not just energy storage, it is now considered to be an endocrine organ, as it produces the resistance hormones, leptin, adiponectin, and resistin.

Leptin and ghrelin are satiety hormones, and regulates energy storage. Leptin decreases the feeling of hunger while ghrelin increases appetite. Leptin levels are lowered during starvation, which, facilitates the stimulation of the hunger center. In the case of sufficient fat storage, it suppresses hunger. Adiponectin is an anti-inflammatory adipokine that regulates insulin sensitivity, and suppresses apoptosis and reduces oxidative stress. Its preventive role in atherosclerosis, obesity, diabetes, liver diseases, kidney diseases, and neurocognitive disorders is suggested in recent studies [8]. Like hormonal control, nervous mechanisms of energy balance are also interlinked. The vagus nerve is a cranial nerve that constitutes the sympathetic component of the autonomic nervous system. The vagus nerve lowers heart rate and sweating; and controls muscle movements of larynx for speech and breathing. It carries axons that increase secretions of the mucous glands of the pharynx, larynx, and other organs in the thorax, and abdomen. The vagus nerve also has sensory functions, and carries sensations from the epiglottis region, thoracic and abdominal viscera, auditory meatus and tympanic membrane, and aorta. The hormones regulating energy storage are involved in many other functions. The orexin system, apart from controlling energy homeostasis and feeding, regulates sleep/wake states, and the maintenance of wakefulness. The orexin system also affects behavior and emotions [9]. The role of ghrelin in promoting growth, and in regeneration mechanisms has been studied. The experiments demonstrate the enhancement of the regenerative potential of the GI epithelium by modulating cell proliferation. Ghrelin hormone also accelerates the regeneration process when hyperthermia is induced suggesting its regenerative potential [10]. Hormones, and nervous controls of energy balance regulate other vital functions. These coordinated functions suggest interlinks between various vital organs, with nutritional status as the common trigger [11].

The complex and interconnected functions of the gut constitutes the enteric nervous system (ENS). This brain in the gut has several added features. The ENS can operate autonomously to regulate motility, and secretions of the gut [12]. Another typical attribute of ENS is the modulation of the response, based on the nutrient composition of the food. The neurotransmitter, serotonin (5-hydroxytryptamine) is found in the central nervous system, platelets, and gut. The enterochromaffin cells of the gut secrete serotonin, which is taken up by platelets. The activities of serotonin in vasoconstriction, depression, memory, sleep, appetite, and temperature control have been known for years. The recent understanding of serotonin explains more functions like gastro-intestinal motility, enteric neurogenesis, mucosal growth and maintenance, intestinal inflammation, osteoblast proliferation, and promotion of hepatic regeneration [13]. Serotonin is an important growth factor of ENS that elaborates the gut–brain relationship.

Many research publications in recent years discuss gut microbiota. In the human body, microorganisms constitute an ecosystem consisting of around 100 trillion bacteria and other microorganisms—almost 10 times the number of the cells of the body. The gut microbiota or intestinal flora has many functions, and is considered to function like an organ. It synthesizes vitamins B and K, provides protection from pathogens, maintains the intestinal barrier, absorbs lipids and polysaccharides, and modulates intestinal motility. The gut microbiota and brain interaction maintains gastrointestinal homeostasis [14]. Microbiota can affect several nervous and cognitive functions. The alteration of healthy flora can lead to a range of disorders like inflammatory gastrointestinal disorders, and obesity [15].

The concept of *gut health* is becoming a point of discussion. More than 6000 papers discuss the various aspects of gut health. More than 6500 papers published so far focus on the association of gut health, and metabolism. The concept is important as a determinant of diseases, as well as its possible role in preventive and therapeutic strategies. The characteristic of microbiota is an important indicator of overall, physiological functions. The detailed study of gut health-related gastrointestinal (GI) functions is now considered important in preventive medicine.

Various components define gut health. Gut health is now considered a new objective in medicine. A healthy gut can ensure good digestion, and absorption of food, prevent GI illness, has normal intestinal microbiota and effective immune status, and can provide a status of wellbeing [16]. Various intestinal, and extra-intestinal diseases are associated with the GI barrier and microbiota. For example, diarrhea, inflammatory bowel disease, celiac disease, allergic diseases, arthritis, obesity, and many autoimmune diseases have a strong relation to gut health.

DIET THERAPY

Different regions and cultures have different dietary practices. Traditionally, people have used local food resources as part of their everyday diet. It is believed that for regular consumption, foods available in neighboring regions are more compatible with people's systems than exotic foods. However, in the process of human development, coupled with migration, travel, tourism, and in general, due to process of globalization, foods, and diets from one region have moved from one country, to other continents. Thus, Chinese and Indian food are popular in the West; Italian cuisine is popular worldwide; the Mediterranean diet is considered to be healthy; and Western fast food is considered to be junk food. Based on traditional experiences and scientific evidence, it is possible to propose a wide range of diets for health promotion, disease prevention, and treatment. Scientists and dieticians have prepared several types of diets for special purposes, or for treating, and preventing specific diseases. A short review of a few of these special diets will be interesting.

The Western dietary pattern is criticized mostly because it contains refined food, its high caloric content, low vegetable content, the excessive use of saturated fatty acids and sugars, and low fiber. The effect of maternal Western diet on neonates was studied in a mice model, in order to understand its mechanism. The milk of such mice was found to contain long chain, and saturated fatty acids in large proportions. This "toxic" milk can trigger inflammation in the mice neonates [17].

Ten trials with a whole grain component, like oats, have reported the lowering of low density lipoproteins (LDL) cholesterol, compared to the control foods. However, experts suggest that the positive findings about whole grain oats should be cautiously studied. A systematic review has indicated that most of the trials on whole grain oats were funded by companies with commercial interests in whole grain. Experts are emphasizing the need for randomized controlled trials on whole grain foods other than oats [18].

A diet rich in fruits, vegetables, low-fat dairy products, low cholesterol, and less sugary products can help to lower blood pressure in individuals with pre-hypertension. The health benefits of this diet were studied in a trial known as the Dietary Approaches to Stop Hypertension (DASH). The DASH diet is introduced by the National Heart, Lung, and Blood Institute of the United States. The diet plan suggests plenty of fruits, vegetables, whole grains; limited use of fat, dairy items, meat, fish, poultry, sugar-sweet foods and beverages, and red meat. The DASH diet plan also aims at providing a balanced diet for general prevention of diseases, apart from its therapeutic effects on hypertension.

A more comprehensive lifestyle modification program that includes exercise and weight loss, together with the DASH diet can significantly improve insulin sensitivity [19]. Another review has concluded that the DASH diet alone can lower blood pressure in overweight individuals. It has shown that a DASH-like diet can significantly protect against Cardiovascular diseases (CVDs), stroke, and cut heart failure (HF) risk by 20–30%. The DASH diet may play an important role in glycemic control in long-term interventions [20].

The Mediterranean diet primarily consists of plant foods, such as fruits and vegetables, whole grains, legumes and nuts. It uses healthy fats, such as olive oil instead of butter, and uses herbs and spices, instead of salt to flavor foods. It restricts the use of red meat to not more than few times a month, and uses fish and poultry at least twice a week. It also recommends moderate drinking of red wine. Several studies have been done in recent years to assess the benefits of the Mediterranean diet [21]. A meta-analysis of 12 studies involving 140,001 individuals was published recently in the *Journal of the American College of Cardiology* has reported a greater than 21% reduced risk for diabetes, and consequent prevention of cardiovascular disease [22]. A randomized trial involving 7447 persons at high cardiovascular risk was undertaken to study the role of the Mediterranean diet in the prevention of cardiovascular disease. Results of this well-controlled trial indicate that the Mediterranean diet, supplemented with extra virgin olive oil, or nuts can reduce the incidence of many cardiovascular events [23]. Eleven relevant, randomized controlled trials involving 402 participants have reported improved glycemic control, and a significant decrease in the glycated-hemoglobin (HbA1c), in subjects on a low-glycemic index [24].

A diet plan suggested by the General Motors Corporation for their employees to shed weight, has now become one of the most talked-about diet regimens on the Internet, and television [25]. However, scientific evidence on this diet plan is lacking. The plan is considered the fastest way to reduce weight, through intensive modifications in a week-long crash course. The plan is also criticized for the associated side effects.

We feel that several regions and cultures across the globe have unique food and dietary practices. Most of them are based on locally available food materials. The environmental compatibility between the food and people must be kept in mind before propagating, or generalizing particular practices to other regions or communities. What is important is the basic principle behind that practice, and not necessarily the actual food materials, which can vary depending on availability in respective geoclimatic regions.

There is little doubt that Mediterranean diet patterns are healthy. Many scientific studies have established their usefulness in preventing cardiovascular diseases. Extra virgin oil may add to their protective effects. However, these studies have been conducted on European, Caucasian communities for whom it is a traditional, natural diet. We cannot assume that similar benefits can be achieved if these dietary practices are embraced by Asian, or African communities. However, every region will have their own fruits and vegetables, and consuming these foods when they are fresh, may give similar benefits. In tropical regions, local fruits like banana, green leafy vegetables, local varieties of berries, and the use of sesame or coconut oil, may be more compatible with the people's systems, affordable, and more beneficial than exotic blueberries, or extra virgin olive oil. Ayurveda advocates a principle that nature does provide all the necessary foods and nutrients in all surroundings, and local foods are the best, and most compatible to people. Therefore, for a change, one can taste different foods, and cuisines from different parts of the world; but for health protection, it is best to use locally available, natural food resources.

DIETARY ADVICE

The routine dietary advice from dieticians, clinicians, and health professionals is very general. Eat fresh fruits and vegetables. Avoiding salt, sugar, sodas, refined flour, saturated fats, butter, oily, fried, spicy, and food stuffs seem to be general mantra. But in moderation everything is needed. Moreover, the environmental and socioeconomic conditions, coupled with cultural practices need to be taken into consideration when offering any dietary advice. Honest scientific advice can be beneficial in the prevention, and control of several diseases. Many professional bodies, associations and councils offer evidence-based, dietary advice. As the scientific knowledge advances, the nature of dietary advice keeps on changing. However, many other forces, and vested interests also play an influential role in making, and revising dietary guidelines. In the process, the situation becomes very confusing for people, and even for professionals.

Recent controversy about the role of fats in diet is a classic case. The American Heart Association's (AHA) diet and lifestyle recommendations advise reducing foods containing partially hydrogenated vegetable oils, and saturated fats. This includes foods like butter. The general advice here is for people to reduce their intake of foodstuffs like butter. There have been good studies supporting this advice. For example, a meta-analysis of 44 trials, involving 18,175 healthy adults, with and without dietary advice related to reduction of salt and fat intake, and with an increase in the intake of fruit, vegetables, and fiber showed improvements in cardiovascular risk factors. Two trials followed people for 10–15 years, and showed a reduced incidence of heart disease, stroke, and heart attack [26]. Based on the guidelines from reputed professional organizations like the AHA, generally it has been accepted that eating saturated fats and butter is to be avoided.

In March 2014, a new meta-analysis was published by scientists from University of Cambridge in the *Annals of Internal Medicine*, which concluded: "Current evidence does not clearly support cardiovascular guidelines that encourage high consumption of polyunsaturated fatty acids, and low consumption of total saturated fats" [27]. This came as a big comfort to the butter industry. Several articles were published in the media, with the message that butter is not bad. On June 12, 2014, *Time Magazine* published very provocative cover story "Eat Butter," with article title "Ending the War on Fat." The first paragraph stated: "For decades, it has been the most vilified nutrient in the American diet. But new science reveals fat isn't what's hurting our health." Interestingly, *Time Magazine* had carried a cover story in March 1984, giving exactly the opposite message—butter was bad. This case is quite representative where one can sense the undercurrents—vested business interests, and the nexus between scientists, media, and industry. Finally, in the process, it is the gullible public that suffers.

Of course, there are well-meaning scientists, media, and business people as well. Following the publication of *Time's* provocative article giving the clean bill of health to butter, the reputable journal, *Science* published a policy-analysis article, where many experts demanded a retraction, and press promotion as damage control. Many scientists pointed out the errors in this study, which compelled the authors of the original article in *Time* to amend their conclusions. Many skeptics believe that this was an attempt to check the growing popularity of olive oil, and an industry attempt to regain markets for butter. This is another case where scientists from reputable institutions have played dubious roles, and seem to have worked in a fashion that is beneficial to certain industries.

In India, cow's milk is traditionally consumed. Ayurveda also recommends fresh milk from naturally grazing cows. Environmentalists and veterinarian scientists are showing concerns about the quality of milk and milk products obtained from hybrid varieties and also use of synthetic hormones and antibiotics just to increase the milk output. The quality of milk from such cows is different than artificially fed cows.

For Indian people, milk from local breeds may be healthier than western breeds like Jersey. Buffalo milk will have different properties as well as the hybrid varieties. Camel milk may be suited for tribal in the desert, goat milk may be suitable for shepherds or yak milk may be suitable for people staying in high altitudes. It is difficult to standardize milk that can be healthy and beneficial to the general global population. The geographical, regional, and cultural practices must be considered seriously while we are globalizing.

Milk is a whole food with numerous nutritive components; especially for infants. However, milk is not as simple as it may appear. Many infants may be immunologically sensitive to milk protein *casein*. Many individuals may have milk sugar *lactose* intolerance. Quality and composition of cow's milk is different than buffalo. Several variants of beta-casein like A1, A2, A3, B, C, etc. are known. Of these A2 from bovine milk is similar to other beta-casein in mammals. The A1 and A2 caseins have different secondary structures, enzymatic hydrolysis products are different. This leads to different physiological properties and impact on digestive, immune, and brain development processes. Anecdotal evidence suggests symptomatic relief in patients with neurological, gastric, and immunological problems, after consuming A2 beta-casein containing milk. Hence, consuming milk containing A2 beta-casein is considered beneficial especially for infants' growth and development [28].

Thus, even a simple thing like milk has region specific physiological effects. Milk from local breeds of cattle may be more beneficial and healthy. Such incidences actually strengthen our hypothesis that traditional knowledge and practices can be the best advisors, even in modern times. This is more valid in the area of food, nutrition, and dietetics. Traditional practices that have stood the test of the time may provide signposts, such as in the case of the Mediterranean diet. However, we feel that it might be better to offer dietary advice in the context of geoclimatic regions, cultures, and traditions.

NUTRIGENOMICS AND PERSONALIZED DIET

Diet is known to have a direct impact on the body. A good diet provides good nutrition, and promotes growth, and maintenance of various cells, tissues, and organs of the body. Several traditions believe that food has impact on behavior. According to yoga philosophy, certain types of meat are known as *rajasik*. They may cause "fight" type response with aggression, anger, annoyance, and irritation. Some foods like milk are known as *satvik*. They cause "flight" type response and can be very soothing, peaceful, and relaxing. Some types of foods, such as butter cause "freeze" type response and are known as *tamasik*. They can be lethargic, sluggish, and sleep inducing (Figure 7.1). Thus since ancient times different types of food was accepted to influence living behavior.

It is generally accepted that "we are what we eat." However, it is now known that diet-related changes go beyond the body and the mind of an individual in this life, but can also affect following generations. These changes are not necessarily limited to, or dependent on, our genome—but go beyond genetics. These are known as diet-induced, epigenetic changes, which are not limited to one generation, but can be inherited by the descendants. Recent research on the effect of diet in the control of gene expression is showing how a pregnant woman's eating behavior influences the long-term health of her children [29].

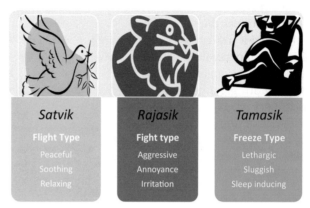

FIGURE 7.1
Diet classification according to yoga.

Several gut hormones are known to enter the brain. Some such hormones, which can influence cognitive ability, are also produced in the brain. Brain-derived neurotrophic factor is known to respond to specific food intake. Molecular mechanisms of interactions between food and cognitive functions need to be studied more, so that specific diets can be advised to increase neural resistance, and improved mental fitness [30].

Everybody has a different genetic makeup, hence every individual may respond to the same diet differently. Diet is also known to affect the expression of genes. A particular diet for one individual can be pathogenic, but may protect another individual by increasing resistance to certain diseases. Nutritional genomics, or nutrigenomics, is a branch of genomic research that studies the association between diet and genes. It can suggest dietary intervention considering nutritional status, nutritional requirement, and genotype of an individual. The personalized approach to diet can be helpful for preventing, and treating many diseases.

An interesting study by Canadian researchers indicates possible mechanisms of diet-induced, gene expression. Healthy volunteers were given a *prudent* diet consisting of fruits, vegetables, whole grain products, and lesser amounts of refined products. Another group was given a typical Western diet consisting of a high intake of processed meats, refined products, desserts, and sweets. The gene expression relevant to immune and inflammatory responses were different in both the diet patterns. The finding suggested that volunteers on the Western diet had a proclivity to cancer, and cardiovascular diseases [31]. It is known that specific nutrients can affect cognitive processes, and emotions. These studies actually support the notion of the three broad categories of diet in Indian culture discussed earlier: *satvik, rajasik,* and *tamasik* types.

Advancing research is completing the circle to move again toward natural products. The rising incidence of diseases produced by the action of free radicals, and inflammatory processes can be prevented with the help of a proper diet that utilizes the rich treasure of natural products. Italian researchers have suggested a strategy for neurodegenerative diseases that is based on an increased use of natural substances to augment neuroprotective pathways [32]. The researchers suggest that the use of natural substances for augmenting health-promoting genes, and for the reduction of the expression of disease-promoting genes.

AYURVEDA CONCEPTS OF NUTRITION

Vedic literature describes various dietary recipes, and their effects on health. The Veda and *Upanishads* describe food consumption practices, and culture. Food is often praised as the supreme spirit, a "vital force," and divine gift from nature. The Veda and *Upanishands* describe various food items, including cereals, pulses, milk and milk products, vegetables, fruits, oils, salt and spices. The Vedas prescribe a code of conduct for food acquisition, cooking, and serving. Various patterns of meals are elaborated in detail. Consumption of food is considered sacred worship. The *Bhagvad Gita* describes properties of healthy food. It suggests that any food should be consumed within 3 h of preparation. A tasteless recipe, decomposed, putrid food, and remnants and unwholesome ingredients should not be consumed. Food during the Vedic period included barley, rice, green gram, black gram, lentils, ghee, and milk as staples. Contrary to the current belief, nonvegetarian food seems to have been prevalent in ancient India. Most of the Ayurveda classics have a special section on nonvegetarian diet.

Ayurveda texts refer to food as the source of energy and vitality. Good food nourishes *Dhatu*, and maintains *Dosha* in a balanced state. Ayurveda describes properties and activities of different foods. Good food should maintain health, should not vitiate *Dosha*, should nourish *Dhatu*, and strengthen the body. Diet is an important aspect of therapeutics. Food can be an etiological factor in disease pathogenesis, and its role in health and disease treatment is vital. Generally, wheat, barley, green-gram, milk, fruits, and green vegetables are examples of "health promoting" food. However, there cannot be an easy or universal definition of good food. The choice of appropriate food may differ according to a person's *prakriti*. Food has temporal, and spatial influences, and the choice of food may vary greatly, depending on chronological cycles, and seasons. Ayurveda has made this easier by offering detailed "dos" and "don'ts," called *pathya* and *apathya*, respectively. The "path" toward health is known as *pathya* (dos) in Sanskrit, and its antonym is *apathya* (don'ts). The success of the treatment largely depends on the patient's adherence to *pathya*, and *apathya*, diet guidance. The emphasis and importance of *pathya* and *apathya* is clear in this quote from an Ayurveda text: "Those who follow proper diet, require no medicine, and those who do not follow a proper diet will not benefit from medicine."

"Proper" diet is a relative, and contextual term. Good food for one person may not be "proper" or useful food for another. An Ayurvedic physician or *Vaidya* examines patients, and tries to establish the link between causative agents, and the manifested disease. The disease process, and pathogenesis is interpreted in terms of *Dosha, Dhatu, Agni,* and other relevant components. Upon considering the nature of pathogenesis, the diet is advised. Every disease has a set of defined dos and don'ts. For example, for an asthmatic patient, consuming curd is contraindicated, however, it can be consumed with asafetida as an additive spice. Thus the food can cater to the specific needs of a particular person. This is one of the important approaches of Ayurveda, which differs from present practices of nutrition, where all individuals are treated in the same manner. Good diet taken at the wrong time, or improper diet, can be responsible for the causation of diseases. For example, drinking excess water during bedtime can make an individual prone to upper-respiratory complaints, such as runny nose, sneezing, coughing, and dyspnea.

Diet and Mind

Ayurveda explains the effects of different types of food and beverages on the mind. Food is considered as a source of complexion, clarity, good voice, longevity, talent, wisdom, happiness, and satisfaction. Even the observance of truth, and obtaining salvation is believed to be possible with the help of proper food. The properties of various food substances are described in Ayurveda. For example, rainwater is considered good for benevolence, pleasantness, and purity. Ghee is supposed to promote memory, intellect, brain function, and mind purity. Many beverages are recommended before and after meals to promote appetite, and digestion, and to provide refreshment, pleasure, satiety, and energy.

The functions of sensory and motor organs depend on proper nutrition. The food is expected to nourish the body, sense organs, mind, brain, and spirit. The food also becomes a culture. In ancient Indian traditions, the meal is considered as a sacred act, worship, and a duty. According to the *Bhagavad Gita,* it is the duty of every earning member of the household to ensure the well-being, and food security of others who cannot afford proper diet.

Ayurveda recognizes the variations, availability, acceptability, customs, and traditions of food. Ayurveda literature on food, nutrition, and dietetics elaborates cultivation, classification, processing, preservation of food, and preparation of a variety of recipes. *Kaiyadev Nighantu* (900 AD), *Bhavpraksha Nighantu,* and *Kshema Kutuhalam* (both 1600 AD) are important texts of Ayurveda dietetics. These classics describe the theory and concepts, food types, variations, processing, dos and don'ts, and prevalent practices and recipes.

Selection of Diet

Ayurveda classifies food materials as dietary, or medicinal substances. Ayurveda suggests specific conditions where food can become medicine, and vice versa. The concept of diet is linked with feelings such as relishing, enjoying, and satisfying, which are also acceptable in terms of modern physiology. In short, food and diet can be considered natural and rewarding, while medicine can be considered compulsory, and punitive.

The selection of diet is based on eight main criteria:

1. Food attributes: Some food items, like black gram, have a heavy property, and are difficult to digest, while some, like green gram, have a light property that are easy on the digestion. These properties of foods need to be considered when advising a diet. For example, patients with poor digestive capacity should be advised to consume foods with light attributes. The concept of attribute, or *guna* is central to the concept of diet. Details about *guna* are explained in the "Primer."

2. Processing: Processing transforms food properties. The same foodstuff can be processed differently in different recipes. For example, milk for an asthmatic patient is boiled with dry ginger. Warm milk is always preferred, and cold milk is avoided. The type and method of processing plays an important role in transformation of the properties, and activities of food. Raw materials undergo various processes, such as soaking in water, roasting, frying, cooking, baking, boiling, broiling, grilling, fermenting, and dipping. Each process affects the properties of raw material, not only changing the taste and form, but making it digestible. For example, boiled food is considered as lighter than fried food. Food processed with water and heat becomes soft, and easily digestible.

3. Compatibility: Every substance affects *Dosha* status. The food combinations should be compatible, and should not increase *Dosha*. Sometimes the combination of useful substances may become incompatible, and harmful. For example, milk and fish combination is considered harmful, as it causes vitiation of all three *Dosha*. Generally incompatible combinations are mentioned as *viruddha aahara*. Frequent consumption of incompatible foods can cause diseases.

4. Quantity: What is the right quantity of food? This depends on many factors, such as requirement, digestion capacity, and *Dosha* status. The food quantity also depends on the type of diet. For example, a person with a strong gastric fire can consume heavy food in larger quantity, but the same might cause indigestion in person with weak gastric fire, or appetite.

5. Habitat: Environmental and climatic condition where food is cultivated and consumed is important. For example, fish are more compatible in coastal areas, but may not be suitable in dry, and arid areas. Wine may be useful in a cold climate, but may be harmful in hot regions.

6. Time: This refers to circadian rhythm, and relates to the stages of disease, age of the individual, and seasons. The food and cuisine should be changed according to the season, and age of the individual, among other factors. Seasonal variations in temperature, rain, humidity, day–night cycle, and other environmental factors affect the body and mind. Seasons also affect digestive capacity, and body strength. Considering the variations, Ayurveda advocates *ritucharya* (seasonal regime), with modifications in diet, lifestyle, and therapies. Though the seasons are different in various continents, we can follow the principles of seasonal regime. Broadly, the summer season has a catabolic effect on the body. In summer, body strength is reduced. The other three seasons have anabolic effects, where the surroundings provide more nourishment to the body. Considering the atmosphere, the six seasons have particular climates, and induce specific effects on *Agni* and *Dosha* that require a specific diet. Ayurveda proposes that meals should be taken only after the previous meal is digested. Ideally, food should be consumed when *Pitta* is at its peak. Skipping meals is considered as a risk factor for obesity, hypertension, and diabetes. According to Ayurveda "one should not eat within 3 h of the previous meal, and should not abstain from eating or drinking for more than 6 h."

7. Consumption: Fresh, hot, unctuous food in proper quantity is always recommended. Such food alleviates *Dosha*, nourishes tissues, improves cognitive functions, and boosts strength, and complexion. There should be no feeling of either fullness or dullness, no discomfort in abdomen, and there should be an easiness in movement. A rest period should follow the meal. Exercise or vigorous activity should be avoided immediately after meals. Sleeping immediately after meals causes an increase in *Kapha*, and can lead to lethargy, weight gain, and reduced brain activity. Growing research on circadian rhythm, and energy regulation is adding new dimensions to the importance of meal timings.

8. Specific precautions: Food properties and the nature of the person should be compatible. Meals should be taken only after previous food is digested, and *Agni* is vibrant again, allowing it to receive food. Incompatible food is to be avoided. The environment should be proper. Hasty, or very slow consumption, is to be avoided.

In short, Ayurvedic advice is based on individual *Prakriti*; the function of *Agni* as a metabolic and digestive regulator; attributes such as *Guna*; taste or *rasa*, and the nature of digested food, which is known as *Vipaka*.

Ayurveda and Metabolism

The concept of *Agni*, or digestive fire, is central to Ayurveda physiology, and therapeutics. For simplicity, it can be thought of as appetite, but it is not merely hunger, or craving. *Agni* involves metabolic, and transformative processes at the physiological, and cellular levels. It encompasses digestion, absorption, and assimilation of food, and drugs. Thus, *Agni* governs all metabolic functions [33]. Maintaining a proper state of *Agni* can lead to health and longevity.

The process of digestion plays a key role in health and disease. *Agni*, in a normal state, can digest food, and convert it into an easily absorbable form that can circulate through microchannels, known as *Srotas*, to provide nourishment, or *Dhatu*, to various cells and tissues. Abnormal *Agni* can lead to disease. Powerful *Agni* accelerates the metabolic rate, while weak *Agni* is considered to be a cause of many diseases. *Agni* regulates the immune system by balancing *Dosha*, and strengthening *Dhatu*. In various autoimmune diseases, Ayurveda advocates improving digestion, and metabolic status through the maintenance of good health of the gut. According to Ayurveda, the gut is considered as *Maha Srotas*, meaning "the large channel." It is not merely the GI tract, or the portion known as the gut, but is considered to be an intelligent system. Thousands of *Srotas*, or microchannels, are connected with *Maha Srotas*, hence, the health of the gut is given substantial importance.

Gut health also depends on microbiota, which can change during various conditions, such as constipation and indigestion. Dietary restriction, avoidance of food, and fasting are considered to be useful in improving digestion, which leads to improved gut health. Recent studies on dietary restriction therapies have highlighted the role of gut health in autoimmune, metabolic, and degenerative diseases. Some studies have shown that obesity can be prevented just by restricting food consumption. However, food restriction alone cannot control major risk factors such as insulin resistance. These results support the emerging view that gut microbiota plays an important role in the etiology, and pathogenesis of metabolic disease. It also suggests that immune system malfunction may lead to metabolic syndrome [34].

The role of *Agni* is pivotal in regulating the body's internal environment. Disturbed *Agni* can lead to *Dosha* vitiation—despite proper diet. Improper digestion can lead to the buildup of toxic substances, known as *Ama*. This is another important concept in Ayurveda. *Ama* literally means "immature" or "undigested" material, which is an undesirable byproduct of digestion. *Ama* is described as a substance with heavy, unctuous, and sticky properties. *Ama* is thought of as a toxic material, which is responsible for pathologies. The role of *Ama* in various molecular, and biochemical processes resulting in metabolic, and inflammatory diseases is important [35] (Figure 7.2).

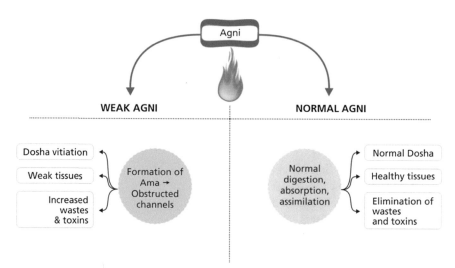

FIGURE 7.2
Agni as regulator of internal environment.

Ayurveda emphasizes the removal, or digestion of *Ama*, and improving *Agni* functions for the maintenance of health. Ayurveda suggests improving *Agni* as treatment of inflammatory diseases. *Panchakarma* therapies aim at improving the gut environment. The treatment of gastrointestinal diseases like diarrhea, irritable bowel syndrome, and amebiasis is done primarily by correction of *Agni*. Scientific exploration of *Agni* and *Ama*, and studying their role in pathogenesis can provide valuable leads in the understanding of metabolic diseases [36].

The concept of gut health is deeply rooted in Ayurveda [37], and has become a new research objective in modern health care [16]. The *Agni* concept can provide important inputs for research into metabolic functions, and bioenergetics. Normal *Agni* maintains a healthy gut environment by ensuring proper digestion, and elimination of excretory products like sweat, urine, and stool as *Mala*. A weaker *Agni* makes the gut environment sluggish—leading to improper digestion, and instead of normal *Mala*, toxic *Ama* is produced. *Ama* is a sticky, heavy substance, which may block the *Srotas*, and trigger inflammatory processes. *Agni*, in its normal state, maintains homeostasis, ensures nourishment, and results in a healthy, long life.

Many substances can be interchangeably used either as drugs, or as food. Many herbs used as spices improve digestion, and the absorption of food—not just its palatability. The use of spices is a specialty of Indian cuisine. Those having potent activity are generally regarded as drugs, while those with nourishing effects are used as food. For example, black pepper is a drug, while barley is a food. But black pepper can be used as spice in diet, and barley water can be used as a medicine. There are six tastes according to Ayurveda. These tastes

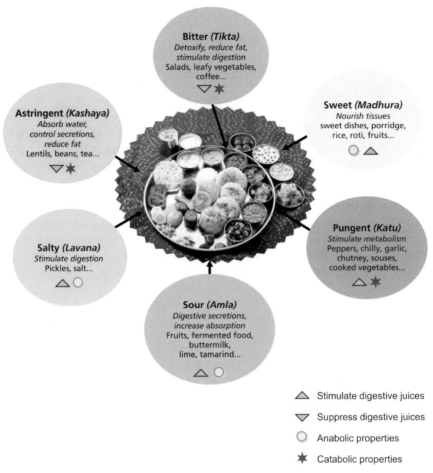

Bitter (Tikta)
*Detoxify, reduce fat,
stimulate digestion*
Salads, leafy vegetables,
coffee...

Astringent (Kashaya)
*Absorb water,
control secretions,
reduce fat*
Lentils, beans, tea...

Sweet (Madhura)
Nourish tissues
sweet dishes, porridge,
rice, roti, fruits...

Salty (Lavana)
Stimulate digestion
Pickles, salt...

Pungent (Katu)
Stimulate metabolism
Peppers, chilly, garlic,
chutney, souses,
cooked vegetables...

Sour (Amla)
*Digestive secretions,
increase absorption*
Fruits, fermented food,
buttermilk,
lime, tamarind...

△ Stimulate digestive juices

▽ Suppress digestive juices

○ Anabolic properties

✳ Catabolic properties

FIGURE 7.3

An ideal meal should include recipes of all six *Rasa*. Actions of *Rasa* can be anabolic or catabolic. Each *Rasa* is required for nourishing tissues and maintaining the balance of *Dosha*.

are not merely those recognized by tongue—the concept of *Rasa* has a deeper meaning. The six *Rasa* include sweet, sour, salty, pungent, bitter, and astringent. Each taste has specific effects on *Dosha* and *Dhatu* (Figure 7.3).

Modern dietetics considers six main types of nutrients as proximate principles. Those are carbohydrates, fats, minerals, protein, micronutrients (vitamin, minerals), and water. A balanced diet is the one which provide required quantities of proximate principles. Similarly, the ideal or balanced diet according to Ayurveda, is the one which provides an appropriate combination of the six tastes or *Rasa*. A balanced diet with all six tastes maintains *Dosha* in a balanced state, leading to good immunity, and a healthy, long life.

DIET, *PRAKRITI*, AND CHRONOBIOLOGY

The concept of personalized diet in Ayurveda is based on *Prakriti*, which gives specific diet advice for health promotion, and disease prevention. A *Prakriti*-specific diet is meant to maintain the balance of *Dosha*.

Kapha Prakriti persons are slow metabolizers; they can tolerate hunger and thirst. They require small amounts of food. They take a long time to digest. They may show weight gain, in spite of small quantity of meals. Their propensity for obesity is high, hence they are advised to fast or consume small meals. They may eat pungent, bitter, and astringent spices with their food, drink hot beverages, and consume dry, and light food.

Pitta Prakriti persons are fast metabolizers, and can digest heavy food. They are prone to ulcers in stomach, and acid peptic diseases; hence, they should control acidity in their food. They prefer frequent munching, also require large amounts of food. They cannot tolerate hunger or thirst. They should avoid spices, and foods with pungent, salty, and sour taste. They should preferably consume food and beverages with cooling properties.

Vata Prakriti persons are generally lean and thin, with dry skin. Their metabolic pattern can be variable, even due to a small change in diet and lifestyle. They may eat ghee, edible oils, and hot food with sweet and sour taste. They eat fast, and consume small amounts. They will not gain much weight, even if they consume a large amount of food. They should consume more oils, and ghee to prevent degenerative diseases, of which they are prone.

What time should meals be taken? The timing of meals is another important aspect of metabolic health. Researchers have studied how the timing of the first meal, breakfast, can have certain effects on the physiological function. The findings suggest that the first meal determines circadian phasing in the liver; the last meal relates to lipid metabolism. This experiment suggests that meal timing can have different physiological effects [38]. Thus, meal timing is an important consideration, especially in the treatment of metabolic diseases. Obese patients are advised to take early meals.

Ayurveda, and many other traditions, advise against eating late night dinners. Jains do not consume any food after sunset. The Ayurvedic concept of micro-channels, or *Srotas* is important to understand as the basis of this practice. *Srotas* in the body are believed to open during sunrise, and start closing during sunset. The microchannels function optimally during daytime, and not during late night. After sunset, the *Srotas* are constricted, and their activity is reduced. If constricted *Srotas* are loaded with food, the balance of *Dosha* is disturbed, and *Ama* is produced. The short-term effects of *Ama* are indigestion, hyper-acidity, headache, giddiness, lethargy, and constipation. The long-term effects of *Ama* are autoimmune, respiratory, and cardiac disease. There are studies on timing of food consumption, and the presentation of acute coronary events. Those who fasted had symptoms mostly in the afternoon [39].

Food and sleep share an important connection. Sleep is an anabolic activity that helps digestion. The parasympathetic system dominates sleep, and metabolic activities. Daytime sleep however, can cause digestion problems, and is considered a risk factor for diabetes. The relationship between the timing of sleep and meals is addressed by Ayurveda. Daytime, post-meal sleep is thought to increase *Kapha*. Therefore, those individuals whose *Kapha* is depleted may benefit from moderate sleep after the daytime meal. However, this kind of advice cannot be generalized. A study in healthy volunteers reports that those who slept during the day had a higher risk for diabetes [40]. Another study suggests that food intake before sleep, or late-night snacks can adversely affect sleep quality [41]. Thus, the timing of food intake and sleep are related, but vary depending on the individual *Prakriti*.

CALORIE RESTRICTION

Yoga therapy and Ayurveda advise fasting to clean the system, and improve its tone and vitality. In the absence of food, *Agni* digests vitiated *Dosha*; this is a basic principle of dietary restriction therapies. These therapies are called *Langhana*, and involve dietary restriction, along with monitoring symptoms of vitiated *Dosha*, *Agni*, and *Dhatu*. The endpoint of *Langhana* is feeling light, increased appetite, proper elimination of wastes, and improved cognitive functions.

The results of clinical trials on fasting support the contention that fasting has protective effects in cases of obesity, hypertension, asthma, rheumatoid arthritis, and that fasting can delay the aging process [42]. The mechanisms of fasting have shown its effects through lipolysis, and the upregulation of gluconeogenesis by the liver in mice. Fasting stimulates gut-derived serotonin upregulation, and prevents glucose uptake by hepatocytes [43]. The neuroprotective effects of fasting are thought to be linked to the activation of sirtuins [44]. Sirtuins are a family of proteins that regulate important, biological pathways such as those of aging, transcription, apoptosis, inflammation, stress resistance, and maintain alertness during calorie restriction. They also control circadian clocks, and mitochondrial biogenesis.

The physiological effects of fasting are receiving much attention against the background of the epidemic of non-communicable diseases such as diabetes, cancer, and autoimmune diseases. It is important to know when not to eat, so as to avoid burdening the digestive system. Intermittent fasting improves the lipid profile, and decreases inflammatory responses. Gene expression related to inflammatory response is reported to be altered with intermittent fasting [45].

Ayurveda advocates fasting as the first, and most important step in the management of any type of fever. Therapeutic fasting is also suggested as a treatment for many diseases, such as rheumatoid arthritis, diabetes, and obesity. The logic of therapeutic fasting is to regain homeostasis, and prevent generation of *Ama*. Therapeutic fasting helps in the digestion of *Ama*, and expedites the excretion of toxins, which relieve the fever.

Many types of fevers are caused by microbial infections. These pathogens produce enzymes known as neuraminidase, which unmasks host's cellular lectins. These lectins can trigger the expression of human leukocyte antigen type II (HLA II). HLA II can then bind to food lectins, and may induce antibody response. Some of these antibodies can attack healthy tissues leading to disease. In this process, if food lectins are not available, further complications can be avoided. Therapeutic fasting is a way to avoid the production of food lectins. It works as a neuraminidase inhibitor, helping to prevent the genesis of auto-immunity during infection, thus removing a root cause of fever [46].

Dietary restriction (DR), or calorie restriction (CR) involve reduced food intake without causing malnutrition. Both CR and DR are often connected with delayed aging, and with the extension of life span, or longevity. Long-term DR, and short-term fasting is known to provide protection against many types of neuronal damage, surgical and ischemia-reperfusion injury, and Parkinson's disease. DR and fasting have been shown to conserve energy, and increase mature B and T-cell numbers in bone marrow [47]. Recent research on monkeys has shown that long-term restriction of the diet significantly improve the survival and longevity [48]. Findings of other studies suggest that CR, or drugs such as rapamycin which mimic the mechanism of CR, may lead to the extension of lifespan [49]. Dietary restriction may improve stress resistance, and metabolic fitness, providing beneficial effects in conditions like surgical stress, inflammation, chemotherapy, and insulin resistance [50].

FASTING AND SPIRITUALITY

In many cultures fasting has been advocated for health promotion, disease prevention, and therapeutic purposes. Fasting is also part of spiritual practices. Buddhist, Jains, Muslims, Hindus, and many other traditions engage in fasting as a ritual. Different traditions have different ways of fasting, in different seasons, and in differing durations. Jains follow *Paryushanparva*. Muslims observe the month of *Ramadan*. Hindus fast during the month of *Shravan*. In many traditional, and spiritual practices fasting is not limited to restrictive or selective intake, or avoidance of food to emphasis that fasting is a way to pursue spiritual goals. In these traditions, fasting is considered to facilitate meditation, awakening, and enlightenment.

A story about the Buddha is interesting. Even after intense fasting and meditation for years, the Buddha was not able to attain his goal of enlightenment. Finally, he gave up a long-duration fast by eating rice pudding, followed by meditation—he then attained enlightenment. Based on this story, an interesting hypothesis has been proposed:

Intense fasting may have inhibited the Buddha's monoamine oxidase (MAO) activity. Eating rice pudding provided dietary tryptophan from carbohydrates. The carbohydrates triggered insulin release, which increased unbound tryptophan, while reducing the levels of competing amino acids at the blood–brain barrier. These effects may have led to a significant influx of tryptophan into the brain, where it was converted to serotonin. In the absence of MAO, serotonin does not degrade, and methyl-transferases convert excess tryptophan, and serotonin into endogenous, psychoactive tryptamines. The endogenous serotonin, and tryptamines are very important neurotransmitters, and many of the psychotropic, and hallucinogenic drugs act through their modulation. The absence of psychoactive substances, and the prolonged fasting may have given the experience its perceived spiritual power [51].

This hypothesis indicates that fasting and meditation techniques may have the ability to alter mental states.

In short, diet is the primary strategy for disease prevention, and health protection. Diet can affect personality, and behavior. Dietary requirements vary from person to person. The interactions between food and microbiota have become an important area of study. The consumption of natural foods, fresh vegetables, and fruits; the restriction of calories; and the mental state of the person positively influences health, and longevity. Guidance from Ayurveda and yoga can play an important role in health care.

If a proper diet is followed, what is the need for medicine? If the diet if not proper, what is the use of medicine?

An ancient Indian proverb

REFERENCES

[1] Kaplan D. The philosophy of food. Berkeley: University of California Press; 2012. Available from: http://www.foodtimeline.org.

[2] http://www.foodtimeline.org.

[3] Dal A. Food in the ancient world from A to Z. London: Routledge; 2003.

[4] Jeffery RW, French SA. Epidemic obesity in the United States: are fast foods and television viewing contributing? Am J Public Health 1998;88(2):277–80.

[5] Helmchen La, Henderson RM. Changes in the distribution of body mass index of white US men, 1890–2000. Ann Hum Biol 2004;31(2):174–81.

[6] Ogden CL, Carroll MD, Kit BK, Flegal KM. Prevalence of obesity in the United States, 2009–2010. NCHS Data Brief 2012(82):1–8.

[7] Hopkins FG. Feeding experiments illustrating the importance of accessory factors in normal dietaries. J Physiol 1912;44(5–6):425–60.

[8] Ye R, Scherer PE. Adiponectin, driver or passenger on the road to insulin sensitivity? Mol Metab 2013;2(3):133–41.

[9] Tsujino N, Sakurai T. Orexin/hypocretin: a neuropeptide at the interface of sleep, energy homeostasis, and reward system. Pharmacol Rev 2009;61(2):162–76.

[10] Kheradmand A, Dezfoulian O, Tarrahi MJ. Ghrelin attenuates heat-induced degenerative effects in the rat testis. Regul Pept 2011;167(1):97–104.

[11] Meier U, Gressner AM. Endocrine regulation of energy metabolism: review of pathobiochemical and clinical chemical aspects of leptin, ghrelin, adiponectin, and resistin. Clin Chem 2004;50(9):1511–25.

[12] Lake JI, Heuckeroth RO. Enteric nervous system development: migration, differentiation, and disease. Am J Physiol Gastrointest Liver Physiol 2013;305(1):G1–24.

[13] Gershon MD. 5-Hydroxytryptamine (serotonin) in the gastrointestinal tract. Curr Opin Endocrinol Diabetes Obes 2013;20(1):14–21.

[14] Mayer EA. Gut feelings: the emerging biology of gut-brain communication. Nat Rev Neurosci 2011;12(8):453–66.

[15] Montiel-Castro AJ, González-Cervantes RM, Bravo-Ruiseco G, Pacheco-López G. The microbiota-gut-brain axis: neurobehavioral correlates, health and sociality. Front Integr Neurosci 2013;7:70. http://dx.doi.org/10.3389/fnint.2013.00070.

[16] Bischoff SC. "Gut health": a new objective in medicine? BMC Med 2011;9:24. http://dx.doi.org/10.1186/1741-7015-9-24.

[17] Du Y, Yang M, Lee S, Behrendt CL, Hooper LV, Saghatelian A, et al. Maternal western diet causes inflammatory milk and TLR2/4-dependent neonatal toxicity. Genes Dev 2012;26(12):1306–11.

[18] Kelly SA, Summerbell CD, Brynes A, Whittaker V, Frost G. Wholegrain cereals for coronary heart disease. Cochrane Database Syst Rev 2007;(2):CD005051.

[19] Hinderliter AL, Babyak MA, Sherwood A, Blumenthal JA. The DASH diet and insulin sensitivity. Curr Hypertens Rep 2011;13(1):67–73.

[20] Shirani F, Salehi-Abargouei A, Azadbakht L. Effects of dietary approaches to stop hypertension (DASH) diet on some risk for developing type 2 diabetes: a systematic review and meta-analysis on controlled clinical trials. Nutrition 2013;29(7–8):939–47.

[21] Rees K, Hartley L, Flowers N, Clarke A, Hooper L, Thorogood M, et al. "Mediterranean" dietary pattern for the primary prevention of cardiovascular disease. Cochrane Database Syst Rev 2013;8:CD009825.

[22] Panagiotakos D, Pitsavos C, Koloverou E, Chrysohoou C, Stefanadis C. Mediterranean diet and diabetes development: a meta-analysis of 12 studies and 140,001 individuals. J Am Coll Cardiol 2014;63(12_S):A1349.

[23] Estruch R, Ros E, Salas-Salvadó J, Covas M-I, Corella D, Arós F, et al. Primary prevention of cardiovascular disease with a mediterranean diet. N Engl J Med 2013;368(14):1279–90.

[24] Thomas D, Elliott EJ. Low glycaemic index, or low glycaemic load, diets for diabetes mellitus. Cochrane Database Syst Rev 2009;(1):CD006296.

[25] http://www.gmdietworks.com.

[26] Rees K, Dyakova M, Ward K, Thorogood M, Brunner E. Dietary advice for reducing cardiovascular risk. Cochrane Database Syst Rev 2013;12:CD002128.

[27] Chowdhury R, Warnakula S, Kunutsor S, Crowe F, Ward HA, Johnson L, et al. Association of dietary, circulating, and supplement fatty acids with coronary risk: a systematic review and meta-analysis. Ann Intern Med 2014;160(6):398–406.

[28] Clarke A, Trivedi M. Bovine beta casein variants: implications to human nutrition and health. Int Conf Food Secur Nutr IPCBEE 2014;67:11–7.

[29] Hunter P. We are what we eat. EMBO Rep 2008;9(5):413–5.

[30] Gómez-Pinilla F. Brain foods: the effects of nutrients on brain function. Nat Rev Neurosci 2008;9(7):568–78.

[31] Bouchard-Mercier A, Paradis A-M, Rudkowska I, Lemieux S, Couture P, Vohl M-C. Associations between dietary patterns and gene expression profiles of healthy men and women: a cross-sectional study. Nutr J 2013;12:24. http://dx.doi.org/10.1186/1475-2891-12-24.

[32] Virmani A, Pinto L, Binienda Z, Ali S. Food, nutrigenomics, and neurodegeneration–neuroprotection by what you eat! Mol Neurobiol 2013;48(2):353–62.

[33] Acharya Y, editor. Caraka Samhita. Varanasi, India: ChaukhambaSurbharati; 1992.

[34] Vijay-Kumar M, Aitken JD, Carvalho FA, Cullender TC, Mwangi S, Srinivasan S, et al. Metabolic syndrome and altered gut microbiota in mice lacking toll-like receptor 5. Science 2010;328(5975):228–31.

[35] Sumantran VN, Tillu G. Cancer, inflammation, and insights from Ayurveda. Evid Based Complement Altern Med 2012;2012:306346. http://dx.doi.org/10.1155/2012/306346.

[36] Chopra A, Saluja M, Tillu G. Diet, Ayurveda and interface with biomedicine. J Ayurveda Integr Med 2010;1(4):243–4.

[37] Chopra A. Ayurvedic medicine and arthritis. Rheum Dis Clin North Am 2000;26(1):133–44.

[38] Wu T, Sun L, ZhuGe F, Guo X, Zhao Z, Tang R, et al. Differential roles of breakfast and supper in rats of a daily three-meal schedule upon circadian regulation and physiology. Chronobiol Int 2011;28(10):890–903.

[39] Al Suwaidi J, Bener A, Gehani AA, Behair S, Al Mohanadi D, Salam A, et al. Does the circadian pattern for acute cardiac events presentation vary with fasting? J Postgr Med 2006;52(1):30–3.

[40] Rehman JU, Brismar K, Holmbäck U, Åkerstedt T, Axelsson J. Sleeping during the day: effects on the 24-h patterns of IGF-binding protein 1, insulin, glucose, cortisol, and growth hormone. Eur J Endocrinol 2010;163(3):383–90.

[41] Crispim CA, Zimberg IZ, Dos Reis BG, Diniz RM, Tufik S, De Mello MT. Relationship between food intake and sleep pattern in healthy individuals. J Clin Sleep Med 2011;7(6):659–64.

[42] Longo VD, Mattson MP. Fasting: molecular mechanisms and clinical applications. Cell Metab 2014;19(2):181–92.

[43] Sumara G, Sumara O, Kim JK, Karsenty G. Gut-derived serotonin is a multifunctional determinant to fasting adaptation. Cell Metab 2012;16(5):588–600.

[44] Kincaid B, Bossy-Wetzel E. Forever young: SIRT3 a shield against mitochondrial meltdown, aging, and neurodegeneration. Front Aging Neurosci 2013;5:48. http://dx.doi.org/10.3389/fnagi.2013.00048.

[45] Azevedo FR De, Ikeoka D, Caramelli B. Effects of intermittent fasting on metabolism in men. Rev Assoc Med Bras 2013;59(2):167–73.

[46] Yarnell E. Proposed biomolecular theory of fasting during fevers due to infection. Altern Med Rev 2001;6(5):482–7.

[47] Shushimita S, De Bruijn MJ, De Bruin RW, IJzermans JN, Hendriks RW, Dor FJ. Dietary restriction and fasting arrest B and T cell development and increase mature B and T cell numbers in bone marrow. PLoS One 2014;9(2):e87772. http://dx.doi.org/10.1371/journal.pone.0087772.

[48] Colman RJ, Beasley TM, Kemnitz JW, Johnson SC, Weindruch R, Anderson RM. Caloric restriction reduces age-related and all-cause mortality in rhesus monkeys. Nat Commun 2014;5:3557. http://dx.doi.org/10.1038/ncomms4557.

[49] Zhao G, Guo S, Somel M, Khaitovich P. Evolution of human longevity uncoupled from caloric restriction mechanisms. PLoS One 2014;9(1):e84117. http://dx.doi.org/10.1371/journal.pone.0084117.

[50] Robertson LT, Mitchell JR. Benefits of short-term dietary restriction in mammals. Exp Gerontol 2013;48(10):1043–8.

[51] Joseph PG. Serotonergic and tryptaminergic overstimulation on refeeding implicated in "enlightenment" experiences. Med Hypotheses 2012;79(5):598–601.

CHAPTER 8

Health Supplements

Hormones, vitamins, stimulants, and depressives are oils upon the creaky machinery of life. The principal item, however, is the machinery.

Martin H. Fischer

NUTRACEUTICALS AND SUPPLEMENTS

Health supplement is a relatively new concept. This is different from nutritional supplements, such as vitamins and minerals, which correct specific deficiency diseases. Health supplements are substances which can be characterized as something between a drug and food. They carry putative claims of specific health benefits. Many times, there is no sufficient, scientific proof to support these claims. Labels are carefully worded to avoid any clash with regulators. In many instances, these labels are vague and misleading, and these products are often aggressively marketed by the industry, with help of attractive, but irrational advertisements. It is a relatively common practice for multilevel marketing strategies, with attractive incentives, to be adopted in order to hook gullible consumers. The health food and supplement industry is rapidly growing, creating huge demand; at the same time, people are left in a state of confusion. Many claims about health supplements are not based on serious research. There are reports of adverse effects of so-called health supplements. Health supplements are promoted as semi "drugs," but they completely bypass the stringent, regulatory process. Therefore, it is important to revisit, review, and analyze the current status of health supplements, and to explore the possibility that traditional knowledge systems, like Ayurveda, can help resolve this situation.

The term nutraceutical covers foods consumed for their health benefits, apart from nutritional effects. It was probably coined in 1989, by Stephen L. DeFelice, founder of Innovation in Medicine in the United States. Dietary supplements and functional foods are part of nutraceuticals.

The Unites States' FDA defines a dietary supplement as "a product intended for ingestion that contains a dietary ingredient intended to add further

201

nutritional value to (supplement) the diet." By this definition, dietary ingredients can include botanicals, vitamins, minerals, amino acids, and other substances such as enzymes, and metabolites. A dietary supplement can be used as a single substance, or in combination with vitamins, minerals, herbs, botanicals, amino acids, or a concentrated metabolite, constituent, or extracts of all these listed. The dietary supplement provides necessary nutrients, and helps in health promotion or in prevention of diseases. The dietary supplement can have various dosage forms—capsules, tablets, powders, and soft gels. In the United States, health supplements are regulated under the Dietary Supplement Health and Education Act (DSHEA). This act has opened the floodgates for the marketing of many supplements, without sufficient scientific rationales and data to support their claims.

The term "functional food" is thought to have been first used in 1980, in Japan. The term includes processed or fortified food—but in its natural form, not in capsules or tablets. Functional food carries the expectation of having effects which prevent or control diseases. There are many definitions of functional food; the European Commission defines it as "a food that beneficially affects one or more target functions in the body, beyond adequate nutritional effects, in a way that is relevant to either an improved state of health, and well-being and/or reduction of risk of disease" [1].

Almost every food has some effect, and the discussion around functional food is becoming complex. Japan has introduced another term, "foods for specified health use" (FOSHU). The FOSHU products are expected to have proven effects on the human body; absence of any safety issues; use of nutritionally appropriate ingredients in acceptable quantities; a guarantee of compatibility with product specifications at the time of consumption; and established quality control methods [2]. Regulations of foods of modified form (dietary supplements) and processed foods (functional foods) are being developed in many countries, in order to prevent the exploitation of patients who might believe dubious claims made by the producers of these foods, and supplements. A wide array of supplements is available with putative health claims. This include nutraceuticals, herbal supplements, antioxidants, vitamins, minerals, lipids, phytochemicals, dietary fats, pre and probiotics, sports supplements, proteins, energy drinks, and dietary fibers (Figure 8.1).

This current regulatory picture illustrates how the development of food and nutrition use is hewing to the drug development path. Many claims about specific substances (like vitamins, antioxidants) or food programs (like the GM diet) are being studied in systematic reviews. The careful analysis of available evidence gives the real picture, which may be contrary to commonly held beliefs.

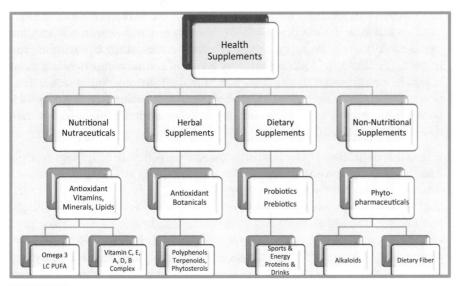

FIGURE 8.1

Various types of health supplements.

NUTRITIONAL AND NONNUTRITIONAL SUPPLEMENTS

We find two extreme approaches toward food: *blame it* and *bank on*. The blame it approach relates particular foods to certain disease, as a causative factor; for example, fat and butter is bad for cholesterol and triglycerides. The bank on approach assumes that certain types of foods or supplements are good for health; for example, special proteins, antioxidants, and omega 3-like supplements are popular for their putative role in health promotion. Examples include amino acid preparations, whey proteins, spirulina, and ginseng, among others. The role of antioxidants in health and disease still remains a debated subject, and will be discussed in the following sections.

Various types of supplements, collectively promoted as nutraceuticals, functional foods, or dietary products are of two broad types: first, the category products that may have actual nutritional value, such as vitamins and minerals. At present, there is a lot of awareness about the importance of eating natural and fresh foods, rather than consuming processed foods. There is growing evidence to support the importance of fresh and natural foods. For instance, to counter vitamin A deficiency, eating fresh carrots is better than taking carotenoid supplementation. Fresh fruits like papaya and vegetables like carrots are much better tolerated, and have been shown to improve vitamin A status in lactating mothers [3]. Just consuming proximate principles like proteins, carbohydrates and fats is not enough for healthy growth and sustenance. We need micronutrients like vitamins and minerals, which play important roles in the regulation of several physiological processes and biochemical pathways.

The discovery of vitamins and minerals as micronutrients was a major landmark in scientific biomedical research. It is known that several diseases are manageable merely by supplementing required quantities of vitamins and minerals. For instance, iodine deficiency is one of the main causes of impaired cognitive development in children. Systematic efforts to develop salt-iodization programs have helped many countries counter iodine deficiency in their populations. According to the WHO, about 56 countries still face iodine deficiency in their populations.

Vitamin A deficiency is the leading cause of preventable blindness in children. It also compromises immunity, and increases the risk of infectious disease. In pregnant women, it may cause night blindness and is associated with an increased risk of maternal mortality. WHO estimates about 250,000 to 500,000 vitamin A-deficient children become blind every year. Iron is essential for building hemoglobin, which is responsible for supplying oxygen to our cells. Iron deficiency anemia can reduce the physical capacity to work. This can lead to serious socio-economic consequences at an individual and community level, which can adversely impact national economies. According to the WHO, almost 30% of the world's population are anemic. The majority of girls and women in developing countries have been found to be anemic.

Therefore, vitamins and mineral are considered as essential micronutrients in the diet, and many supplementation programs have been developed and successfully implemented. For example, the iodization of salt has shown remarkable benefits in controlling iodine deficiency. However, rather than exploring natural and dietary sources, supplementation has been pushed by market forces. It must be remembered that the current practice of medicine is supposed to be evidence based. This means for every intervention, therapy, or strategy, there should be a sufficient scientific base. Most reliable scientific evidence comes from published papers, systematic reviews, and meta-analyses.

Of late, many experts have criticized the irrational supplementation of vitamins and minerals. Several studies, systematic reviews, and meta-analyses indicate that there is not sufficient evidence to support generalized vitamin supplementation. A recent editorial from the *Annals of Internal Medicine*, titled "Enough Is Enough: Stop Wasting Money on Vitamin and Mineral Supplements" is a representative [4]. It must be remembered that while vitamins and minerals are necessary as micronutrients, their unnecessary or over-consumption may show untoward effects.

Thus, dietary supplementation of micronutrients like vitamins, minerals, and antioxidants is a double-edged sword, and must be used very carefully. However, because of less stringent regulatory controls, coupled with the pursuit of profit, today vitamin, mineral, and antioxidant supplementation has become a multibillion-dollar industry.

Against this background, it will be interesting to probe a bit more into the putative role of antioxidants, vitamins, and minerals in health and disease.

ANTIOXIDANTS

The most common mechanism, and reason behind the beneficial effects, of the consumption of fruits and vegetables seems to be through an ability, when consumed to counter a type of reactive oxygen species—known as free radicals—through a scavenging, or antioxidant property. Free radical damage is responsible for oxidative stress and may lead to damage at cellular and molecular levels—including mutilation of DNA, enzymes, proteins, lipids, and alterations in biochemical processes. Oxidative stress may cause cell damage resulting in chronic diseases such as cancer, and oxidative damage in relation to degeneration, and aging—as will be discussed in Chapter 10. However, it may be pertinent to discuss the dietary benefits of the scavenging of free radicals or reactive oxygen species, and reactive nitrogen species, by food supplements consisting of high quantities of antioxidants. The antioxidant theory of disease prevention has gone through several cycles of agreement and disagreement.

Oxygen radicals, and other reactive oxygen/nitrogen species, are involved in several human diseases. These substances are known as free radicals. These free radicals can harm chromosomes, genes, proteins, cells, tissues, and metabolism. Removing or scavenging free radicals is considered as a therapeutic and preventive strategy. Antioxidants are supposed to be free radical scavengers. Therefore, many supplements and nutraceuticals claim to contain antioxidants. However, the role of free radicals in the progression of diseases is not very clear. They seem to be more important in conditions like neurodegenerative diseases and cancers. Reducing oxidative damage may decrease the risk of development of certain diseases. However, supplements of antioxidants may not necessarily decrease oxidative damage. There is limited evidence supporting the benefits of antioxidant supplementation, and data showing the benefits of antioxidant supplementation is not consistent.

The term *antioxidant paradox* is discussed against the background of several experiments and trials. Though reactive oxygen species play a major role in the pathogenesis of several diseases, large doses of dietary antioxidant supplements in these diseases studies demonstrated little or no preventative, or therapeutic effect. This phenomenon is known as the antioxidant paradox [5]. It is accepted that oxygen radicals, and other reactive oxygen species play an important role in several human diseases. However, large amounts of anti-oxidants as dietary supplements has not demonstrated much preventative or therapeutic benefit. This paradoxical situation still continues and the putative, beneficial role of antioxidant supplements remains inconclusive.

The role of reactive oxygen species in the origin and/or progression of most human diseases is still unclear. They probably play an important role in the pathogenesis of diseases such as cancer and neurodegenerative conditions. The endogenous antioxidant mechanisms in the human body are complex, inter-locking, and regulated. The total antioxidant capacity of the body is seemingly not responsive to high doses of exogenous antioxidant supplements; the extent of oxidative damage is rarely altered. It is suggested that careful regulation and modulation of endogenous antioxidant levels by supplying weak pro-oxi-dants may be more useful for treatment and prevention of disease. Rather than banking on exogenous antioxidant supplements, scientists are advising eating plenty of fruits, grains, and vegetables—coupled with lifestyle modifications like avoiding unnecessary weight gain, refraining from habits like smoking, and doing regular exercise, all of which may trigger beneficial adaptation and minimize oxidative damage.

Many efforts, intentionally or otherwise, are aimed at trying to find a "magic pill," which can maintain health, control diseases, and even replace diet! The health-supplements culture is transforming the wholesome natural *diet* into a capsuled *drug*. This is a mechanical approach, which considers energy requirement as a mathematical calculation. Diet is not something that can be mechanically considered as the input and output of calories, especially when the effects of diet on the brain, cognitive functions, and emotions are known. Human perceptions about diet have a long history. We need to go beyond blame it and bank on approaches, and discover a *balanced*, individualized approach.

Many systematic reviews, and meta-analysis studies, based on several well-designed, randomized clinical trials have not been able to provide convincing evidence that antioxidants such as beta-carotene, vitamin A, vitamin C, vitamin E, and selenium, or their combinations prevent gastroin-testinal cancers [6]. It will be interesting to discuss a few prominent exam-ples of antioxidant phytochemicals and vitamins.

VITAMINS

Many vitamins especially vitamins A, E, D, C, and B complex are consid-ered *antioxidants*. The role of antioxidants in health promotion and disease prevention and treatment has been explored for many years. However, in a recent systematic review of 78 randomized clinical trials totaling 296,707 participants showed no evidence to support the usefulness of antioxidant supplements for primary or secondary prevention of diseases. Data from 52 trials involving 80,807 participants with gastrointestinal, cardiovascular, neurological, ocular, dermatological, rheumatoid, renal, and endocrinological diseases were analyzed in this massive review. The authors observed that higher doses of beta-carotene, vitamin A, and vitamin E, in fact may have led to increased mortality. The increased risk of mortality was not associated with the use of vitamin C or selenium. The authors conclude that the current evidence does not support the use of antioxidant vitamin supplements, either in the general population, or in patients with various diseases [7].

Still, antioxidant vitamins, and phytochemicals are extensively used for putative benefits in many age-related, degenerative conditions, such as cataracts, which are influenced by oxidative stress. Various antioxidant vitamin supplementations have been used to prevent the formation or slow the progression of cataracts. Since the results from observational studies have been inconsistent, a systematic review was attempted using the results of nine trials involving 117,272 adults, of over 35 years of age, where supplementation with the antioxidant vitamins, beta-carotene, vitamin C, and vitamin E was compared to an inactive placebo. These well-controlled trials were conducted in Australia, Finland, India, Italy, the United Kingdom, and the United States. The doses of antioxidants given in these trials were higher than the recommended daily allowances. A systematic review of these trials failed to show any meaningful evidence in favor of antioxidant vitamins' effect on the incidence of extraction, or progression of cataracts, or on the loss of visual acuity. The results of this systematic review were so convincing that the authors boldly advised, "We do not recommend any further studies to examine the role of antioxidant vitamins." Even apart from any benefits, this supplementation seems to compromise the basic principle in therapeutics "cause no harm," since a considerable number of participants developed yellowing of the skin due to excess carotenoids [8]. In some cases—including age-related, macular degeneration—benefits from antioxidant vitamin and mineral supplements of a moderate delay in progression of the disease, have been reported. But these studies are small in number and were conducted on well-nourished populations, making it difficult to reach any generalized conclusion. Although generally regarded as safe, vitamin supplements may have harmful effects. A systematic review of the evidence on the harms of vitamin supplements is needed [9].

Oxidative stress is also responsible for liver damage, hence antioxidant vitamins are used to counter hepatic toxicity and prevent liver disorders due to alcoholism, autoimmune disorders, hepatitis B or hepatitis C, and liver cirrhosis. A systematic review of 20 randomized clinical trials, involving 1225 participants who were given beta-carotene, vitamin A, vitamin C, and vitamin E as antioxidant supplements, found no evidence to support or refute the use of antioxidant supplements in patients with liver disease. The analysis has indicated that antioxidant supplements may increase liver enzyme activity [8].

Vitamin B Complex

Vitamins of the B complex groups includes thiamine (B1), riboflavin (B2), niacin (B3), pantothenic acid (B5), pyridoxine (B6), biotin (B7), folic acid (B9), and cyanocobalamin (B12). The B complex is involved in various metabolic and cellular functions. They are used for treating various diseases. In peripheral neuropathy, however, there are not enough studies or evidence to determine its beneficial or harmful effects. A systematic review of 13 trials, involving 741 participants with diabetes and alcoholic peripheral neuropathy showed a very marginal benefit with vitamin B1, as compared to a placebo. Another study indicated that vitamin B complex in a higher dose for four weeks has a better effect in reducing pain and other clinical problems. While vitamin B is generally well-tolerated, larger studies are required to establish the therapeutic benefits in treating neuropathy, and such conditions [21].

Cardiovascular disease (CVD) is the number one cause of death worldwide. Ischemic heart disease, stroke, and congestive heart failure are the most common causes of cardiovascular morbidity and mortality. In addition to the known risks of diabetes, high blood pressure, smoking, and impaired lipid profile, now increased homocysteine levels are also considered as one of the key risk factors. Homocysteine is an amino acid, and its blood levels are known to be influenced by cyanocobalamin (B12), folic acid (B9), and pyridoxine (B6). High total plasma homocysteine levels are associated with an increased risk of atherosclerotic diseases. Therefore, vitamin B complex supplementation is thought to reduce the risk of myocardial infarction, stroke, and angina pectoris. However, a systematic review of 12 randomized clinical trials, involving 47,429 participants found no evidence to support the homocysteine-lowering effect of vitamins B6, B9, or B12 supplementation, as compared with a placebo or standard care. Vitamin B complex supplementation had no meaningful effect on the prevention of myocardial infarction, stroke; nor did it reduce total mortality in participants at risk [22].

Vitamin B6 is a water-soluble vitamin which helps in the development of the nervous system. B6 is present in many foods, including meat, poultry, fish, vegetables, and bananas. Vitamin B6 may have a role in the prevention of pre-eclampsia, and in preterm birth. Vitamin B6 is considered to be helpful for reducing nausea in pregnancy. Overdosing with B6 is reported to be associated with numbness and difficulty in walking. The results of five trials involving 1646 women were analyzed in a systematic review to assess the usefulness of B6 supplementation during pregnancy. The aim of this review was to find if vitamin B6 can reduce the chances of pre-eclampsia and preterm birth. The systematic review did not find much evidence in support of the benefits of vitamin B6 supplementation during pregnancy. In fact, this systematic review concluded that too much vitamin B6 can be harmful [23]. Thus, it has not been shown that a popular supplementation like vitamin B complex, which is commonly used as a tonic, provides any substantial health benefits.

Vitamin C

Vitamin C (ascorbic acid) was actually the first substance tested systematically in clinical trial by James Lind. This is an important nutrient, that is involved in several enzymatic reactions, such as the conversion of cholesterol to bile acids, activation of vitamin B, the development of connective tissues, wound healing, phagocytosis; and it has antioxidant activity. After its first trial in 1747, there have been more than 2600 trials conducted on vitamin C in the last 260 years. The investigations on the efficacy of vitamin C as a treatment, ranged from the common cold to cancer. Despite these efforts the evidence for the use of vitamin C as a treatment is still not conclusive for many diseases.

As a routine practice, vitamin C is prescribed in viral fevers, common colds, and other infections. However, the meta-analysis of 29 therapeutic trials (with 11,306 patients) on vitamin C indicates no consistent effect on the duration or severity of colds [24]. The conclusive evidence for the prophylactic use of vitamin C for the prevention of pneumonia, or asthma is also lacking [25]. The vitamin has been administered during pregnancy with the indication to reduce risk of stillbirth, perinatal death, low birth weight, and intrauterine growth restriction. Analysis of five trials with 766 participating women does not conclusively suggest the role of vitamin C in preventing these pregnancy outcomes. However, the risk of preterm birth was noted in the analysis [26].

The role of vitamin C in cancer prevention and treatment has long been debated. Vitamin C acts as pro-oxidant (toxic) to tumor cells and induces apoptosis [27]. However, convincing evidence for its efficacy in preventing gastrointestinal and lung cancers has not been found [6] [28]. The role of vitamin C is being explored in the alleviation of adverse effects of chemotherapy and radiation treatment.

Vitamin E

For several years it was not clear if vitamin E functions merely as an antioxidant or has any other specific role. Earlier animal studies during the 1950s have shown that major requirement of vitamin E can be met by taking any biologically active, nontocopherol antioxidants. Such antioxidants were shown to prevent vitamin E deficiency symptoms [10]. Many earlier studies have indicated the beneficial effects of vitamin E, on several conditions, especially degenerative conditions. Interestingly, unlike many other vitamins that have specific functions and the deficiency of which can lead to specific, serious conditions, vitamin E is not associated with any specific disease or disorder. General symptoms of vitamin E deficiency include muscle weakness, loss of muscle mass, abnormal eye movements, vision problems, and unsteady walking, among other symptoms. These symptoms can also be due to several other factors. Placing Vitamin E into the category of Vitamins might be a forced conclusion. It might best be categorized as an antioxidant.

Given the evidence, the role of vitamin E beyond mere antioxidant has not been established. Still, vitamin E supplementation is rampant in the cosmetic industry, as well as among the elderly. In 2007, Traber and Atkinson proposed the hypothesis that vitamin E is antioxidant and nothing more [11]. The importance of antioxidant function is to maintain the bioactivity, and integrity of long-chain, polyunsaturated fatty acids in cell membranes. This would also mean that instead of relatively expensive vitamin E, one could use cheaper and easily available antioxidants, which might do the same job.

Still, many laboratories, and animal and epidemiological studies have pointed toward a possible beneficial role for vitamin E in the prevention of CVD, and in the treatment of degenerative conditions like Alzheimer's disease (AD). For instance, the Cambridge Heart Antioxidant Study, known as CHAOS, has concluded that in patients with coronary atherosclerosis, alpha tocopherol treatment can reduce the rate of myocardial infarction [12]. However, currently, very limited evidence exists in humans to support the routine use of vitamin E. A systematic review of three studies showed limited benefit of vitamin E in mild cognitive disorder, and AD. On the contrary, there is evidence implicating vitamin E with potentially serious side effects and increased mortality. The authors of this review, in no uncertain terms, conclude that vitamin E should not be used in the treatment of cognitive disorders and AD [13].

Probably, as a reaction to this conclusion, from 2012 to 2014, numerous articles have been published supporting the uses of vitamins in degenerative conditions like AD. In a controlled, clinical trial on patients with mild to moderate AD the administration of 2000 IU/d of alpha tocopherol slowed functional decline as compared to a placebo—suggesting the beneficial effects of vitamin E in patients with mild to moderate AD [14]. The reputable journal, *JAMA*, preferred to write an editorial highlighting the results of this trial. A few other medical journals, like *Neurology Today*, published supporting articles, although in guarded tones, highlighting this study [15]. The available evidence does not support routine use of vitamin E supplementation by intravenous route in high doses [16].

Vitamin A

In the developing world, many pregnant women are reported to be vitamin A deficient. During pregnancy, additional vitamin A is required. The benefits of vitamin A supplementation to children after six months of age is known to help reduce death and adverse effects on health. A study of potential benefits of vitamin A supplementation on 51,446 newborns showed a significant reduction in infant deaths at six months of age [19].

The role of vitamin A in preventing acute lower respiratory tract infections (LRTI) was studied by analyzing the results of 10 studies, including 33,179 participants. Eight studies found that vitamin A has no significant effect on the incidence or prevalence of acute LRTI. On the contrary, vitamin A caused an increased incidence of acute LRTI in one study. In two other studies vitamin A was found to be associated with increased symptoms of cough and rapid breathing. A few studies also reported that vitamin A significantly reduced the incidence of acute LRTI in children with poor nutritional status. However, vitamin A was reported to increase the incidence of LRTI in healthy children. This review concludes that despite its benefits in preventing diarrhea, vitamin A should not be given to all children, because it has a limited role in preventing acute LRTIs. The study also concluded that the beneficial effects appear mainly in undernutrition populations [20].

Vitamin D

Vitamin D is synthesized in the skin as vitamin D_3 (cholecalciferol) or is obtained from dietary sources or supplements as vitamin D_3, or vitamin D_2 (ergocalciferol). Many earlier studies have suggested that vitamin D may play a role in reducing occurrences of cancer, and CVD such as heart attack or stroke. A systematic review has analyzed data of 56 randomized, placebo controlled trials involving a total of 95,286 participants receiving vitamin D or placebo. The results of this analysis suggest that vitamin D_3 may reduce mortality in elderly people. Vitamin D_3 also seemed to decrease cancer mortality, showing a reduction in mortality of 4 per 1000 people, treated for 5 to 7 years. The study also reveals adverse effects of vitamin D, such as renal stone formation and elevated blood levels of calcium [17].

Scientists from the Imperial College, London, and the University of Edinburgh have done an umbrella review of 107 systematic literature reviews, 74 meta-analyses of observational studies of plasma vitamin D concentrations, and 87 meta-analyses of randomized controlled trials of vitamin D supplementation. They analyzed the relationship between vitamin D, and a wide range of skeletal, malignant, cardiovascular, autoimmune, infectious, metabolic, and other diseases. The beneficial effects of vitamin D supplementation were significant only for birth weight; with probable benefits to prevent dental caries in children; and to those with chronic kidney disease requiring dialysis. After such a huge effort of review of reviews and meta-analyses, the scientists concluded that current evidence does not support the argument that only vitamin D supplementation increases bone mineral density, or reduces the risk of fractures, or falls in older people [18].

MINERALS

Calcium

Calcium is an important mineral, which is needed to build and maintain strong bones. Calcium is vital for bone health especially in older adults for antifracture benefits. Calcium is also needed for proper functioning of heart, muscles, and nerves. Vitamin D is needed for absorption of calcium. The value of calcium supplementation, especially among those who are facing dietary shortage, is well recognized. However, a generalized advice of calcium supplementation has been challenged by many experts. Calcium supplements are known to cause mild gastrointestinal symptoms and have increased the risk of nephrolithiasis. It is also known that calcium supplements can increase the risk of both coronary and cerebrovascular events. A recent study by Lewis et al. known as Calcium Intake Fracture Outcomes Study (CAIFOS) published in the Journal of Bone and Mineral Research 2014 reported no significant benefits of calcium supplement to cardiovascular risk as compared with placebo. After several randomized studies and meta-analysis, scientists have recommended that the clinicians should promote ways and means for patients to receive adequate dietary calcium to achieve the recommended daily intake [44].

Sodium and Potassium

Sodium and potassium are important electrolytes in the body, with a wide range of physiological functions, including its role in homeostasis. The American Heart Association states that sodium is the number one enemy in the battle against high blood pressure. The beneficial role of potassium in controlling hypertension is becoming increasingly clear. However, the right balance of both sodium and potassium is essential for normal health. The majority of Americans consume a lot of salt, giving almost two times the amount of sodium than the recommended levels. In comparison, potassium, which is in fruits and vegetables, is consumed less. This results in an imbalance of sodium and potassium, causing a risk of many disease conditions, including cardiovascular events.

The reduction in sodium intake is general diet advice given to hypertensive patients. A systematic analysis of 167 studies indicate that a reduction of sodium resulted in just a 1% decrease in blood pressure in normotensives; a 3.5% decrease in hypertensive. A significant increase in plasma renin, plasma aldosterone, plasma adrenaline, and plasma noradrenaline; a 2.5% increase in cholesterol; and a 7% increase in triglyceride was also observed [42]. A review of 52 publications from January 1, 1990, to January 31, 2013 demonstrates that high salt intake definitely increases blood pressure. In addition, it also plays a role in endothelial dysfunction, cardiovascular structure and function, albuminuria and kidney disease progression, and cardiovascular morbidity and mortality, in the general population.

Conversely, increased dietary potassium intake can attenuate these effects. Studies have shown a linkage of potassium supplementation with a reduction in stroke rates and CVD risk. By moderate supplementation of potassium, over-weight, obese adults, and aging populations have shown greater sensitivity to the effects of a reduced salt intake. A diet that includes modest salt restriction, with increased potassium intake serves as a strategy to prevent, or control hyperten-sion, and decrease cardiovascular morbidity and mortality. Thus, the body of evi-dence supports the conclusion that a reduction in sodium intake and increased dietary potassium may prevent kidney disease, stroke, and CVD [43].

Selenium

Selenium is an essential trace element. It gives antioxidant protection, and helps redox-regulation in humans. Selenium supplements are frequently used by cancer patients. Since many adverse effects of radiotherapy and chemotherapy are linked to oxidative processes, it has been claimed that selenium reduces the side effects of cancer therapy. A systematic review of three clinical studies involving a total of 162 cancer patients could not find clear evidence to support the use of selenium supplements to reduce the side effects of chemotherapy, radiotherapy, or surgery. Although no adverse effects were reported in the studies, selenium intoxication due to overdosing has been reported in literature. More research is needed in order to ascertain the usefulness and the correct doses of selenium supplements when administered to counter the side effects of cancer therapy [40].

In another review, the effects of selenium supplements on healthy adults when used to prevent the occurrence of CVD has been studied. An analysis of 12 randomized trials consisting of 19,715 healthy adults receiving selenium sup-plements or a placebo were compared. This study has concluded that selenium supplements taken by healthy adults may not lead to prevention of major CVDs. Moreover, the study indicated an increased risk of developing type 2 diabetes, and suggests that selenium supplementation may be unnecessary for well-nourished communities [41].

Other minerals like iron and zinc have been used in many supplements. We have not covered these minerals in this section, because they do not merely act as antioxidants; and sufficient evidence is available about their role in pre-venting and treating certain deficiency disorders. For example, calcium supple-mentation may be beneficial in certain stages of the life cycle, such as during the growth period, pregnancy, and in treating conditions such as osteoporosis. Iron supplementation is beneficial in overcoming anemia.

LIPIDS, OILS, AND FATS

Lipids are available in nature as different oils and fats. They are key compo-nents of the cell membrane and have a substantial role in various metabolic processes. Many supplements containing fish oil, flax seed and other vegetable oils, essential oils, and animal and milk fat are available in the market. Many of them are used as antioxidants, growth promoters, and skin and hair tonics.

Fish oil has attracted considerable attention recently because of the cardioprotective effects of the n-3 fatty acids contained in fish oil, including eicosapentaenoic acid (EPA), and docosahexaenoic acid (DHA). The American Heart Association initiated recommendation of fish and/or fish oil supplements for patients with heart disease in 2002. As a result, use of fish oil supplements substantially increased in 2006. However, a recent meta-analysis has reported that omega-3 fatty acid supplementation was not associated with a lower risk of major CVD events [32].

Studies on α-linolenic acid (ALA), a plant-derived omega-3 fatty acid, and CVD risk have generated inconsistent results. In a meta-analysis summarizing the evidence regarding the relationship of ALA and CVD, 27 original studies including 251,049 individuals, and 15,327 CVD events were analyzed. While in observational studies, higher ALA exposure is seen to be associated with a moderate lowering of CVD risk, large, randomized clinical trials are needed for definitive conclusions [33]. Twenty-three randomized, controlled trials including 1075 participants, receiving a mean dose of 3.5 g/d of omega-3 polyunsaturated fatty acids (PUFA) , did not show a significant change in cholesterol, HbA1c, fasting glucose, fasting insulin, or body weight. In type 2 diabetes, omega-3 PUFA supplementation is shown to lower triglycerides and VLDL cholesterol. However, a raise in LDL cholesterol was observed. Moreover, there was no statistically significant effect on glycemic control or fasting insulin [34].

Six studies, with a total of 1039 patients suggest a marginal, significant benefit of omega-3 therapy for the maintenance of remission of Crohn's disease. Evidence from two large, high quality studies suggests that omega-3 fatty acids are probably ineffective for maintenance of remission in CVD. Omega-3 fatty acids appear to be safe, although they may cause diarrhea and upper-gastrointestinal tract symptoms [35].

A systematic analysis of two trials, with a total of 37 children with autism spectrum disorder (ASD), did not show much evidence in favor of omega-3 fatty acids supplementation. Given the paucity of rigorous studies in this area, more evidence needs to be generated from large, well-conducted, randomized controlled trials that examine both high and low functioning individuals with ASD with a longer follow-up period [36].

A systematic review of 13 trials with 1011 participants has revealed little evidence that PUFA supplementation provides any benefit for the symptoms of ADHD in children and adolescents. The majority of data showed no benefit of PUFA supplementation [37].An analysis of six randomized, controlled trials involving 1280 participants found no significant difference in children's neurodevelopment or visual acuity. However, in two studies LCPUFA supplementation was found to be associated with increased head circumference. Thus, current evidence to support or refute LCPUFA supplementation to breastfeeding mothers to improve infant growth, and development is not convincing [38].

A systematic review of 11 trials, involving 331 participants did not corroborate the effectiveness of a cholesterol-lowering diet for hyper-cholesterolaemia. Experts have suggested a large number of randomized controlled trials in order to investigate the effectiveness of omega-3 fatty acids, plant sterols, stanols, and soya protein for use in a cholesterol-lowering diet [39].

PHYTOCHEMICALS

Nature has marvelously provided a rich diversity of phytochemicals. Hundreds of phytochemicals, vitamins, and minerals as antioxidant supplements are also shown to be beneficial in preventing several diseases, including cancer, and a cluster of metabolic and degenerative diseases. Some of the important types of beneficial phytochemicals include polyphenols, terpenoids, organo-sulphur compounds, and phytosterols. For instance, more than 8000 different polyphenols are present in edible plants, which are part of the human diet [29]. Polyphenols include phenolic acids, such as gallic acid, ellagic acid, anacardic acid, and caffeic acid; flavonoids like cynidin, delphinidin, luteolin, and epicatechin; resveratrol; and curcuminoids. Terpenoids include carotenoids like beta-carotene, lutein, lycopene, ginsenosides, and ursolic acid. Organosulphur compounds include isothiocynates, sulforaphane, alicin, and mercaptans etc [30]. Many essential lipids like omega-3 fatty acids also have a wide range of beneficial effects.

Most edible fruits and vegetables have several antioxidant phytochemicals with the power to promote health and prevent diseases. Many studies have shown that consumption of fruits and vegetables help in health promotion and disease prevention. Various types of phytochemicals, vitamins, and minerals are naturally present in fruits and vegetables.

Instead of supplements consisting of extracts or actives, it is advisable to eat diets rich in plant products like grains, nuts, fruits, and vegetables to help maintain human health. Diets rich in vegetables and fruits are associated with a reduced risk of many diseases, including neurodegenerative disorders. Some phytochemicals like polyphenols, flavonoids, and anthocyanin are powerful antioxidants. They may have evolved as secondary metabolites to counter insects and other predators. These chemicals at low doses may stimulate adaptive neural stress responses, also known as hormesis. These pathways include cell-survival signaling kinases, the transcription factors, and histone deacetyl-ases of the sirtuin family. It is hypothesized that phytochemicals such as res-veratrol, and curcumin might protect neurons against injury by stimulating the production of antioxidant enzymes, neurotrophic factors, and proteins, which can help cells to withstand neuronal stress [31].

Most of these phytochemicals have a wide range of target pathways, and pri-marily work as antioxidants. They also have anti-inflammatory, antiprolifer-ative, and immunomodulatory activities. For instance, ellagic acid found in pomegranates and berries, resveratrol found in grape seeds and curcumin from turmeric, all have been shown to beneficially influence cell cycle, NF-kB signaling, p53 signaling, and hormone signaling.

ANTIOXIDANTS AND DISEASE PREVENTION

Antioxidant phytochemicals are found in fruits, vegetables, nuts, and legumes. Some of the most popular phytochemicals include carotenes, resveratrol, curcumin, anthocynidines, lutein, lycopene, phytosterols, tocopherols, and omega-3 fatty acids. Many phytochemicals are used to prevent disease like can-cer, CVDs, a cluster of diseases in metabolic syndrome, cognitive, and degen-erative disorders.

While a few stand-alone, or poorly controlled studies may justify their use, suf-ficient evidence to support their putative benefits in diseases like cancers does not exist. For example, a systematic analysis of 43 controlled trials was carried out to assess the beneficial and harmful effects of antioxidant supplements. After careful analysis, the authors concluded that there was no convincing evi-dence to justify the use of antioxidant supplements to prevent gastrointestinal cancers. On the contrary, antioxidant supplements seem to increase overall mortality [45]. Efforts to study the association between green tea consumption and various types of cancers have had interesting results. This involved patients with cancers of the digestive tract; gynecological cancer, including breast can-cer; urological cancer, including prostate cancer; lung cancer; and cancer of the oral cavity. The evidence to support the usefulness of green tea to reduce the risk of cancer was not conclusive. This means that the beneficial effects of drinking green tea remains unproven in cancer prevention, but this practice appears to be safe at moderate, regular, and habitual use [46].

It is known that people with chronic kidney disease (CKD) have a high risk of developing heart disease and risk of dying prematurely. It is believed that people with CKD often have higher oxidative stress, leading to an increased rate of disease progression. A study to evaluate the putative beneficial effects of antioxidant therapy for patients with CKD did not show much benefit in reducing the risk of heart disease, or death. However, there was some evidence to suggest that people on dialysis may benefit from antioxidant treatment, and that these therapies could reduce the risk of kidney disease becoming worse. The authors of this systematic review have concluded that the current evidence about the role of antioxidants to prevent risk of heart disease in patients with CKD is not sufficient. They have advised more systematic studies to explore antioxidant therapy benefits for patients with CKD [47].

Glutamine is an amino acid popularly used as supplement to enhance muscle growth in weightlifting, bodybuilding, and endurance. Glutamine is also used to expedite healing after surgeries. A combination of glutamine and antioxidants is used as a supplement in critically ill patients. A randomized, clinical trial was conducted to study the effect of glutamine and antioxidants. This multicentric trial involved 1223 critically ill patients, with multiorgan failure on mechanical ventilation; they were given glutamine, antioxidants, or a placebo. The results indicate no evidence to support the clinical benefits either of glutamine, or antioxidants. Contrary to expectation, the study reported that the glutamine group actually showed an increased mortality among patients [48].

PROBIOTICS

In 1907, Élie Metchnikoff, professor at the Pasteur Institute in Paris, suggested the possibility of replacing harmful microbes with useful microbes by modifying the gut flora. Metchnikoff proposed the hypothesis that the aging process is influenced by toxic substances produced by proteolytic bacteria in the large intestine. Live forms of beneficial microbes, for example, lactic acid bacteria, are present in many fermented foodstuffs like curd, buttermilk, and cheese. Use of such foods to promote health through the modification of gut microbial flora is known as probiotic. The actual meaning of probiotic is something that is "good for life," The term *prebiotics* is used for a nondigestible food ingredients, which can stimulate the growth of good bacteria in the colon. The term *synbiotic* is used when a product contains both probiotics and prebiotics. Several prebiotic, probiotic, and synbiotic products are available as health supplements in the world market [49].

The WHO's "Guidelines for the Evaluation of Probiotics in Food," have given recommendations for the evaluation and validation of the probiotic health claim. The International Scientific Association for Probiotics and Prebiotics (ISAPP) has formed an expert panel to discuss the field of probiotics. The WHO definition of a probiotic as "live microorganisms, which when administered in adequate amounts, confer a health benefit on the host" has been reinforced by this panel. This panel has suggested that the FDA should not regulate probiotics along the same lines as investigational, new drug applications. The panel felt that such an action can hinder the conduct and increase the cost of safe, much-needed research on probiotics, through unnecessary bureaucratic requirements.

The panel also opined that the drug approach to probiotic research can add to the cost and delay of much needed efficacy studies. The panel recognized the need to improve communication to the public and health-care professionals about the benefits of probiotics—which benefits can be expected for the general population. The panel also acknowledges that robust evidence must be provided for benefits associated with specific strains [50].

Probiotics are used in many gastro-intestinal conditions. A meta-analysis of 82 randomized, controlled trials involving 11,811 patients has shown the beneficial effects of probiotics in reducing antibiotic-associated diarrhea [51]. A systematic review of 19 trials, involving 1650 patients with irritable bowel syndrome, has also shown the beneficial effects of probiotics; however, more research is needed to identify the most effective microbial species, or strains [52,53]. A systematic review of 13 trials, involving 1347 patients, has shown some benefits of probiotics on functional constipation, with a low risk of adverse events, although the results were found to be inconsistent [54].

Probiotics seem to be the emerging market in health supplements. However, more scientific evidence of probiotics' health benefits and safety for intended used is required. Altering gut microbial flora can have a variety of effects—some can be good, some can be bad, and many are still unknown. Therefore, the assurance of safety of these products is necessary. It is indicated that through interference with commensal microflora, probiotics can result in adverse effects like sepsis, fungemia, and GI ischemia. Generally, critically ill patients in intensive care units, critically sick infants, postoperative and hospitalized patients, and patients with immune-compromised complexity were the most at-risk populations. Therefore, although overwhelming evidence suggests that probiotics are safe, it is crucial to carefully consider the risk-benefit ratio before prescribing them to critically ill patients [55].

DIETARY FIBER

Diet consists of several other ingredients apart from primary metabolites. The dietary fiber, tough, and not digested in the gut has several metabolic functions. The fiber bind nutrients, synthesizes new metabolites, and modulates nutrient absorption through processes like fermentation. The type and quantity of fiber affect the microbial composition of the gut. The role of fiber in gut health was ignored for many years, due to a focus on so-called active ingredients. Dietary fiber has attracted the attention of researchers. Fiber helps in prevention of colon cancer [56]. The higher the intake of fiber, the lower the risk is for both CVD, and coronary heart disease [57]. A recently published meta-analysis explores the relationship of fiber intake, and the risk of CVD. This analysis is based on 22 cohort studies, consisting of data of 23 years, including 32,730

CVD patients. The data suggest that consumption of fiber before and after myocardial infarction was responsible for lowering cardiovascular mortality [58]. The evidence for dietary fiber can be considered as conclusive, and needs translation into practice for better health indicators.

Increasing evidence suggesting the protective role of fiber is also boosting the use of fiber supplements. The use of artificial, or *purified* fiber supplements poses new questions about fiber–drug interactions. For example, levodopa elimination was slower with the administration of *Plantago ovata* husk. A fiber known as glucomannan has been found to interact with glibenclamide, reducing its absorption. Another fibrous substance, guar gum, has not affected the absorption of glipizide [59].

The subject of natural and artificial fiber also calls for serious research along the lines of Frederick Hopkins' position supporting a *holistic* diet, rather than *purifying* the foods. The fiber case reminds us to rethink the trend of delivering diet through capsules. The diet is not just *fulfillment* of nutrients. It has several personal, social, and cultural dimensions; and some role in mental satisfaction. Even the food for astronauts during long space travel is prepared with care and variety so that the travelers feel almost like they are eating at home [60].

Several other types of health supplements and products are available in the market and the consumers are often mislead, misinformed and exploited through attractive advertisements. Especially the online teleshops are aggressively promoting various kinds of herbal teas for reducing weight, fat burner supplements to counter obesity, special oils and scalp applications for growing hair. Special products from milk and soy, protein supplements for building muscles; products, different kinds of energy drinks, aphrodisiacs, mother's tonics, baby foods, infant formula and many such categories are freely available in market. Many such health supplements do not have scientific evidence supporting putative claims. Many claims are magical and are actually illegal. We are not covering such health supplements in this book.

HIGH PROFITS—LOW EVIDENCE

The industry of nutraceutical and health food supplementation is very profitable. Quite similar to the influence of the pharmaceutical industry, the nutraceutical industry is also using all types of marketing, and business tactics to influence world polity, professionals, and regulators. In addition, through attractive advertisements, they also lure patients who are looking for new options. As a result, the use of many types of nutraceuticals and dietary vitamin and mineral supplements has increased substantially over the past 20 years.

In general, the use of dietary supplements increases with age. Scientists from Brigham and Women's Hospital, and Harvard Medical School completed an interesting study to track trend of dietary supplement use for 20 years involving over 125,000 people. This systematic, longitudinal study on a large population has reported very interesting results. According to this study, the percentage of participants who used at least one health supplement, including multivitamins, was considerably increased from 1986 to 2006.

Longitudinal trends of the use of multivitamins, vitamin D, folic acid, calcium, magnesium, and fish oil supplements has continued to increase since 1990. The use of vitamin D supplements increased primarily in the 2000s. It is widely recognized that vitamin D plays an important role in calcium absorption for bone health, especially in postmenopausal women. Despite fortification of milk with vitamin D, one third of the US population had vitamin D inadequacy, or deficiency in 2001–2006.

Vitamin D deficiency is common among the elderly population mainly because of limited sun exposure, malabsorption, and lower dietary intake. Vitamin D deficiency is known to be a potential risk factor for various chronic diseases, such as hypertension, CVD, and cancer. An increased number of the population who are postmenopausal may have contributed to the increase in the use of vitamin D supplements for the prevention, and treatment of osteoporosis during this study's period.

The use of folic acid supplements increased in 1996, coinciding with the United States' Food and Drug Administration's requirement to fortify grain products with folic acid for prevention of neural tube defects. Its putative protection against chronic diseases, such as CVD, and cancer may have increased the voluntary use of folic acid supplements. Attempts to promote folate fortification in grain flours are being systematically undertaken in many countries, including India. Obviously, this may open a large market for the food, and supplement business. However, except in the prevention of neural tube defects, many other health benefits of folic acid fortification have not yet been proved. Moreover, few reports indicate that excess folate is likely to promote tumor initiation [61].

Among health supplements, iron was the only one that showed a consistent decrease in use. Iron is important for the growth and development of infants and children, and in the prevention of iron-deficiency anemia. Since the 1990s, iron deficiency has been relatively uncommon, and the prevalence of iron-deficiency anemia decreased in the general population in the United States. It is known that excessive iron intake from supplements may reduce the absorption of zinc, and a few pharmaceutical drugs. It can also cause gastrointestinal distress, increased risk for infection, colorectal cancer, and CVD. These may be a few of the reasons for the decrease in iron supplementation during the 20-year period.

No more is it a mystery as to why some supplementation suddenly spreads like wildfire. In previous years, it was antioxidants, vitamins, and minerals, in general. These days, medical doctors contribute to the propagation of specific supplements. For example, there was sudden spurt in vitamin B12 and vitamin D3 use in the past few years. This does not happen automatically, or because of new scientific findings, or the availability of superior evidence. Often, these increases are the result of systematic, promotional campaigns on the part of industries, smartly hyping the use of particular supplements; and such use spreads all over the world. We have discussed the evidence base for many common vitamins, and antioxidants in earlier sections. To further strengthen our argument, we present a few more examples.

As recently as 2014, the United States' Preventive Services Task Force (USPSTF), which is a body of high-level experts, considered the subject of health supplements, in totality. After carefully reviewing all available scientific evidence, the USPSTF has concluded that "the current evidence is insufficient to assess the balance of benefits and harms of multivitamins for the prevention of cardiovascular disease or cancer." The USPSTF recommends against beta-carotene, and vitamin E supplements for the prevention of CVD or cancer [62]. Not just the USPSTF, but most of the reputed scientific, and professional bodies have opined in favor of natural food, and a balanced diet. In no uncertain terms, they have cautioned against rampant, yet avoidable, health supplementation. Here are a few examples which should suffice in knowing the truth:

- An independent consensus panel sponsored by the National Institutes of Health concluded that "the present evidence is insufficient to recommend for, or against multivitamins to prevent chronic disease."
- The Academy of Nutrition and Dietetics, in a 2009 position statement, admitted that "although multivitamin supplements may be useful in meeting the recommended levels of some nutrients, there is no evidence that they are effective in preventing chronic disease."
- The American Cancer Society found that "current evidence does not support the use of dietary supplements for the prevention of cancer."
- In 2007, The American Institute for Cancer Research determined that "dietary supplements are not recommended for cancer prevention," and recommended a balanced diet, with a variety of foods rather than supplements.
- The American Heart Association recommends that "healthy persons receive adequate nutrients by eating a variety of foods, rather than supplementation."
- The American Academy of Family Physicians' clinical recommendations are also consistent with the USPSTF recommendations.

With so much of evidence against them, most of the health supplements are no different than placebos, mumbo jumbo, and superstitions to which gullible patients often fall prey. However, this approach of giving health supplements like drugs will not be sustainable. Moreover, it is very unethical. Still, globally, the health supplement industry is growing in leaps and bounds. Several questions emerge from this analysis:

> Why does a particular health supplement suddenly become important?
> Why does their popularity gradually wane?
> What benefits do these supplements actually offer to people?
> Are they at least safe? Are they really necessary?
> Why this is happening, despite many advisories from experts?

Actually, most of these questions have straightforward answers. However, often vested interests, and the nexus with doctors, regulators, and industry make these questions confusing, and complex. But just to cast the reality of the situation in a straight and simple manner—it is about business. It is much easier to do business in the arena of food supplements, than in the highly regulated pharmaceutical industry. With less stringent norms and regulations, health food supplements have become the choice of many mainstream pharma companies, which are facing major setbacks due to the innovation deficit in drug discovery.

In the process, gullible people waste hard-earned money paid to receive doubtful health benefits. Except in very genuine cases, such as specific deficiency diseases in endemic areas and in vulnerable populations, the generalized indiscriminate use of most health supplements is devoid of sufficient scientific evidence. When we make this statement, we do not undermine the importance of vitamins, minerals, antioxidants, and other health supplements. Vitamin C has a role in preventing diseases like scurvy. Vitamin A undoubtedly plays an important role in preventing night blindness. Calcium and vitamin D can treat rickets, and prevent osteoporosis. Iron is necessary for the treatment of anemia. Folate is crucial to prevent neural tube defects. There are definite, clinical conditions where health supplements can play a vital role. To some extent, they can help to prevent many diseases. But we are worried about the way in which today's health supplement industry is moving. We feel greatly concerned because of the rampant commercialization; poor quality of products; inadequate knowledge about dose, kinetics, and drug–food interactions; virtually no studies related to their absorption, distribution, metabolism, excretion, and toxicity (ADMET); and grossly insufficient scientific evidence to support health claims.

The better way is to improve the quality of nutrition through the consumption of good food. Eating good food is an enjoyable experience. It is not equivalent to taking extracted nutrients in a capsule. Instead of spending creative energies in developing proprietary protein milk shakes, popping multivitamin pills, or submitting to shots of vitamin D, it is advisable to educate, and empower people about how to make tasty protein recipes, the importance of physical activity, and exposure to sunlight.

NUTRITION—NOT IN A CAPSULE

The problems related to the medicalization of society were briefly discussed in Chapters 1 and 3. Currently, and as a practical effect, the aggressive use of pharmaceutical drugs has led people to consume drugs like foods. The amount of tablets, capsules, and liquids consumed by the average, elderly American is so large that there may not be much space left in which to consume food. It is astonishing to see the large number of different medications an individual must consume, especially during advanced age. Over 37% of elderly Americans are reported to use more than five prescription drugs. In the United States, prescription drug spending doubled in just one decade—reaching a whopping $234 billion in sales. The Centers for Disease Control (CDC) data sheet has reported a worrisome increase in the most commonly used drugs, such as asthma medicines for children, central nervous system stimulants for adolescents, antidepressants for middle-aged adults, and cholesterol-lowering drugs for older Americans [63]. It is important to note that many of these drugs have untoward effects.

As a natural reaction to this dominance of pharmaceutical drugs, a large number of consumers are trying many alternatives. The use of health supplements, and nutraceuticals seems to be a new wave. Unfortunately, this is also driven by industry interests, and is not necessarily based on the needs of the consumer. The global nutraceutical product market is showing a steep rise, and is estimated to have reached over $200 billion in sales. Many large pharmaceutical companies like Pfizer and Abbot are now manufacturing nutraceutical products, and health supplements.

Today, the industry is trying to convert food and nutrition into a package of tablets, capsules, granules, and drinks. As compared to the highly regulated pharmaceutical drug market, nutraceutical development requires significantly lower investment, has looser regulatory controls, and higher profitability. However, as we have seen from several examples in this chapter, many health supplements, and vitamins do not have sufficient scientific evidence to support their putative health claims. Consuming natural, regional, seasonal foods, and fruits and vegetables is more economical, and healthier.

There is a fundamental difference between food, diet, nutrition, and medicines, drugs, and pharmaceuticals. The former are a natural part of our life and health, while the latter are created by us to treat diseases. It is now well-known that many diseases are preventable, and even treatable, purely through diet, and lifestyle modifications. Diet is our everyday food, and it cannot be reduced to a dosage form—like a tablet, capsule, or syrup.

We need to revisit traditional knowledge, and practices to regain control of our health. Our health is really in our hands. We can adopt healthy dietary practices, and lifestyles. Time-tested, traditional knowledge, and practices can serve as important sources. Already, this awareness is palpable among national leadership institutions. The Division of Diabetes Translation of the Centers for Disease Control and Prevention, in the United States, has launched the Traditional Foods Project, with the objective of studying traditional foods for the prevention of type 2 diabetes. This initiative focuses on community efforts to reclaim traditional foods, and increase physical activity. The project encourages policy changes for availability, and access to local, traditional foods, and forms of exercise [64].

These are encouraging trends, where people can be informed and empowered to take care of their own health. Health supplements, and nutraceutical products cannot be of much use for attaining, and maintaining health. The health supplement industry should not go on the same path of the drug industry— replacing the *medicalization* with the *supplementalization* of society.

The health supplement industry should remember Hippocrates' advice: "Let the food be your medicine" and *not* "Let the medicine be your food."

REFERENCES

[1] ftp://ftp.cordis.europa.eu/pub/fp7/kbbe/docs/functional-foods_en.pdf.

[2] http://www.mhlw.go.jp/english/topics/foodsafety/fhc/02.html.

[3] Ncube TN, Greiner T, Malaba LC, Gebre-Medhin M. Supplementing lactating women with puréed papaya and grated carrots improved vitamin A status in a placebo-controlled trial. J Nutr 2001;131(5):1497–502.

[4] Guallar E, Stranges S, Mulrow C, Appel LJ, Miller ER. Enough is enough: stop wasting money on vitamin and mineral supplements. Ann Intern Med 2013;159(12):850–1.

[5] Halliwell B. The antioxidant paradox: less paradoxical now? Br J Clin Pharmacol 2013;75(3): 637–44.

[6] Bjelakovic G, Nikolova D, Gluud LL, Simonetti RG, Gluud C. Antioxidant supplements for prevention of mortality in healthy participants and patients with various diseases. Cochrane Database Syst Rev 2008;(2):CD007176.

[7] Bjelakovic G, Nikolova D, Gluud L, Simonetti R, Gluud C. Antioxidant supplements for prevention of mortality in healthy participants and patients with various diseases. Cochrane Database Syst Rev 2012;3:CD007176.

[8] Mathew MC, Ervin A-M, Tao J, Davis RM. Antioxidant vitamin supplementation for preventing and slowing the progression of age-related cataract. Cochrane Database Syst Rev 2012;6:CD004567.

[9] Evans JR. Antioxidant vitamin and mineral supplements for slowing the progression of age-related macular degeneration. Cochrane Database Syst Rev 2006;(2):CD000254.

[10] Machlin L, Gordon R, Meisky K. The effect of antioxidants on vitamin E deficiency symptoms and production of liver "peroxide" in the chicken. J Nutr 1959;67(2):333–43.

[11] Traber MG, Atkinson J. Vitamin E, antioxidant and nothing more. Free Radic Biol Med 2007;43(1):4–15.

[12] Stephens NG, Parsons A, Schofield PM, Kelly F, Cheeseman K, Mitchinson MJ, et al. Randomised controlled trial of vitamin E in patients with coronary disease: Cambridge Heart Antioxidant Study (CHAOS). Lancet 1996;347(9004):781–6.

[13] Farina N, Isaac MG, Clark AR, Rusted J, Tabet N. Vitamin E for Alzheimer's dementia and mild cognitive impairment. Cochrane Database Syst Rev 2012;11:CD002854.

[14] Dysken MW, Sano M, Asthana S, Vertrees JE, Pallaki M, Llorente M, et al. Effect of vitamin E and memantine on functional decline in Alzheimer disease: the TEAM-AD VA cooperative randomized trial. JAMA 2014;311(1):33–44.

[15] Fitzgerald S. Vitamin E has modest benefit for early AD. Neurol Today 2014;14(4):1–8.

[16] Brion LP, Bell EF, Raghuveer TS. Vitamin E supplementation for prevention of morbidity and mortality in preterm infants. Cochrane Database Syst Rev 2003;(4):CD003665.

[17] Bjelakovic G, Gluud LL, Nikolova D, Whitfield K, Wetterslev J, Simonetti RG, et al. Vitamin D supplementation for prevention of mortality in adults. Cochrane Database Syst Rev 2014;1:CD007470.

[18] Theodoratou E, Tzoulaki I, Zgaga L, Ioannidis JPA. Vitamin D and multiple health outcomes: umbrella review of systematic reviews and meta-analyses of observational studies and randomised trials. BMJ 2014;348:g2035. http://dx.doi.org/10.1136/bmj.g2035.

[19] Haider BA, Bhutta ZA. Neonatal vitamin A supplementation for the prevention of mortality and morbidity in term neonates in developing countries. Cochrane Database Syst Rev 2011;(10):CD006980.

[20] Chen H, Zhuo Q, Yuan W, Wang J, Wu T. Vitamin A for preventing acute lower respiratory tract infections in children up to seven years of age. Cochrane Database Syst Rev 2008;(1):CD006090.

[21] Ang CD, Alviar MJ, Dans AL, Bautista-Velez GG, Villaruz-Sulit MV, Tan JJ, et al. Vitamin B for treating peripheral neuropathy. Cochrane Database Syst Rev 2008;(3):CD004573.

[22] Marti-Carvajal AJ, Sola I, Lathyris D, Karakitsiou DE, Simancas-Racines D. Homocysteine-lowering interventions for preventing cardiovascular events. Cochrane Database Syst Rev 2013;(1):CD006612.

[23] Thaver D, Saeed MA, Bhutta ZA. Pyridoxine (vitamin B6) supplementation in pregnancy. Cochrane Database Syst Rev 2006;(2):CD000179.

[24] Hemilä H, Chalker E. Vitamin C for preventing and treating the common cold. Cochrane Database Syst Rev 2013;1:CD000980.

[25] Milan SJ, Hart A, Wilkinson M. Vitamin C for asthma and exercise-induced bronchoconstriction. Cochrane Database Syst Rev 2013;10:CD010391.

[26] Rumbold A, Crowther Caroline A. Vitamin C supplementation in pregnancy. Cochrane Database Syst Rev 2005;(1):CD004072.

[27] Park S. The effects of high concentrations of vitamin C on cancer cells. Nutrients 2013;5(9):3496–505.

[28] Cortés-Jofré M, Rueda JR, Corsini-Muñoz G, Fonseca-Cortés C, Caraballoso M, Bonfill Cosp X. Drugs for preventing lung cancer in healthy people. Cochrane Database Syst Rev 2012;10:CD002141.

[29] Fraga CG, Galleano M, Verstraeten SV, Oteiza PI. Basic biochemical mechanisms behind the health benefits of polyphenols. Mol Aspects Med 2010;31(6):435–45.

[30] González-Vallinas M, González-Castejón M, Rodríguez-Casado A, Ramírez de Molina A. Dietary phytochemicals in cancer prevention and therapy: a complementary approach with promising perspectives. Nutr Rev 2013;71(9):585–99.

[31] Mattson MP, Cheng A. Neurohormetic phytochemicals: low-dose toxins that induce adaptive neuronal stress responses. Trends Neurosci 2006;29(11):632–9.

[32] Enns J, Yeganeh A, Zarychanski R, Abou-Setta AM, Friesen C, Zahradka P, et al. The impact of omega-3 polyunsaturated fatty acid supplementation on the incidence of cardiovascular events and complications in peripheral arterial disease. BMC Cardiovasc Disord 2014;14(70). http://dx.doi.org/10.1186/1471-2261-14-70.

[33] Pan A, Chen M, Chowdhury R, Wu JHY, Sun Q, Campos H, et al. α-Linolenic acid and risk of cardiovascular disease: a systematic review and meta-analysis. Am J Clin Nutr 2012;96(6):1262–73.

[34] Hartweg J, Perera R, Montori V, Dinneen S, Neil HAW, Farmer A. Omega-3 polyunsaturated fatty acids (PUFA) for type 2 diabetes mellitus. Cochrane Database Syst Rev 2008;(1):CD003205.

[35] Turner D, Zlotkin SH, Shah PS, Griffiths AM. Omega 3 fatty acids (fish oil) for maintenance of remission in Crohn's disease. Cochrane Database Syst Rev 2009;(1):CD006320.

[36] James S, Montgomery P, Williams K. Omega-3 fatty acids supplementation for autism spectrum disorders (ASD). Cochrane Database Syst Rev 2011;(11):CD007992.

[37] Gillies D, Sinn JK, Lad SS, Leach MJ, Ross MJ. Polyunsaturated fatty acids (PUFA) for attention deficit hyperactivity disorder (ADHD) in children and adolescents. Cochrane Database Syst Rev 2012;7:CD007986.

[38] Delgado-Noguera MF, Calvache JA, Bonfill Cosp X. Supplementation with long chain polyunsaturated fatty acids (LCPUFA) to breastfeeding mothers for improving child growth and development. Cochrane Database Syst Rev 2010;(12):CD007901.

[39] Shafiq N, Singh M, Kaur S, Khosla P, Malhotra S. Dietary treatment for familial hypercholesterolaemia. Cochrane Database Syst Rev 2010;(1):CD001918.

[40] Dennert G, Horneber M. Selenium for alleviating the side effects of chemotherapy, radiotherapy and surgery in cancer patients. Cochrane Database Syst Rev 2006;3:CD005037.

[41] Rees K, Hartley L, Day C, Flowers N, Clarke A, Stranges S. Selenium supplementation for the primary prevention of cardiovascular disease. Cochrane Database Syst Rev 2013;1:CD009671.

[42] Jürgens G, Graudal NA. Effects of low sodium diet versus high sodium diet on blood pressure, renin, aldosterone, catecholamines, cholesterols, and triglyceride. Cochrane Database Syst Rev 2004;(1):CD004022.

[43] Aaron KJ, Sanders PW. Role of dietary salt and potassium intake in cardiovascular health and disease: a review of the evidence. Mayo Clin Proc 2013;88(9):987–95.

[44] Bauer DC. The calcium supplement controversy: now what? J Bone Miner Res 2014;29(3):531–3.

[45] Bjelakovic G, Nikolova D, Simonetti RG, Gluud C. Antioxidant supplements for preventing gastrointestinal cancers. Cochrane Database Syst Rev 2008;(3):CD004183.

[46] Boehm K, Borrelli F, Ernst E, Habacher G, Hung SK, Milazzo S, et al. Green tea (*Camellia sinensis*) for the prevention of cancer. Cochrane Database Syst Rev 2009;(3):CD005004.

[47] Jun M, Venkataraman V, Razavian M, Cooper B, Zoungas S, Ninomiya T, et al. Antioxidants for chronic kidney disease. Cochrane Database Syst Rev 2012;10:CD008176.

[48] Heyland D, Muscedere J, Wischmeyer PE, Cook D, Jones G, Albert M, et al. A randomized trial of glutamine and antioxidants in critically ill patients. N Engl J Med 2013;368(16):1489–97.

[49] Schrezenmeir J, de Vrese M. Probiotics, prebiotics, and synbiotics–approaching a definition. Am J Clin Nutr 2001;73(2 Suppl):361S–4S.

[50] Hill C, Gurner F, Reid G, Gibson GR, Merenstein DJ, Pot B, et al. Expert consensus document: the International Scientific Association for Probiotics and Prebiotics consensus statement on the scope and appropriate use of the term probiotic. Nat Rev Gastroenterol Hepatol 2014;11(8):506–14.

[51] Hempel S, Newberry SJ, Maher AR, Wang Z, Miles JNV, Shanman R, et al. Probiotics for the prevention and treatment of antibiotic-associated diarrhea: a systematic review and meta-analysis. JAMA 2012;307(18):1959–69.

[52] Moayyedi P, Ford AC, Talley NJ, Cremonini F, Foxx-Orenstein AE, Brandt LJ, et al. The efficacy of probiotics in the treatment of irritable bowel syndrome: a systematic review. Gut 2010;59(3):325–32.

[53] Ford AC, Quigley EMM, Lacy BE, Lembo AJ, Saito Ya, Schiller LR, et al. Efficacy of prebiotics, probiotics, and synbiotics in irritable bowel syndrome and chronic idiopathic constipation: systematic review and meta-analysis. Am J Gastroenterol 2014;109(10):1547–61.

[54] Dimidi E, Christodoulides S, Fragkos KC, Scott SM, Whelan K. The effect of probiotics on functional constipation in adults: a systematic review and meta-analysis of randomized controlled trials. Proc Nutr Soc 2014;100(4):1075–84.

[55] Didari T, Solki S, Mozaffari S, Nikfar S, Abdollahi M. A systematic review of the safety of probiotics. Expert Opin Drug Saf 2014;13(2):227–39.

[56] Zeng H, Lazarova DL, Bordonaro M. Mechanisms linking dietary fiber, gut microbiota and colon cancer prevention. World J Gastrointest Oncol 2014;6(2):41–51.

[57] Threapleton DE, Greenwood DC, Evans CEL, Cleghorn CL, Nykjaer C, Woodhead C, et al. Dietary fibre intake and risk of cardiovascular disease: systematic review and meta-analysis. BMJ 2013;347:f6879. http://dx.doi.org/10.1136/bmj.f6879.

[58] Li S, Flint A, Pai JK, Forman JP, Hu FB, Willett WC, et al. Dietary fiber intake and mortality among survivors of myocardial infarction: prospective cohort study. BMJ 2014;348:g2659. http://dx.doi.org/10.1136/bmj.g2659.

[59] González Canga A, Fernández Martínez N, Sahagún Prieto AM, García Vieitez JJ, Díez Liébana MJ, Díez Láiz R, et al. Dietary fiber and its interaction with drugs. Nutr Hosp 2010;25(4):535–9.

[60] Food for space flight. Available from: http://spaceflight.nasa.gov/shuttle/reference/factsheets/food.html.

[61] Deghan Manshadi S, Ishiguro L, Sohn K-J, Medline A, Renlund R, Croxford R, et al. Folic acid supplementation promotes mammary tumor progression in a rat model. PLoS One 2014;9:e84635. http://dx.doi.org/10.1371/journal.pone.0084635.

[62] Moyer VA., U.S. Preventive Services Task Force. Vitamin, mineral, and multivitamin supplements for the primary prevention of cardiovascular disease and cancer: U.S. preventive services task force recommendation statement. Ann Intern Med 2014;160(8):558–64.

[63] Gu Q, Dillon CF, Burt VL. Prescription drug use continues to increase: U.S. prescription drug data for 2007-2008. NCHS Data Brief 2010;(42):1–8.

[64] http://www.cdc.gov/diabetes/projects/ndwp/traditional-foods.htm.

CHAPTER 9

Drug Discovery and Ayurveda

At the end of times, the merchants of the word will deceive the nations of the world through their pharmacia.

Revelations 18:23

HISTORY OF DRUGS

Today, drugs and pharmaceuticals have become an essential commodity. They have played significant roles in the prevention and treatment of several diseases; the result is better life expectancy. However, we seem to believe that drugs can give us health and wellness. Today, many drugs are failing due to untoward or toxic effects, and the quest for better and safer drugs continues even more aggressively. We are getting more and more dependent upon and have become practically intoxicated with drugs. Still, this multitrillion dollar pharmaceutical industry is facing multiple challenges, especially from the last decade. The industry is under great pressure to reach desired levels of innovation, profitability, and productivity. This sector seems to be engaged in a desperate attempt for survival—resulting in closures, mergers, and acquisitions on one side, and being confronted with economical, and ethical challenges on the other.

Wiki at wikiversity.org defines drugs as "chemical substances used or intended to be used to modify or explore the physiological condition or pathological state for the benefit of the recipient." The *Oxford English Dictionary* defines a drug as "a medicine or other substance which has a physiological effect when ingested or otherwise introduced into the body." It also extends this to "a substance taken for its narcotic or stimulant effects, often illegally." On other hand, the term *medicine* is defined as "a drug or other preparation for the treatment or prevention of disease." Thus, typically the term *drug* is used to mean a substance which may have medicinal or poisonous properties. Historically, drugs have been developed from poisonous sources. Thus, some of the early discovered drugs like morphine, cocaine, tubocurarine, codeine, quinine, colchicine, belladonna, and many others actually originated from potentially poisonous plants. This is the case particularly because it is easier

229

to identify poisonous characteristics of substances, than it is to recognize medicinal properties [1]. There is still confusion between what a drug is, and what medicine is. Current efforts are in the direction of *drug discovery* and not *medicine discovery*. A few examples in support of this contention are discussed in later sections.

Because of an overemphasis on synthetic drugs and market forces, traditional medicine and home remedies are being ignored. Simple home remedies, which could take care of many common ailments, are getting disparaged, disintegrated, and are on the verge of extinction in many parts of the world.

The history of using drugs in some form or other for treating diseases and alleviating symptoms dates back to ancient times, and includes the use of many ludicrous therapies. In Chapter 2 (Evolution of Medicine), we discussed how modern medicine has gradually developed over the years by scientific and observational efforts of scientists. The history of drug development reveals that most of the early discoveries resulted from traditional knowledge, community experiences, and serendipity. The discovery of early drugs seems to arise from mostly poison sources.

The pioneering effort of the systematic classification of various botanical, animal, and mineral sources is available in *Charaka Samhita* and *Sushruta Samhita*. Details about their actions on body systems, along with their role in promoting health, and treating diseases are available in these two classics. Descriptions are available about which parts (root, stem, leaves, flowers seeds) are to be used for which symptoms; whether singly or in combination with other materials; in which form (liquid, tablets, linctus); and along with methods of processing. Detailed prescription about indications, dosage, precautions, and dietary advice is also available. Ayurveda also describes various drugs for disease treatment, health protection, regenerative therapies, and palliative care. Many drugs useful to maintain health, enhance the capacities of sense organs, and to attain longevity are elaborately mentioned in Ayurveda. In early times, exploration was undertaken to find various sources for rejuvenation and longevity-promoting drugs (elixirs). In Ayurveda, many drugs of herbomineral origin, made through sophisticated processes of detoxifying metals like mercury, were developed; this branch of Ayurveda is known as *Rasashastra*.

The Greek physician, Galen (AD 129–200) devised the first modern pharmacopoeia, describing the appearance, properties, and uses of many plants of his time. The foundations of the modern pharmaceutical industry were laid when techniques to produce synthetic replacements for many of the natural medicines were developed. Natural products chemistry, as a branch, seems to have begun in1804, when a German pharmacist, Friedrich Sertürner, isolated morphine from opium. He was the first chemist to isolate an alkaloid from a botanical source. Morphine was obtained from *Papaver somniferum*, the poppy plant, properties of which were known for over 5000 years.

The seeds of synthetic drugs were sown in 1874, when salicylic acid was isolated from willow bark. This is the precursor of the synthetic drug popularly known as aspirin. Pharmaceutical research took a major leap forward when organic synthetic chemistry was developed. During that time, pharmacologists, microbiologists, and biochemists began to unravel the chemistry of natural processes in humans, animals, plants, and microorganisms. This led to the identification of many key chemical molecules and provided more opportunities to develop novel compounds. Many new drugs emerged to treat infections, infestations, cancers, ulcers, and heart and blood pressure conditions. Many drugs were developed through random screening of thousands of synthetic chemicals. Many also came in as serendipity, discovered as a result of the sharp-eyed observations of physicians and scientists. Examples of such drugs include sulfonamides, isoniazid, antipsychotic drugs, antihistamines, and penicillin.

The emergence of the modern pharmaceutical industry is the outcome of different activities that developed potent, single molecules with highly selective activity for different pathological conditions. In many cases, synthetic drugs improved on nature. For example, a new range of local anesthetics derived from cocaine avoided its dangerous effects on blood pressure. Chloroquine is much less toxic than naturally occurring quinine.

Drug discovery moved away from serendipity to rational drug design when American Chemists Hitchings and Elion worked on DNA-based antimetabolites, leading to analogs of purines as anticancer agents. Watson and Crick's discovery of the DNA structure, and a better understanding of the processes of replication, transcription, and translation helped in the discovery of antiviral drugs. Recombinant DNA technology, molecular cloning, and other developments in molecular biology helped advance the progression of biologicals as new therapeutic agents. The chemical dominance of biologicals and vaccines began to evolve. The discovery of polymerase chain reaction, combinatorial chemistry (CC), high-throughput screening (HTS), molecular modeling, and bioinformatics have substantially contributed to the new era of genomic-based drug discovery.

The spectacular successes in medicinal chemistry and the development of synthetic drugs resulted in natural product chemistry becoming less popular; it was virtually abandoned by major drug companies. While drug discovery was becoming more specific, and commercially lucrative, efforts to develop new drugs for the treatment of communicable diseases of the poor, like malaria, trypanosomiasis, filariasis, tuberculosis, schistosomiasis, leishmaniasis, and amebiasis came to a near standstill. Although botanical medicines were produced in many countries, the clinical efficacy of these medicines was not usually evaluated. The composition of these complex herbal mixtures was only crudely analyzed. Thus, herbal medicines came to be associated with anecdotes, or "old wives tales," and quack medicine; and exploitation of sick, the desperate, and the gullible. Sadly, herbal medicines continue to be poor in quality control, both for materials and clinical efficacy.

Today, pharmaceutical scientists are experiencing difficulty in identifying new lead structures, templates, and scaffolds in the finite world of chemical diversity. A number of synthetic drugs have adverse and unacceptable side effects. There have been impressive successes with botanical medicines, most notably, quinghaosu and artemisinin from Chinese medicine. Considerable research on pharmacognosy, chemistry, pharmacology, and clinical therapeutics has been carried out on Ayurvedic medicinal plants. Numerous molecules have come out of the Ayurvedic experiential base, examples include *Rauwolfia* alkaloids for treating hypertension, psoralens for vitiligo, Holarrhena alkaloids for amebiasis, guggulsterons as hypolipidemic agents, *M. pruriens* for Parkinson's disease, piperidines as bioavailability enhancers, baccosides in mental retention, picrosides for hepatic protection, Phyllanthins as antivirals, curcumines for inflammation, and withanolides and many other steroidal lactones and glycosides as immunomodulators [2].

The mass screening of plants for discovering new drug leads is very expensive and inefficient. There is evidence that traditional knowledge can offer leads for the treatment of HIV/AIDS and cancer [3]. It is estimated that over 100 new, natural product-based leads are in clinical development [4]. About 60% of anticancer and 75% of anti-infective drugs approved from 1981 to 2002 have natural origins [5]. Many experts feel that it would be cheaper and more efficient to study plants described in ancient texts [6]. Many active compounds from traditional medicine sources could serve as good starting compounds and scaffolds for rational drug design. Most of these compounds are part of routinely used, traditional medicines, and hence their tolerance and safety are relatively better known than any other chemical entities that are new for human use [7]. CC approaches based on natural product scaffolds are being used to create screening libraries of closely resembling drug-like compounds [8,9]. Due to these advantages, the natural and traditional medicines are reemerging as promising new leads to boost new drug discovery.

DRUG DISCOVERY PROCESS

Modern drug discovery has significantly contributed to improving our quality of life, and has been responsible for curing, irradiating, and controlling several diseases and disorders. While many drugs may have untoward effects, the benefits from drugs are far-reaching, and it may be unreasonable to severely criticize modern drugs as killers; in most cases, fatalities occur due to the misuse of painkillers, tranquilizers, opioids, and other such drugs [10].

Drugs and pharmaceuticals are essentially chemical or biological materials. A process of identifying potential substances and converting them into drugs and pharmaceuticals is long, investment-intensive, and complex. This process is generally known as drug discovery. The discovery pipeline involves many steps; basic research, discovery of target and leads, validation, drug candidates, synthesis, drug development, drug manufacture, and drug distribution.

The first step is exploratory, which involves several years of scientific research to understand the biochemical and pathophysiological processes of specific, medical conditions for which a drug is required. The exploratory steps involve basic research, which can give indications of specific receptor targets, which can be modulated. Once the targets are identified, the next step is to find promising lead compounds, which can bind and interact with the target receptors. This binding and interaction is quite similar to any enzyme action, where there is a high degree of specificity—like a lock and key arrangement. The challenging task is to find the most appropriate key as the lead compound or candidate drug that can modulate the target in the desired direction.

This discovery of candidate drug molecules leads to the next step of screening potential compounds, using suitable assay systems. Currently, various powerful and rapid technologies are available. In the past, the various structural analogs of lead molecules were synthesized manually, and so there were many limiting factors. Now, CC synthesis and structure libraries can turn up hundreds of chemically similar compounds within a short time. The high-throughput assays can rapidly test activities of these hundreds of compounds, in an automated fashion. This helps in identifying the best potential candidates, which may have a high degree of specificity and affinity to the target. Often these studies can also be done *in silico*, using suitable software and bioinformatics tools.

Selected lead compounds follow a rigorous path through in silico, in vitro, and in vivo assays in the process of optimization. Out of hundreds of selected leads, only a few may emerge as potential drug candidates. This is followed by the preclinical phase, which involves cell-based and animal-based assays to screen the safety and activity of drug candidates. At this stage a lot of pharmaceutical development work also happens involving the formulation of dosage form, delivery system, stability, pharmacokinetics, bioavailability, pharmacodynamics, and dosing studies. At this stage, the industry may decide whether the candidate drug merits further study. If the decision is positive, then the industry has to file an investigational drug application with the regulatory authority, such as the Food and Drug Administration (FDA). Once the safety and efficacy is established, and the regulatory compliances are made, the candidate

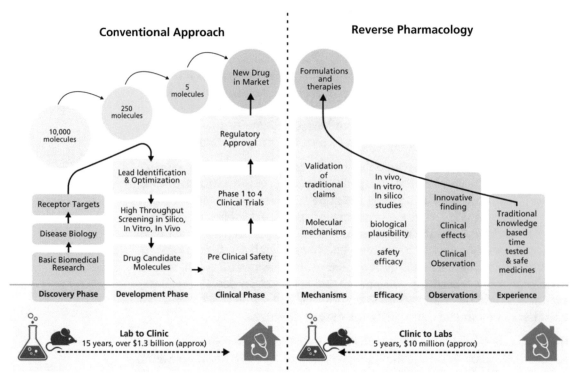

FIGURE 9.1
Conventional drug discovery process and reverse pharmacology approach.

drugs may become eligible for clinical studies involving phase 1 to phase 4 clinical trials. It is generally estimated that out of 10,000 compounds screened, about 250 may enter into the preclinical phase, about 2–5 may enter the clinical phase, and if lucky, one drug molecule may get approval from the regulator. This is a very long, drawn-out, investment-intensive process involving an investment of 10–15 years, and around $1.5 billion (Figure 9.1).

In addition to the complexity of the discovery process, the discovery space size has become too large for efficient data mining. It is becoming exceedingly difficult to explore and validate new targets, leads, and candidate drugs (Figure 9.2). According to theoretical estimations about 10^{60} small molecules exist in the world of which about 21 million have been studied for commercial uses. Over 10,000 ligand-binding domains are known. The therapeutic target database of National University of Singapore contains about 2015 targets of which only 286 are in clinical trials. Out of about 18,000 drug candidate molecules only about 1540 are approved.

It was hoped that with the use of high-throughput technologies, supported by genomics, computational sciences, and bioinformatics the drug discovery process would be faster, more economical, and safer. Unfortunately,

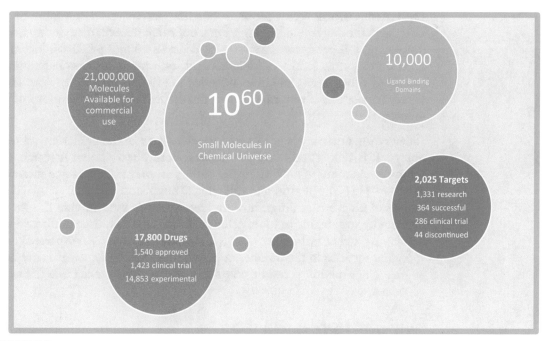

FIGURE 9.2
Drug discovery space.

this hope has not become a reality. In fact, the number of new drugs entering the market through these technologies has been reduced. Moreover, the number of drugs being withdrawn after launching in the markets is increasing. There is certainly something fundamentally wrong in the approaches for drug discovery.

DRUG DISCOVERY IMPASSE

Drug discovery and development has undergone several transitions during the last few years. The advances in biomedical sciences, biotechnology, vaccinology, computer-aided drug design, genomics, and molecular technologies have brought mind-boggling changes, leading to more powerful, and a broader spectrum of chemical and biological drugs for many diseases. Still today, the rate of new drugs discovered has dropped, and many existing drugs continue to be withdrawn by regulators for safety reasons. The chemical drugs are rapidly being replaced with biologicals. With the advent of pharmacogenomics and systems biology concepts, pharmaceutical medicine is becoming more personalized. Industry leaders and pharmaceutical and biomedical scientists are desperately looking for novel approaches, new ideas, and innovations to disrupt this discovery bottleneck. Undoubtedly, the pharmaceutical sector is facing a severe innovation deficit.

The pharmaceutical sector seems to be going in circles—doing more of the same, with the determined, but irrational hope that this will disrupt the current impasse [11]. Several strategies have been discussed and practiced, including open innovations, industry–academia and industry–industry collaborations, and many other recent initiatives intended to expedite lead generation. These efforts are nothing if not indicative of the desperation and aspirations of the industry [12].

Today's drug discovery is not limited to serendipity or the availability of technology. It is rightly indicated by scientists from Pfizer Global Research, Drs Schmid and Smith, that the strategies of the past may not guarantee success in the future [13]. The pharmaceutical industry has historically seen an incredible growth of blockbuster drugs; however, recent trends indicate that this model may not be sustainable in future. Due to the average cost, time of discovering, developing and launching new drugs is steadily increasing. An expected corresponding increase in the number of newer, safer, and better drugs is not happening. The situation is fast deteriorating, and analysts predict that the worst is yet to come.

WHY ARE DRUGS FAILING?

It may sound strange but this is the reality. Very few exceptional drugs, such as aspirin, have been able to withstand the real test of time. Large numbers of drugs fail and are withdrawn from the market for several reasons [14]. Drugs fail most of the time because of toxicity, untoward events, and other serious safety-related issues. Drugs like antibiotics also fail, either because of microbes developing resistance, or simply because the target on which the drug is acting is modulated, making that drug redundant. Drugs acting on single targets as agonists or antagonists are likely to produce direct and/or indirect cascading effects in the whole physiological system; because the body actually constitutes a metanetwork of targets, interacting with each other to regulate complex, biological, metabolic processes [15]. Systems biology principles indicate that a seemingly simple drug, acting on a specific target can actually trigger a complex reaction somewhere else, quite similar to the butterfly effect. Also, drugs die if they are ineffective. Interestingly, drugs are also killed by pharmaceutical companies for business reasons. When new generation drugs are developed, older ones are withdrawn. As a principle, drug discovery scientists would like to kill a drug as early as possible in the pipeline, because carrying a bad drug to the end of the pipeline increases the risks and costs.

Many drugs are terminated by regulatory authorities such as the FDA—even after they are approved. This is worrying, especially because the regulatory process to approve any drug is indeed a very stringent, and drawn-out process, in which the safety and efficacy of any new drug is assured, by a series of chemical, biological, and clinical investigations. Still, starting with thalidomide in

1950s, until today, hundreds of drugs have been killed by regulators and were withdrawn from markets.

A representative list of such drugs withdrawn from the market is available on Wikipedia [16]. This massacre of drugs has resulted in significant financial loss, sending the whole sector into a traumatic, panic situation. Clearly, market-driven approaches to drug discovery have not been successful or sustainable. Moreover, there seems to be a substantial decline in pharmaceutical ethics.

To address the current challenges, discovery impasse and innovation deficit faced by the industry, a new branch known as polypharmacology has emerged. Polypharmacology attempts to study drugs' interactions with multiple targets. It uses computational power and supercomputer-based virtual HTS for docking studies to improve efficiency of discovery process. Polypharmacology also attempts repurposing existing drug molecules for different therapeutic conditions. However, these efforts require some guidance for selecting the right type of targets and new scaffolds of drug molecules. We feel that traditional knowledge can play a vital role in this process of repurposing existing drugs [17].

INSPIRATION FROM TRADITIONS

Any drug, whether chemical, botanical, or biological, will have inherent limitations if it is focused only on a single target. The high specificity to a specific target can actually turn out to be a limitation. It is important to address multiple targets emanating from a syndrome-related metabolic cascade, so that holistic management can be effectively achieved. Therefore, it is necessary to shift the strategy from one that focuses on a single target, new chemical entity—the *drug*, to one of a multiple-target, synergistic *formulation* discovery approach.

Traditional knowledge-inspired discovery attempts have shown promising potential in several chronic diseases such as cancer, diabetes, and arthritis through the modulation of multiple targets. It is possible that polybotanical complex formulations from traditional medicines, such as Ayurveda and Traditional Chinese Medicine, have similar rationales traditional knowledge systems, such as Indian Ayurveda and Traditional Chinese Medicine can be extremely valuable in identifying real, *medicinal* resources.

In the Indian subcontinent, documentation and use of medicinal plants started during the Vedic period, of which documentation of more than 100 medicinal plants is found. *Charaka Samhita* and *Sushruta Samhita* contain detailed descriptions of over 800 medicinal herbs and over 8000 formulations. Ayurveda also deals with healthy lifestyle, health protection and sustenance, disease prevention, and diagnosis and treatment. The legacy of the epoch-spanning practice of Ayurveda is the development and use of time-tested home remedies for common ailments, which continue as local health traditions.

The Ayurvedic drug delivery system is unique, because it upholds the concept of a *target organ* or *target therapeutic function*, rather than active principles, and a single or a few biochemical reactions by a given herb. In Ayurveda, the belief is that drugs do not act against diseases only because of their physical nature or properties. They function in a specific time, and after reaching their target tissue. It is only then that the drug makes a specific, final therapeutic action. Ayurveda treatment aims at the alleviation of symptoms, maintenance of *Dosha* balance—at the same time ensuring that no harm is caused to the patient. Ayurvedic formulations consist of rational combinations of natural materials meant for multitargeting, rather than single-targeting. Such traditional wisdom may play an important role in guiding evidence-based scientific research on herbal medicines.

We hypothesize that a strategic mind-set shift from *drug discovery*, to *formulation discovery*, to *therapeutics discovery* will open up entirely new avenues to integrative health and medicine. A few examples of Ayurveda-inspired approaches to natural product drug discovery are discussed in the following sections.

The traditional medicine-inspired drug discovery and development has two main approaches. In the first approach, classical, traditional medicines can be standardized using modern methods. Here, their original nature is not changed. This approach is relatively straightforward, because these medicines have been used for many years in treating patients. Therefore, their safety and efficacy have been experienced. Here modern science's role is to ensure quality and reproducibility.

The second approach is concerned with using traditional knowledge for natural product chemistry. This may lead to the isolation of new chemical entities, which can be developed as new drugs for use in modern medicine. Today, this approach is very relevant to the pharmaceutical industry.

The third approach can be termed *formulation discovery*. Normally, drug discovery follows the one gene/one target/one drug approach, however, a multitarget, multi-ingredient formulation approach may be the smarter approach. In our view, drug discovery need not always be confined to the discovery of a single molecule. We are dealing with polygenic syndromes and not just isolated diseases—multitarget approaches are necessary [18]. Due to the diversity of structures, herbal extracts can deal with multiple targets simultaneously, and may have synergistic effects. Therefore, the development of standardized, synergetic, safe, and effective herbal formulations, with sufficient and robust scientific evidence can offer a faster and much more economical alternative.

Many of the formulations of Ayurveda and Traditional and Complementary Medicine (T&CM) have remained unchanged for thousands of years. These formulations are in use, even now. As a general rule, no drug can remain in the market for such a long time if it is not effective, or if it has any overly harmful side effects. The very fact that people have been using these Ayurvedic and

T&CM remedies for such a long time bolsters the belief in their safety and efficacy. It is possible to systematically document the use of these formulations by following pharmacoepidemiological methods. This can be valuable evidence to support their putative safety and efficacy [19].

REVERSE PHARMACOLOGY

The reverse pharmacology described here relates to reversing the *laboratory-to-clinic* process of discovery, to one of *clinics-to-laboratories* [20]. This is known as reverse pharmacology, which is defined as the science of integrating documented clinical experiences and experiential observations into leads, through transdisciplinary exploratory studies, and further developing these into drug candidates through robust preclinical and clinical research [21]. In this process, *safety* remains the most important starting point, and efficacy becomes a matter of validation (Figure 9.1).

Reserpine, an alkaloid isolated from *Rauwolfia serpentina*, also known as Indian snakeroot, was a major discovery made by using the reverse pharmacology approach. In 1931, Indian chemists Sen and Bose convincingly demonstrated the antihypertensive and tranquillizing effects of the plant. They also observed unique side effects, such as depression, extra pyramidal syndrome, gynecomastia, and other side effects [22]. Later, in 1949, Rustom Jal Vakil, who pioneered the development of cardiology in India, published a trial of *Rauwolfia* on patients with essential hypertension, which reported a reduction in systolic and diastolic blood pressure [23]. It took decades to delineate the mechanisms of these side effects. This effort led to a watershed for new antidepressants, anti-Parkinson's disease drugs, and prolactin-reducing drugs [24]. Reserpine as an antihypertensive alkaloid became available in the market as a result of work carried out by Ciba–Geigy. This was probably the first time the principles of reverse pharmacology were systematically used for focused, fast-track drug discovery based on Ayurveda knowledge.

Reverse pharmacology was practiced for several years at Ciba–Geigy and Podar Ayurveda Hospital, Mumbai. Some promising work was undertaken two to three decades ago through a composite drug research program jointly conducted by the Indian Council of Medical Research and India's governmental Council for Scientific and Industrial Research (CSIR). Taking the lead from Ayurveda, a cholesterol-lowering drug, Guggulipid was developed from *Commiphora mukul* [25]. The Drug Controller General of India approved the drug for marketing in 1986. Guggulipid is being manufactured and marketed by Cipla Ltd, Mumbai under the brand name Guglip; however, the availability of authentic raw material has remained a limiting factor. A memory enhancer derived from *Bacopa monnieri* and developed by the Central Drug Research Institute is also available in the market.

Another example of the reverse pharmacology approach is a drug developed from M. *pruriens* seeds for the treatment of Parkinson's disease [26,27]. A commercial product, Zandopa, manufactured by Zandu is claimed to be standardized, safe, effective, economical, and to be derived from a natural source, and which can effectively replace synthetic L-DOPA formulations in patients who comply with its dosage regimen.

Regrettably, most of the efforts in reverse pharmacology approach have remained academic, and have not been pursued to any great extent in order to understand their advanced molecular mechanisms. Most likely, because of inadequate industry involvement during the development cycle, the potential of such efforts to become globally successful products was not optimally explored. The government of India's CSIR under the national network project known as the New Millennium Indian Technology Leadership Initiative (NMITLI), attempted to bridge this gap by bringing industry and academia together right from the beginning, in order to undertake herbal drug development projects on psoriasis, osteoarthritis (OA), hepatitis, and diabetes [28,29].

PHYTOPHARMACEUTICALS

Recently, the Indian government published a draft amendment to the Drugs and Cosmetics Act, and its Rules, by defining phytopharmaceuticals as botanical-based drugs. For the evaluation and marketing of plant-based substances as drugs, the Schedule mandates that scientific data must back quality, safety, and efficacy claims. This will create a new category, similar to synthetic, chemical drugs currently covered under Schedule Y of the Drugs and Cosmetics Act. This initiative is expected to give a boost to scientific research-based drug development from traditional medicine. Previously, the approval of Guggulu tablets developed by Indian scientists took almost a decade, and was approved as a special case by the India's Drug Controller General. These new regulations, which put phytopharmaceuticals into a new category, are necessary, because many provisions for synthetic drugs are not appropriate or relevant for botanical-based products or phytopharmaceuticals. D.B.A. Narayana, a scientist formerly with Unilever, played a significant role in convincing the Indian governmental authorities of the need for, and in spurring the creation of this new category of phytopharmaceuticals [30].

Experts feel that this new category of phytopharmaceuticals will be helpful in facilitating traditional knowledge-inspired drug discovery. In this effort, many Ayurvedic medicines can be put through extensive evaluation, using biomedical sciences. This will encourage the use of scientifically developed herbal products by modern medical practitioners—not only in India, but at the global level.

CLASSICAL AYURVEDIC FORMULATIONS

Ayurveda pharmacology and therapeutics offer many sophisticated formulations, which have been used for hundreds of years. India's Traditional Knowledge Digital Library (TKDL, http://www.tkdl.res.in) has compiled Indian traditional knowledge; it is available in the public domain in the form of several documents and authoritative texts. The TKDL covers more than 290,000 transcripts from 150 AYUSH texts that cover 36,000 classical Ayurveda formulations. Out of these 36,000, the commonly used combinations of Ayurvedic practice number about 2000. Out of these, around 500 are estimated to be manufactured by various Ayurvedic drug manufactures. About 100 formulations are very popular and are being used even at the community level as over-the-counter products. Some of these drugs continue to be used for preventive and primary health care in Indian homes. There are a few classic combinations, which are used as generic drugs. Here, we provide a few representative examples of such formulations.

Triphala

Triphala is a name of a traditional formulation which consists three fruits, known as myrobalans. The first ingredient is *Emblica officinale*, commonly known as Indian gooseberry in English, or *amala* in the vernacular. The second and third ingredients are *Terminalia chebula*, commonly known as *harada* in the vernacular, and *Terminalia belerica*, known as *behada* in the vernacular. These myrobalan fruits with seeds removed are normally used as a mixture consisting of an equal proportion of all three. This combination has synergistic properties. Triphala is an effective medicine for all three *Dosha*. It is used as a general health promoter, for cosmetics purpose to improve skin and hair quality, and for diabetic wound management. It is considered as good as *Rasayana*, which facilitates nourishment to all *Dhatu*. Triphala is the drug of choice for the treatment of several diseases, especially those of metabolism, dental and skin conditions, and for wound treatment. It has a very good effect on the health of the eyes, and is thought to delay degenerative changes, such as cataracts [31] (Figure 9.3).

Triphala is the most common ingredient of over 400 formulations indicated for diabetes-like conditions. The possible mechanism of Triphala in diabetes is through lipid peroxide inhibition, free radical scavenging [32]. A controlled clinical trial on a combination of *Triphala* and Guggulu (*Commiphora mukul*) has indicated its efficacy in treating obesity. This formulation significantly reduced weight and other anthropometric variables. The study also reported significant reduction in LDL cholesterol, and triglyceride levels, with increased HDL levels [33]. Its activity in urinary tract infections with multidrug-resistant bacteria has also been reported [34]. Recent research on *Triphala* has shown its activity as an anticaries agent, root canal irritant,

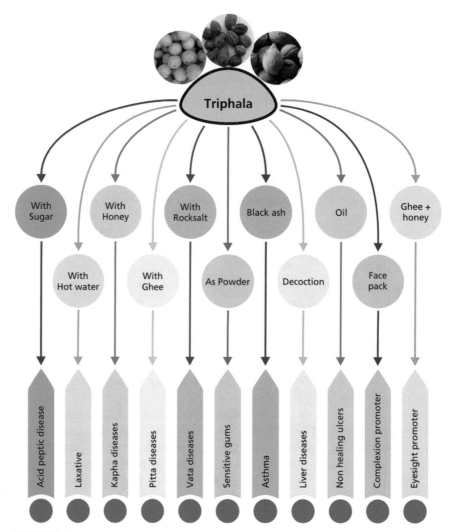

FIGURE 9.3
Multidimensional effects of Triphala.

and in the protection against, and control of gum infections [35]. *Triphala* mouthwash was shown to reduce dental plaque and gingivitis as compared to chlorhexidine [36]. In a phase 1 clinical trial on healthy volunteers, its immunostimulatory effects on cytotoxic T cells and natural killer cells have been reported [37]. *Triphala* is shown to induce apoptosis in tumors cells of the human pancreas, in both cellular and in vivo models [38]. *Triphala* is the most common drug available in almost every home in India that practices herbal remedy as a primary care.

Trikatu

Trikatu is a combination of three acrid substances: *Piper nigrum* known as black pepper, *Piper longum* known as long pepper, and *Zingiber officinale* known as ginger. This combination is used in more than 1500 Ayurveda formulations. *Trikatu* is used as a drug of choice in various diseases of *Vata* and *Kapha*. It enhances appetite, is a digestive and a fat metabolizer, and is effective in respiratory, gastrointestinal, and liver diseases—especially in obstructive jaundice. A good deal of work has been done on *Trikatu* as a bioavailability enhancer, because it facilitates the absorption of drugs from the gastrointestinal tract, and protects them from being metabolized in the liver [39]. However, in animal studies, the coadministration of *Trikatu* did not show much activity in enhancing the bioavailability of drugs like rifampicin, diclofenac, and pefloxacin [40,41,42]. However, an active ingredient present in long pepper, piperine, combined with rifampicin, effectively inhibited rifampicin-resistant mutant strains [43]. Piperine also caused an increase in the bioavailability of drugs like phenytoin, propranolol, and theophylline in healthy volunteers [44]. Taking inspiration from Ayurvedic practice and its knowledge of *Trikatu*, scientists have developed the revolutionary category of *bioenhancers* [45]. This is one of the best examples of reverse pharmacology approach.

Chyavanprash

Chyavanprash is a popular rejuvenating and immunomodulator formulation. It has been used since Charaka's time to increase vitality and longevity. It can be used to treat several maladies, such as respiratory complaints, frequent illness episodes, immunocompromised state, stress, general debility after chronic illness, infertility, and degenerative diseases. The main ingredient of this formulation is Indian gooseberry or *amalaki* fruits. This complex formulation is made from *amalaki* as a base ingredient boiled in decoctions of selected medicinal herbs. The addition of raw sugar, and pure Indian ghee, or clarified butter, makes this a tasty, nutritious, and health-promoting, jelly-like formulation. *Chyavanprash* is one of the most popular Ayurvedic formulations, with a market share of about half a billion Indian rupees annually. Numerous brands of *Chyavanprash* are available in market. Despite this, research on *Chyavanprash* is sparse. There is no good clinical trial published either on its putative benefits for preventive health or on any other therapeutic indications. A few studies have reported its antioxidant properties and procholinergic action, as well as in increasing learning ability and learning–retention capacity in animals [46]. A study has also reported its activity in reducing cisplatin-induced acute nephrotoxicity [47]. This formulation certainly deserves more attention from the scientific community.

Asava and *Arishta*

Asava and *arishta* are fermented products prepared according to classical procedures, using infusions and decoctions, respectively. Actually, these can be categorized as medicinal wines. They are prepared to augment the efficacy and improve the shelf life and potency of various herbal juices, infusions, and decoctions. These preparations are manufactured through a slow fermentation process, using natural yeast found in the flowers of *Woodfordia fruticosa* or *Madhuka indica*. Herbal juices, decoctions, and infusions are mixed with raw sugar, and other herbal ingredients like *trikatu*, and allowed to ferment naturally in wooden barrels. It can take approximately 6 months to generate 5–10% ethyl alcohol in the solution. The natural process of alcohol generation is self-limiting and stops at a certain concentration. *Asava* and *arishta* are used in several conditions for rapid action. Their medicinal properties are supposed to increase as they mature over time in the wooden vats. Properties of *asava* and *arishta* are known to differ from original juices or decoctions. They are easy to assimilate, and can be used as a vehicle, along with other medicines to improve bioavailability. According to Ayurveda, they have hot and penetrating properties, which may help to reach the *Srotas* or microchannels to relieve obstructive pathology. These formulations are preferred in respiratory, skin, metabolic, digestive, reproductive, and neuromuscular diseases. These are contraindicated in bleeding disorders and ulcers—especially among persons of *Pitta Prakriti*. There is a good case to be made for integrating wine technology with *asava* and *arishta*, so that a range of medicinal wines can be prepared for health promotion and prevention of diseases such as diabetes, atherosclerosis, ischemic diseases, dementia, and Alzheimer disease.

Bhasma

Bhasma are unique to Ayurveda pharmaceutics. They hold a prominent position in an Ayurveda clinic. An Ayurveda physician generally uses various *bhasma*, prepared from metals like iron, zinc, silver, copper, gold, and mercury; minerals like chalcopyrite, mica, calcium, potassium; and gemstones like ruby, topaz, and diamond. *Bhasma* have very specific and potent actions on specific tissues. For example, the neuroprotective effects of silver *bhasma*, and the bone-strengthening effects of calcium *bhasma* are documented in ancient Ayurveda texts.

Although, the Charaka, Sushruta, and Vagbhata do not mention many metal preparations, the *bhasma* tradition is evident from the eighth century AD. *Bhasma* preparation is a very complex process and involves a series of trituration treatments, with specific herbal juices, followed by baking at very high temperatures to the point of incineration. In this process, even heavy metals may lose their original properties, and the final product is a complex, herbomineral ultrafine powder. The *bhasma*–nanomedicine link is getting stronger with advancing research [48]. *Bhasma* prepared using classical, traditional processes are safe if used in proper doses and for correct indications. This is a very sophisticated treatment, and only experienced, specialist practitioners can best use it. *Bhasma* are like prescription drugs, which must be taken only under the medical care of a specialist. The so-called *bhasma* toxicity is likely to be related to poor manufacturing processes, contamination, and lack of stringent implementation of regulations [49]. Recently, scientists have studied an organometallic derivative of mercury *bhasma* known as *rasasindura*. It was shown to improve median life span and starvation resistance in *Drosophila melanogaster* [50]. The chemistry and biology of *bhasma* is certainly a potential and challenging area for biomedical research.

Ayurveda gives thousands of rational formulations, which can be used for different clinical conditions. We feel that a deeper understanding of the art of these formulations may help us to design and develop new, multitargeted, synergistic formulations. Single molecule-based drug discovery can and should move on to polyherbal formulations discovery.

New, rational, and innovative herbal formulations can be developed in accordance with stringent regulatory requirements on par with any modern drug. We hope that the conventional skepticism against herbals will slowly wane, if quality-assured, standardized, polyherbal formulations are developed. Here, we present two case studies, where such herbal formulations were successfully developed, under less time restraints, and necessitating fewer resources than conventional drug development.

EVIDENCE-BASED FORMULATIONS

The Ayurvedic formulary gives thousands of multi-ingredient preparations, and an excellent rationale for such formulations can be found in the Ayurvedic classics. One such attempt to design a multi-ingredient formulation (Artrex) for the treatment of rheumatoid and osteoarthritis has been successfully completed, and the formulation has been tested in a well-designed, randomized, double-blind, placebo-controlled clinical trial. This formulation gives therapeutic benefits in acute conditions of pain and inflammation, and it also addresses immunopathological interventions required for the long-term management of slow, progressive, degenerative diseases, such as rheumatoid arthritis (RA) [51,52]. It has ingredients with analgesic and anti-inflammatory activities similar to nonsteroidal anti-inflammatory drugs (NSAIDs), and also includes ingredients with immunomodulatory, anabolic, disease-modifying, and free radical scavenging activities. Thus, the formulation as a whole acts as a combination of NSAIDs and disease-modifying antirheumatic drugs (DMARDs) [53]. The product was codeveloped with Bioved Pharmaceuticals, and has been patented in India and in the United States. It is available in the market in a few counties [54].

The NMITLI project was a team effort comprising six research institutes, five clinical centers, and six industry collaborators. The clinical team developed integrative protocols and appropriate research methodologies for evidence-based Ayurveda and botanical drug development [55]. Two formulations for OA, which performed better than a placebo, or glucosamine in exploratory trials, were then taken up for further mechanistic studies [56]. All the formulations prepared for clinical trials were manufactured and labeled generally in accordance with the FDA guidance to industry for botanical drugs. Most of the required tests were performed in a process starting from the data of raw material, botanical identification, chemical profile, and DNA analysis, and evaluating the stability of the finished products [57,58,59,60,61]. *In vitro* studies using suitable cell and tissue culture models on these formulations revealed significant chondroprotection, proteoglycan release, nitric oxide release, aggrecan release, and hyaluronidase inhibition as markers in an explant model of OA cartilage damage [62,63,64,65].

This team effort led to the design of synergistic polyherbal formulations that were found to be safe and devoid of any genotoxicity or mutagenic activity. Short-listed formulations entered a series of randomized clinical trials, comparing these formulations with the known drugs, glucosamine and celecoxib. Thus, this project was completed in 5 years, with an expenditure of over 2 million USD. This treatment may ultimately cost just 25 USD a month for patients, and have much better therapeutic benefits, including chondroprotection, which no other modern drug offers. Finally, one best formulation was selected, leading to one Indian and one Patent Cooperation Treaty application—with a dossier of the data required for possible regulatory submissions [66]. Currently, CSIR is in the process of identifying a suitable industrial partner for further development, optimization, manufacturing, registrations, and marketing.

INNOVATIVE FORMULATION DISCOVERY

The pharmaceutical sector is going in circles, doing more of the same, while still hoping to disrupt the current impasse. Powerful, high-throughput technologies are available, and the sector requires high-throughput thinking to bring about disruptive, game-changing innovations. In a closely interconnected matrix of molecules, cells, tissues, organs, and complex physiological systems, it may well be very naïve to continue to believe that addressing one target, with one chemical or biological entity—as a drug—will offer sustained health benefits.

Presumably, any drug—whether chemical, botanical, or biological—will have inherent limitations, if it is only focused on a single target. It is important to address multiple targets from a syndrome-related, metabolic cascade perspective, so that holistic management can be effectively achieved. The *magic bullet* approach may have been relevant in the twentieth century when infectious diseases were predominant. In the twenty-first century, we are facing major epidemic of lifestyle diseases like diabetes, asthma, and metabolic syndrome where multiple genes and multiple targets are involved. These complex conditions cannot be treated with a single-target approach. Therefore, it is necessary to move from the single-target, new chemical entity—or *drug*—to a multiple-target, synergistic *formulation* discovery approach (Figure 9.4).

It is possible that polybotanical complex formulations from traditional medicines such as Ayurveda and Traditional Chinese Medicines may have similar rationales. Such traditional knowledge-inspired discovery attempts have shown promising potential in several chronic diseases, such as cancer, diabetes, and arthritis—through the modulation of multiple targets. Several single- or multi-botanical formulas are in wide use, globally, however, their rationales and scientific evidence for their pharmacodynamic actions remain insufficient [14].

The current notion of *formulations* is confined to formulating dosage forms and drug delivery systems. Here, an active drug is combined with different excipients, adjuvants, and additives, so that easily consumable, palatable, convenient dosage forms like tablets, capsules, syrups, ointments, and injections can be produced. In our opinion, this formulation is merely to ensure the correct delivery of active ingredients. In such a formulation, active drug and nonactive excipients are used. We are suggesting that the formulation can be designed in such a way that many active ingredients, extracts, or substances can be combined—based on an array of targets in disease conditions. Such synergistic

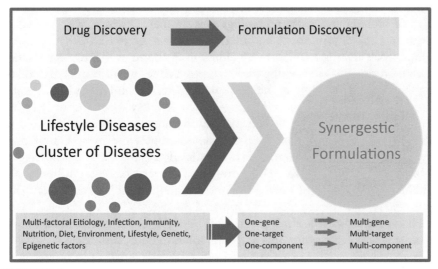

FIGURE 9.4
Magic bullets are not relevant in the twenty-first century.

formulation discovery is a very sophisticated process. We feel that such sophisticated, rational, and time-tested formulations are already available in traditional systems like Ayurveda and T&CM. We need to standardize and test those using conventional preclinical and clinical methods along the lines of reverse pharmacology. Such evidence-based, innovative, multitarget formulations discovery can address issues related to drug toxicity, resistance, and redundancy of the current, single-chemical entities—or drugs.

NETWORK PHARMACOLOGY

Current drug discovery attempts are being made to design ligands with maxim selectivity to act on specific drug targets. Like enzyme action, the drug–receptor relation is considered to be highly specific—like a "lock and key." However, just as one master key can open many locks, and many drugs may modulate multiple proteins, rather than acting upon single targets. As compared with multitarget drugs, highly selective, single-target compounds may actually have less clinical efficacy.

Many drugs used in specialties like oncology, psychiatry, and even newer anti-infective drugs, have effects on multiple targets. To better understand the underlying complex biological and pharmacological processes for chronic diseases like asthma, cancer, and neurodegenerative diseases, it is important to know the network of different pathways where the drugs are likely to act. In designing a new generation of drugs, which can modulate multiple biological targets, a network pharmacology approach is required [67].

When the single-target-based drug discovery approach was evolving, knowledge of biology and pathophysiology was limited. Diseases were looked at more in terms of symptoms, and therapeutics were targeted mainly to relieve symptoms. This triggered discovery efforts in direction to identify specific targets and receptors. Modulating these receptors either by blocking or stimulating the respective biological pathways with specific chemical entities was the discovery strategy. These specific drugs have been very successful in relieving specific symptoms. For example, hypertension can be controlled by beta-blockers, calcium channel blockers, and angiotensin-converting enzyme inhibitors. Hyperglycemia can be treated with drugs stimulating insulin secretion; hypercholesterolemia can be treated with drugs inhibiting cholesterol synthesis, pain and inflammation can be treated with prostaglandin inhibitors and so on. This quick fix, reductionist approach dominated discovery science for several years. It exists even now. This strategy may still be valid for infectious diseases and treatment of specific symptoms like pain, inflammation, and edema.

However, chronic, complex diseases, or syndromes, where multiple targets are involved, this approach might not be applicable. Many metabolic networks collectively are responsible for pathogenesis of complex syndromes and clusters of disease. Lifestyle disorders like obesity, diabetes, cardiovascular diseases, and cancers require a multitargeted approach. Therefore, knowledge of network pharmacology is important for rational formulation discovery.

Complex formulations used in traditional systems like Ayurveda and T&CM offer these benefits. The possible synergistic effects of multicomponents, multichannels, and multitargets seem to be a better proposition for the treatment of chronic, complex, multifactorial diseases. Establishing the rationale behind the use of complex formulations used in traditional medicine is a major challenge.

Pharmacology networks can be constructed for various diseases using systems biology with bioinformatics methods [68]. The idea of T&CM formulation was hypothesized to be consistent with a network pharmacology approach. Network pharmacology, and cheminformatics approaches are useful for multitarget formulations, and also in understanding their mechanisms of action. Network pharmacology provides a new way to understand the interrelationship between complex diseases and drug interventions through the network target paradigm [69]. The constructed network system can pinpoint the main active components and their corresponding targets, which can be helpful for the therapeutic applications of complex formulations of traditional medicine [70].

The recent advances in *omics* technologies and computational and systems biology have provided new insights in identifying multiple target modulations [71]. For example, curcumin is reported to regulate multiple cell signaling pathways and targets, such as cyclooxygenase-2, tumor necrosis factor (TNF-α), epidermal growth factor, human epidermal growth factor receptor 2, vascular endothelial cell growth factor, proteosome, and activate apoptosis, downregulate cell survival gene products, and upregulate tumor suppressor protein—p53, cyclin-dependent kinase inhibitors showing potential in the disease conditions such as cancer, diabetes, and Alzheimer's disease [72].

Panax ginseng has been reported to modulate multiple targets in type 2 diabetes mellitus, and insulin resistance such as insulin receptor substrate-1, c-Jun NH2-terminal kinase, 5′ adenosine monophosphate-activated protein kinase, phosphatidylinositol 3-kinase (PI3K), Akt/PKB, Thr308, Ser473, and glucose transporter type 4 modulating the uptake, and disposal of glucose in adipocytes [73].

Network pharmacology has been effectively used for natural product anticancer drug discovery. Since many natural products have low toxicity, and a capacity for better absorption, and metabolism, they can play an important role in the discovery of multitarget formulations. Dr Xiaojie Xu and others from Peking University in China have carried out an interesting study on natural product compounds from *Holarrhena antidysenterica* and *Hypericum perforatum* to analyze interactions with targets of diseases. Based on these interactions, they constructed network of interactions between natural products and several

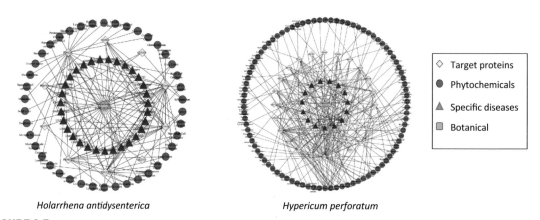

| Holarrhena antidysenterica | Hypericum perforatum |

◇ Target proteins
● Phytochemicals
▲ Specific diseases
▢ Botanical

FIGURE 9.5

The drug–target–disease pharmacology network. *Adapted from: Luo F, Gu J, Chen L, Xu X. Systems pharmacology strategies for anticancer drug discovery based on natural products.* Mol BioSyst *2014;10(7):1912–17.*

cancer target proteins. The network pharmacology analysis indicated that *H. antidysenterica* and *H. perforatum* had potential therapeutic effects against complex diseases, such as prostate cancer, breast cancer, and brain cancer. Such studies provide new insights to understand the pharmacological properties of herbal medicines and help in drug discovery for complex diseases like cancer (Figure 9.5) [74].

Thus, the traditional knowledge coupled with network pharmacology may come in handy to offer an array of new scaffolds, and multitarget, synergistic formulation discovery. Dr Aiping LU from Hong Kong Baptist University and Jian Li from Beijing University of Chinese Medicine with other researchers have conceptualized an interesting model to repurpose old Traditional Chinese Medicine (TCM) drugs based on polypharmacology network and bioinformatics. If validated, this approach can be very useful for new drug and formulation discovery. A few more examples of research on TCM formulations in this direction will be interesting.

FINDINGS FROM TCM

New methods are needed for molecular target/pathway identification to demonstrate scientific evidence in support of traditional use and claims. Some pioneering research to understand rationales and mechanisms of action of multi-ingredient formulations has been attempted by scientists from the National Center for Toxicological Research of the United States' FDA, in collaboration with reputable institutions from China and the United States. This group has used sophisticated technology, including microarrays and bioinformatics tools to examine TCM-induced changes in gene expression. They used the Connectivity Map (CMAP) database, which contains microarray expression data from cultured cell lines (such as human breast cancer cell line, MCF-7), which were treated with bioactive, small molecules with known mechanisms of action, so that the activity of the test formulations could be compared. The current version of the CMAP database contains more than 7000 expression profiles representing treatments from 1309 compounds, and has been used for the discovery of functional connections between drugs, genes, and diseases through the common gene expression changes on the same cell lines. Using this technology, drugs affecting common molecular pathways can be identified, and the putative mechanisms of action of unknown drugs can be explored. This approach has been used to study the effect and mechanism of a TCM herbal formulation for women's health. This sophisticated study shows that treatment with a herbal formulation, and its four, individual, herbal ingredients resulted in an increased antioxidant response; phytoestrogenic activity was consistent with the claimed use for women's diseases.

The CMAP profiles of herbal formulations showed similarities to known chemopreventive agents like withaferin A and resveratrol. This study justified use of withaferin A as a nontoxic, chemopreventive agent. Most importantly, it demonstrated the feasibility of combining microarray gene expression profiling with CMAP mining to discover mechanisms of actions, and to identify new health benefits of herbal formulations [75]. It is suggested that such a genomic approach should be integrated with a traditional chromatography-based fingerprinting method and metabolomics to obtain a more complete understanding of polyherbal formulations.

Another study was done by Chinese researchers on a TCM formulation containing arsenic, known as Realgar–Indigo naturalis formula, to evaluate the putative efficacy in treating human acute promyelocytic leukemia (APL). This study reports that the traditional formulation shows synergistic effect both in murine APL model in vivo and in APL cell differentiation in vitro. These scientists also suggest that the traditional principle of designing a formula, which uses some adjuvant components, may facilitate the delivery of the active moieties to the disease site in the body [76].

An example from TCM clinical practice for the treatment of RA is similar to Ayurveda concept. According to TCM, patients of RA can be classified into two main patterns—cold and hot. The gene expression profiles of typical cold- and hot-pattern RA patients showed different protein–protein interaction-related networks. Thus, based on genomic data, the molecular networks of the cold and hot patterns could be compared with disease-related network analysis available in the public domain. This can provide a molecular network of the RA cold and hot patterns. For new drug discovery, the cold or hot patterns in RA can be used to construct molecular networks, which can be merged with the herbal formula pharmacological network. This can help to discover the best potential herbal formula candidates. Mapping to a drug–target network can help with the

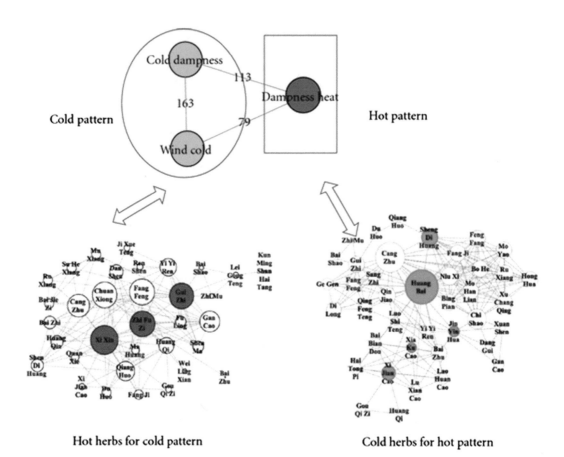

FIGURE 9.6

Formulations for hot and cold patterns as per TCM. The common herbal formula and combinations for the treatment of rheumatoid arthritis with TCM cold and heat pattern was obtained by text mining. *Reproduced with kind permission: Jian Li, Cheng Lu, Miao Jiang, et al. Traditional Chinese Medicine-Based Network Pharmacology could lead to new multicompound drug discovery.* Evid Based Complement Altern Med *2012:11. Article ID 149762.* http://dx.doi.org/10.1155/2012/149762.

prioritization of compounds, while mapping to other biological networks can identify interesting target pairs, and their associated compounds in the context of biological systems. This approach, proposed by Chinese researchers Jian Li et al., could be a useful way to build up and investigate the pharmacological network for multiple compound, new drug candidates [77] (Figure 9.6).

Traditional polyherbal formulations are complex systems, which make them difficult to evaluate for pharmacodynamics action. A modular approach has been proposed by Dr Jianglong Song and colleagues from Chinese Academy of Sciences, Beijing using a computational complex network technique, which can provide molecular mechanisms. This approach first constructs homologous networks, followed by the identification of primary pharmacological

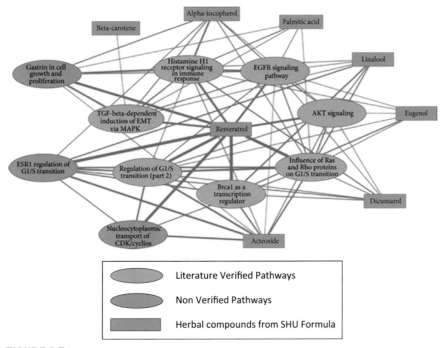

FIGURE 9.7

Molecular mechanism of TCM formulation. An illustration of Shu-feng-jie-du (SHU) formula intervening the influenza development through multiple pathways. *Adapted with kind permission from: Jianglong Song, Fangbo Zhang, Shihuan Tang, et al. A module analysis approach to investigate molecular mechanism of TCM Formula: a trial on Shu-feng-jie-du Formula.* Evid Based Complement Altern Med *2013;2013:14. Article ID 731370.* http://dx.doi.org/10.1155/2013/731370.

units and pathway analysis. This approach was used to study the traditional Chinese formulation known as Shu-feng-jie-du, which is used to treat influenza infection [78] (Figure 9.7).

INTEGRATIVE APPROACHES

The studies, we discussed earlier, give hope that even complex, polyherbal formulations can be standardized with a reasonably good understanding of the mechanisms of action. These complex formulations can be studied for the pharmacodynamic activities of individual ingredients as well. The putative synergistic role of different components in the formulations can also be studied, using modern methods of chemistry, biomedical analysis, and computational and molecular biology.

It will be beneficial if Chinese and Indian scientists share their knowledge, and collaborate—exchanging mutual facilities, and expertise. Such a confluence of two great traditions can certainly expedite the process of rational formulation discovery. This integration of wisdom, ideas, and technology will help modern science to rediscover drug discovery, and take it to greater heights.

Thus, we feel that integrative approaches are needed for drug discovery and also for development. This integration can be at various levels. Firstly, for preparing rational formulations, integrating appropriate ingredients so that chronic, polygenic, difficult-to-treat diseases, and complex syndromes can be effectively treated. Secondly, to deal with mapping multiple targets of disease, in multicomponent, synergistic formulations. Thirdly, to align and integrate formulations with the discovery of appropriate therapeutics. We feel that it is necessary to conceptualize *therapeutic discovery*, which will involve holistic consideration for treating a diseased person. This will involve integration of diet, detox processes, and lifestyle modifications, along with the prescription of appropriate, personalized formulations. We feel that such a personalized *therapeutic discovery* as an integrative package is the future of medicine.

In short, the potential of natural products, ethnopharmacology, and Ayurveda-inspired drug discovery is largely unexplored. For instance, more than 8 million living species exist on this planet. Botanical Gardens Conservation International estimates that there are 400,000 plant species. Out of which, 50,000 are edible plant species. Medicinal plant species are estimated to be about 20,000. According to Canada's Department of Agriculture, around 13,000 species of plants have been used as traditional medicines for at least a century. Data from the United States suggest that in the past decade around 1800 medicinal plant species were commercially available. Currently, research on herbal drugs has not crossed even 1000 scientific articles. There are only two botanical drugs approved by US FDA. This shows the vast biological resource available for bioprospecting and drug discovery. Concerted efforts to improve our understanding of botanical materials are needed in order to fast-forward the drug discovery process, which is facing a severe innovation deficit. This may give impetus to traditional knowledge-inspired new drugs and formulations, which are effective, safe, affordable, and available for the global community.

We hope that traditional knowledge-inspired approach may be a consolation to the pharmaceutical sector. This approach may offer innovative leads and materials to motivate pharmaceutical companies to improve their efficiency and productivity, without compromising ethics, and by not deceiving people in their desperation for profits. We hope that the term Pharmacia does not get associated with sorcery, and the merchants of the world do not deceive the nations as was warned by the biblical revelations.

We end this chapter with a Vedic incitement "Everything found in the universe has utility. There is no shortage of resources in nature, but what is rare is the *talented strategist* who can make the best use of available resources, who can innovate ethical, and viable options in the best interest of people."

REFERENCES

[1] Patwardhan B, Vaidya ADB. Natural products drug discovery: accelerating the clinical candidate development using reverse pharmacology approaches. Indian J Exp Biol 2010;48(3): 220–7.

[2] Patwardhan B. Ayurveda: the "Designer" medicine: a review of ethnopharmacology and bioprospecting research. Indian Drugs 2000;37(5):213–27.

[3] Cragg GM, Boyd MR, Cardellina JH, Newman DJ, Snader KM, McCloud TG. Ethnobotany and drug discovery: the experience of the US National Cancer Institute. Ciba Found Symp 1994;185:178–90; discussion 190–6.

[4] Harvey AL. Natural products in drug discovery. Drug Discov Today 2008;13(19-20):894–901.

[5] Gupta R, Gabrielsen B, Ferguson SM. Nature's medicines: traditional knowledge and intellectual property management. Case studies from the National Institutes of Health (NIH), USA. Curr Drug Discov Technol 2005;2(4):203–19.

[6] Holland BK. Prospecting for drugs in ancient texts. Nature 1994;369(6483):702.

[7] Patwardhan B, Vaidya ADB, Chorghade M, Joshi SP. Reverse pharmacology and systems approaches for drug discovery and development. Current Bioactive Compounds 2008;4(4):201–212.

[8] Verdine GL. The combinatorial chemistry of nature. Nature 1996;384(6604 Suppl):11–3.

[9] Koehn FE, Carter GT. The evolving role of natural products in drug discovery. Nat Rev Drug Discov 2005;4(3):206–20.

[10] Harmon K. Prescription drug deaths increase dramatically. Sci Am 2010. Available from: http://www.scientificamerican.com/article/prescription-drug-deaths/

[11] Patwardhan B. The new pharmacognosy. Comb Chem Highthroughput Screen 2014;17(2):97.

[12] Simpson PB, Reichman M. Opening the lead generation toolbox. Nat Rev Drug Discov 2013;13(1):3–4.

[13] Schmid EF, Smith DA. Is pharmaceutical R&D just a game of chance or can strategy make a difference? Drug Discov Today 2004;9(1):18–26.

[14] Patwardhan B. Death of drugs and rebirth of health care: Indian response to discovery impasse. In: Chaguturu R, Editor. Collaborative Innovation in Drug Discovery. John Wiley & Sons, Inc; 2014. p. 173–94.

[15] Pujol A, Mosca R, Farrés J, Aloy P. Unveiling the role of network and systems biology in drug discovery. Trends Pharmacol Sci 2010;31(3):115–23.

[16] List of withdrawn drugs available at http://en.wikipedia.org/wiki/List_of_withdrawn_drugs.

[17] Ellingson SR, Smith JC, Baudry J. Polypharmacology and supercomputer-based docking: opportunities and challenges. Mol Simul 2014;40(10–11):848–54.

[18] Zimmermann GR, Lehár J, Keith CT. Multi-target therapeutics: when the whole is greater than the sum of the parts. Drug Discov Today 2007;12(1-2):34–42.

[19] Vaidya R, Vaidya A, Patwardhan B, Tillu G, Rao Y. Ayurvedic pharmacoepidemiology - a proposed new discipline. JAPI 2003;51:528.

[20] Vaidya ADB. Reverse pharmacological correlates of Ayurvedic drug actions. Indian J Pharmacol 2006;38(5):311–5.

[21] Vaidya ADB, Devasagayam TPA. Current status of herbal drugs in India: an overview. J Clin Biochem Nutr 2007;41(1):1–11.

[22] Sen G, Bose K. *Rauwolfia serpentina*, a new Indian drug for insanity and hypertension. Indian Med World 1931;21:194–201.

[23] Vakil RJ. A clinical trial of *Rauwolfia serpentina* in essential hypertension. Br Hear J 1949;11(4):350–5.

[24] Svensson T. Effects of chronic treatment with tricyclic antidepressant drugs on identified brain noradrenergic and serotonergic neurons. Acta Psychiatr Scand Suppl 1980;280:121–3.

[25] Satyavati G. Gum guggul (*Commiphora mukul*)-the success story of an ancient insight leading to a modern discovery. Ind J Med Res 1988;87:327–35.

[26] Manyam BV, Dhanasekaran M, Hare TA. Neuroprotective effects of the antiparkinson drug *Mucuna pruriens*. Phyther Res 2004;18(9):706–12.

[27] Vaidya RA, Sheth AR, Aloorkar SD, Rege NR, Bagadia VN, Devi PK, et al. The inhibitory effect of the cowhage plant *Mucuna pruriens* bark and L-dopa on chlorpromazine-induced hyperprolactinemia. Neurol India 1978;26(4):177–8.

[28] Padma TV. Ayurveda. Nature 2005;436(7050):486.

[29] Patwardhan B, Mashelkar R. Traditional medicine-inspired approaches to drug discovery: can Ayurveda show the way forward? Drug Discov Today 2009;14(15-16):804–11.

[30] Narayana DA, Katiyar C. Draft amendment to drugs and cosmetics rules to license science based botanicals, phytopharmaceuticals as drugs in India. J Ayurveda Integr Med 2013;4(4):245–6.

[31] Gupta SK, Kalaiselvan V, Srivastava S, Agrawal SS, Saxena R. Evaluation of anticataract potential of Triphala in selenite-induced cataract: In vitro and in vivo studies. J Ayurveda Integr Med 2010;1(4):280–6.

[32] Sabu MC, Kuttan R. Anti-diabetic activity of medicinal plants and its relationship with their antioxidant property. J Ethnopharmacol 2002;81(2):155–60.

[33] Paranjpe P, Patki P, Patwardhan B. Ayurvedic treatment of obesity: a randomised double-blind, placebo-controlled clinical trial. J Ethnopharmacol 1990;29(1):1–11.

[34] Bag A, Bhattacharyya SK, Pal NK. Antibacterial potential of hydroalcoholic extracts of Triphala components against multidrug-resistant uropathogenic bacteria - a preliminary report. Indian J Exp Biol 2013;51(9):709–14.

[35] Prakash S, Shelke AU. Role of Triphala in dentistry. J Indian SocPeriodontol 2014;18(2):132–5.

[36] Bhattacharjee R, Nekkanti S, Kumar NG, Kapuria K, Acharya S, Pentapati KC. Efficacy of Triphala mouth rinse (aqueous extracts) on dental plaque and gingivitis in children. J Investig Dent 2014. http://dx.doi.org/10.1111/jicd.12094.

[37] Phetkate P, Kummalue T, U-Pratya Y, Kietinun S. Significant increase in cytotoxic T lymphocytes and natural killer cells by Triphala: a clinical phase i study. Evid Based Complement Altern Med 2012;2012:239856. http://dx.doi.org/10.1155/2012/239856.

[38] Shi Y, Sahu RP, Srivastava SK. Triphala inhibits both in vitro and in vivo xenograft growth of pancreatic tumor cells by inducing apoptosis. BMC Cancer 2008;8:294.

[39] Atal CK, Zutshi U, Rao PG. Scientific evidence on the role of Ayurvedic herbals on bioavailability of drugs. J Ethnopharmacol 1981;4(2):229–32.

[40] Karan RS, Bhargava VK, Garg SK. Effect of Trikatu, an Ayurvedic prescription, on the pharmacokinetic profile of rifampicin in rabbits. J Ethnopharmacol 1999;64(3):259–64.

[41] Lala LG, D'Mello PM, Naik SR. Pharmacokinetic and pharmacodynamic studies on interaction of "trikatu" with diclofenac sodium. J Ethnopharmacol 2004;91(2-3):277–80.

[42] Dama MS, Varshneya C, Dardi MS, Katoch VC. Effect of trikatu pretreatment on the pharmacokinetics of pefloxacin administered orally in mountain Gaddi goats. J Vet Sci 2008;9(1):25–9.

[43] Sharma S, Kumar M, Sharma S, Nargotra A, Koul S, Khan IA. Piperine as an inhibitor of Rv1258c, a putative multidrug efflux pump of Mycobacterium tuberculosis. J Antimicrob Chemother 2010;65(8):1694–701.

[44] Hu Z, Yang X, Ho PCL, Sui YC, Heng PWS, Chan E, et al. Herb-drug interactions: a literature review. Drugs 2005;65(9):1239–82.

[45] Atal N, Bedi KL. Bioenhancers: revolutionary concept to market. J Ayurveda Integr Med 2010;1(2):96–9.

[46] Bansal N, Parle M. Beneficial effect of chyawanprash on cognitive function in aged mice. Pharm Biol 2011;49(1):2–8.

[47] Aditya M, Nair CKK. Ayurvedic formulations as therapeutic radioprotectors: preclinical studies on Brahma Rasayana and Chyavanaprash. Curr Sci 2013;104(7):959–66.

[48] Pal D, Sahu CK, Haldar A. *Bhasma*: the ancient indian nanomedicine. J Adv Pharm Technol Res 2014;5(1):4–12.

[49] Patwardhan B, Warude D, Tillu G. Heavy metals and Ayurveda. Curr Sci 2005;88(10):1535–6.

[50] Dwivedi V, Anandan EM, Mony RS, Muraleedharan TS, Valiathan MS, Mutsuddi M, et al. In vivo effects of traditional Ayurvedic formulations in *Drosophila melanogaster* model relate with therapeutic applications. PLoS One 2012;7(5):e37113. http://dx.doi.org/10.1371/journal.pone.0037113.

[51] Chopra A, Lavin P, Patwardhan B, Chitre DA. 32-week randomized, placebo-controlled clinical evaluation of RA-11, an Ayurvedic drug, on osteoarthritis of the knees. J Clin Rheumatol 2004;10(5):236–45.

[52] Chopra A, Lavin P, Patwardhan B, Chitre D. Randomized double blind trial of an Ayurvedic plant derived formulation for treatment of rheumatoid arthritis. J Rheumatol 2000;27(6):1365–72.

[53] Chopra A. Efficacy of Ayurvedic formulation in rheumatoid arthritis and osteoarthritis. American College of Rheumatology scientific meetings, Florida 1996, and Asia Pacific League against Rheumatism, Singapore 1997.

[54] Patwardhan B. Method of treating musculoskeletal disorders and a novel composition therefore. US Patent 5494668; 1996.

[55] Chopra A, Saluja M, Tillu G. Ayurveda-modern medicine interface: a critical appraisal of studies of Ayurvedic medicines to treat osteoarthritis and rheumatoid arthritis. J Ayurveda Integr Med 2010;1(3):190–8.

[56] Chopra A, Saluja M, Tillu G, Venugopalan A, Sarmukaddam S, Raut AK, et al. A randomized controlled exploratory evaluation of standardized Ayurvedic formulations in symptomatic osteoarthritis knees: a Government of India NMITLI project. Evid Based Complement Altern Med 2011;2011:724291. http://dx.doi.org/10.1155/2011/724291.

[57] Joshi K, Chavan P, Warude D, Patwardhan B. Molecular markers in herbal drug technology. Curr Sci 2004;87(2):159–65.

[58] Chavan P, Warude D, Joshi K, Patwardhan B. Development of SCAR (sequence-characterized amplified region) markers as a complementary tool for identification of ginger (Zingiber officinale Roscoe) from crude drugs and multicomponent formulations. Biotechnol Appl Biochem 2008;50(Pt 1):61–9.

[59] Shinde VM, Dhalwal K, Mahadik KR, Joshi KS, Patwardhan BK. RAPD analysis for determination of components in herbal medicine. Evid Based Complement Altern Med 2007;4(Suppl 1):21–3.

[60] Chavan P, Joshi K, Patwardhan B. DNA microarrays in herbal drug research. Evid Based Complement Altern Med 2006;3(4):447–57.

[61] Dnyaneshwar W, Preeti C, Kalpana J, Bhushan P. Development and application of RAPD-SCAR marker for identification of Phyllanthus emblica LINN. Biol Pharm Bull 2006;29(11):2313–6.

[62] Sumantran VN, Kulkarni A, Chandwaskar R, Harsulkar A, Patwardhan B, Chopra A, et al. Chondroprotective potential of fruit extracts of Phyllanthus emblica in osteoarthritis. Evid Based Complement Altern Med 2008;5(3):329–35.

[63] Sumantran VN, Chandwaskar R, Joshi AK, Boddul S, Patwardhan B, Chopra A, et al. The relationship between chondroprotective and antiinflammatory effects of *Withania somnifera* root and glucosamine sulphate on human osteoarthritic cartilage in vitro. Phyther Res 2008;22(10):1342–8.

[64] Sumantran VN, Joshi AK, Boddul S, Koppikar SJ, Warude D, Patwardhan B, et al. Antiarthritic activity of a standardized, multiherbal, Ayurvedic formulation containing *Boswellia serrata*: In vitro studies on knee cartilage from osteoarthritis patients. Phyther 2011;25(9):1375–80.

[65] Sumantran VN, Kulkarni A, Boddul S, Chinchwade T, Koppikar SJ, Harsulkar A, et al. Chondroprotective potential of root extracts of *Withania somnifera* in osteoarthritis. J Biosci 2007;32(2):299–307.

[66] Chopra A. Validating safety & efficacy of Ayurvedic derived botanical formulations: a clinical arthritis model of NMITLI. In: 5th Oxford International Conference on the Science of Botanicals (ICSB). USA: University of Mississippi at Oxford; 2006.

[67] Hopkins A. Network pharmacology. Nat Biotechnol 2007;25(10):1110-1.

[68] Cho D-Y, Kim Y-A, Przytycka TM. Chapter 5: network biology approach to complex diseases. PLoS Comput Biol 2012;8(12):e1002820. http://dx.doi.org/10.1371/journal.pcbi.1002820.

[69] Li S. Network target: a starting point for traditional Chinese medicine network pharmacology. Zhongguo Zhong Yao Za Zhi. 2011;36(15):2017–20.

[70] Shi SH, Cai YP, Cai XJ, Zheng XY, Cao DS, Ye FQ, et al. A network pharmacology approach to understanding the mechanisms of action of traditional medicine: Bushenhuoxue formula for treatment of chronic kidney disease. PLoS One 2014;9(3):e89123. http://dx.doi.org/10.1371/journal.pone.0089123.

[71] Gu J, Gui Y, Chen L, Yuan G, Lu HZ, Xu X. Use of natural products as chemical library for drug discovery and network pharmacology. PLoS One 2013;8(4):e62839. http://dx.doi.org/10.1371/journal.pone.0062839.

[72] Gupta SC, Patchva S, Aggarwal BB. Therapeutic roles of curcumin: lessons learned from clinical trials. AAPS J 2013;15(1):195–218.

[73] Zhang Z, Li X, Lv W, Yang Y, Gao H, Yang J, et al. Ginsenoside Re reduces insulin resistance through inhibition of c-Jun NH2-terminal kinase and nuclear factor-kappaB. Mol Endocrinol 2008;22(1):186–95.

[74] Luo F, Gu J, Chen L, Xu X. Systems pharmacology strategies for anticancer drug discovery based on natural products. Mol Biosyst 2014;10(7):1912–7.

[75] Wen Z, Wang Z, Wang S, Ravula R, Yang L, Xu J, et al. Discovery of molecular mechanisms of traditional Chinese medicinal formula Si-Wu-Tang using gene expression microarray and connectivity map. PLoS One 2011;6(3):e18278. http://dx.doi.org/10.1371/journal.pone.0018278.

[76] Wang L, Zhou G-B, Liu P, Song J-H, Liang Y, Yan X-J, et al. Dissection of mechanisms of Chinese medicinal formula Realgar-Indigo naturalis as an effective treatment for promyelocytic leukemia. Proc Natl Acad Sci USA 2008;105(12):4826–31.

[77] Li J, Lu C, Jiang M, Niu X, Guo H, Li L, et al. Traditional chinese medicine-based network pharmacology could lead to new multicompound drug discovery. Evid Based Complementary Altern Med 2012;2012:149762. http://dx.doi.org/10.1155/2012/149762.

[78] Song J, Zhang F, Tang S, Liu X, Gao Y, Lu P, et al. A module analysis approach to investigate molecular mechanism of TCM formula: a trial on Shu-feng-jie-du formula. Evid Based Complement Altern Med 2013;2013:731370. http://dx.doi.org/10.1155/2013/731370.

CHAPTER 10

Longevity, Rejuvenation, and *Rasayana*

I don't want to achieve immortality through my work. I want to achieve
it through not dying.

Woody Allen

LONGEVITY AND HEALTHY AGING

Aging *per se* is a natural and normal stage of growth. It is not a disease. Aging
can bring on some disabilities, and diseases, which can be avoided if one
attains healthy aging. It is a human tendency to aspire to live longer. Longevity is living longer, which is result of healthy aging. Longevity is based
on several factors including genetics, lifestyle, and environmental factors. The
optimal life span for a human is considered to be 100 years, although there
are few exceptional individuals who may live longer. Many studies indicate
that about one third of the human life span depends on genetics, and the
remainder can be attributed to lifestyle and environmental factors, many of
which depend on individual choices. About 130 genes are considered to be
in some way associated with human longevity. Recent studies have reported
that even modest physical exercise and leisure time can extend life expectancy
by 4–5 years. Studies have indicated that there is a genetic inheritance among
long-lived individuals. Families with a history of exceptional longevity may
inherit biological factors, which can modulate aging processes, and disease
susceptibility [1].

Aging as such, is not a disease, but the aging process can pose vulnerability
to many diseases. Diseases and disabilities of old age are emerging as major
problems. The diseases associated with aging are probably the outcome of several factors, including nutrition, lifestyle, and environmental exposure during
the earlier stages of development from childhood. Therefore, disease prevention strategies to avoid problems linked to old age should be given importance.
Admittedly, modern medicine has no definitive cures for many chronic and difficult-to-treat diseases. Many drugs have been tried, but except for condemning a
patient to a life sentence of cortisone therapy, there is nothing much to offer. As
regards to disease prevention, modern medicine accepts the common wisdom

259

Integrative Approaches for Health. http://dx.doi.org/10.1016/B978-0-12-801282-6.00010-3

of avoiding risk factors, exercising, and eating nutritious foods. There is no definite regimen which will prevent age-related disabilities, and health problems.

We propose a simple equation concerning disease, aging, and longevity:

Aging—(disabilities + disease) = Healthy aging + longevity.

Thus prevention and timely treatment of age-related deformities and diseases may help us to move toward healthy aging and longevity. We feel that in such a situation, traditional knowledge systems like Ayurveda and Yoga have much to offer. Physical and mental exercise, relaxation, meditation techniques from Yoga, and the interventions of Ayurveda through *Rasayana* and *Jara Chikitsa*, can aid in the prevention of the disabilities of aging, and in the promotion of rejuvenation—with the help of some unique therapies, which claim to restore the vitality of aging cells. The meditation and relaxation techniques of Yoga can offer better mental health. Details of this will be discussed in later sections.

POPULATION AGING

During this decade, the advances in biomedical sciences, therapeutics, diagnostics, and surgery have resulted in improved quality of health care. As a result, the average life expectancy has significantly increased in many parts of the world. Longevity is seen as an obvious, and happy outcome. The World Population Aging 2013 report of the United Nations estimate that, by 2050, global population of those over 60 years of age will top 22%. This means that the elderly population will total over two billion. Generally, people are able to live longer, but not necessarily healthier. The population aging phenomenon is happening much faster than at any time in the past. For example, in a country like France, it took over 100 years to double the aging population, but now it is estimated that countries such as China and Brazil will reach a similar demographic transition in just 25 years. The earlier benchmark of 60 years as elderly is advancing; today, more than ever before, people are likely to live for 80 or 90 years. By 2047, the percentage of older people is projected to exceed the number of children for the first time. This means that the world is aging. It is known that women have a life span of 6–8 years longer than men. The number of people who are aged 80 years or older will increase fourfold by 2050. This is bound to impose formidable challenges in terms of health services, which countries like Japan are already facing. The developing countries have already started feeling the heat.

Several socioeconomic impacts are bound to be felt because of this demographic transition. Higher rates of suicides, especially in old age, present a paradoxical situation where some people wish to live longer, while many others are desperate to end their life. In many countries euthanasia has been legalized, and people from other countries are demanding this right. Of course, a wide range of ethical considerations are involved in such decisions.

As populations are aging, and people are living longer, the prevalence of non-communicable diseases and disability is seen to be increasing. Currently, according to WHO statistics, more than 45% of old people have one or more chronic disabilities, which make the aged group more vulnerable. This could be due to cardiovascular, cerebrovascular, musculoskeletal diseases, metabolic diseases like diabetes, osteoporosis, and various cancers—especially of the prostate and colon. Even in poor countries the majority of older people die of diseases like heart disease, cancer, and diabetes, rather than from infectious and parasitic diseases. The number of frail elderly who are not able to look after themselves is expected to increase four times in the developing countries by 2050. The risk of dementia rises with age. It is estimated that about 30% of people over 85 years-old face cognitive decline, and may not have access to affordable, required care. These consequences of demographic transitions need to be addressed carefully.

BIOLOGICAL PROCESS OF AGING

Aging is a natural, progressive, biological process of changes responsible for increased susceptibility to disease and degeneration, which finally leads to death. This process is common to all living forms where aging and death is universal. Indian philosophy offers a transcendental perspective on the life cycle, from birth to death. Thanatology is a science of death, which deals with physical and biological changes, including postmortem, and forensic sciences [2]. Indian philosophy provides an art of accepting death gracefully, and with dignity.

The aging process is controlled by genetic and epigenetic factors, which are influenced by environmental conditions. This can lead to differences in intra- and interspecies life spans. Various theories associated with the nature of the aging process include encoding of aging in DNA, progressive deterioration in protein synthesis, cross-linkage of macromolecules, auto immune damage, and free radical reaction damage. Reasonable scientific consensus is in favor of the free radical theory of aging [3]. Both plants and animals require oxygen to sustain life. The free radical reactions are the result of the biochemical reduction of oxygen to water in photosynthesis and respiration.

The aging process involves reactive oxygen species (ROS) as free radical reactions. ROS include superoxide anion, peroxide, nitrous oxide, and hydroxyl radicals. These ROS are mainly responsible for various damaging and degenerative conditions responsible for aging. These free radical reactions are also responsible for age-related deterioration of the cardiovascular, central nervous, and immune systems. Aging in humans is a multidimensional process. It is known that physical, mental, and social developments can happen even at later stages in life. Aging also reflects cultural and societal conventions, and needs to be considered in a holistic fashion. The phenomenon of aging is not restricted only to the visible changes of biological aging.

Crucial hallmarks of aging have been very lucidly explained in a review by Spanish researchers, López-Otín et al. Their scheme enumerates nine hallmarks: genomic instability, telomere attrition, epigenetic alterations, loss of proteostasis, deregulated nutrient sensing, mitochondrial dysfunction, cellular senescence, stem cell exhaustion, and altered intercellular communication [4] (Figure 10.1).

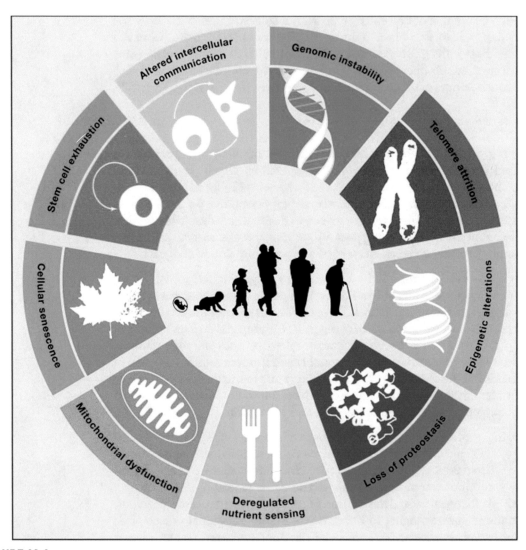

FIGURE 10.1

Mechanisms of Aging. *Reprinted from Publication The Hallmarks of Aging published in Cell June 6, 2013;153(6):1194–1217, with permission from Elsevier.*

Scientists think that the healthy life span can be increased if we are able to control free radical-induced damage. This can be attempted by modifications in diet and lifestyle, including body weight control and diets, which can minimize free radical reactions in the body. Dietary interventions consisting of antioxidants have been shown to reduce free radical reaction damage leading to the increased life span of experimental animals like mice, rats, fruit flies, nematodes, and rotifers.

Antioxidants have been shown to counter cancers and also to enhance humoral and cell-mediated immune responses. In Chapter 7 (Food and Diet), we discussed how diets high in antioxidants such as α-tocopherol, ascorbic acid, and selenium, and low in polyunsaturated lipids, can decrease free radical reaction damage. Recently, eminent scientist, James Watson, has proposed a hypothesis that the root cause of many diseases, such as diabetes and cancer, may be linked to oxidative damage caused by free radicals of ROS or reactive nitrogen species (RNS).

About 3.5 billion years ago, through free radical reaction, life was formed from amino acids, nucleotides, and other basic chemicals. These reactions were induced mainly by ionizing radiation from the sun, and later, from photosynthesis and respiration. Thus, free radical reactions have been a natural part of life since the beginning. Approximately 1.3 billion years ago, when the atmospheric oxygen increased, the anaerobes drastically reduced. For about 500 million years, life was mostly confined to the sea; therefore, it was protected from UV radiations from the sun. With a further increase in atmospheric oxygen, evolution accelerated, and primates appeared about 65 million years ago; humans appeared about 4–5 million years ago. The life span of a species was related to their ability to counter damaging free radical reactions. The free radical theory of aging was first proposed in 1954. Although this theory is widely accepted, evidence is available both for and against it.

For instance, *Proteus anguinus* (salamander) has exceptional longevity without any special defenses against oxidative damage. A few studies on mice with their major antioxidant genes knocked out did show the expected elevation in oxidative damage, without impact on their life span. Such findings raise fundamental questions about the usefulness of the free radical damage theory for understanding the aging process, and variations in life span and life histories [5,6].

Aging, Advanced Glycation End Product and Lipofuscin

The accumulation of some of the unwanted intracellular garbage or waste materials, like advanced glycation end product (AGE) and lipofuscin (LF), have been shown to be directly related to cellular degeneration during old age; and are linked to several age-related diseases [7]. Lipofuscin is a yellowish-brown, granular pigment composed of highly oxidized and cross-linked, complex lipoproteins. LF is composed of insoluble and nondegradable residues of lysosomal digestion, which is found in various tissues and cells, including the heart, liver, kidney, retina, adrenals, and nerves and ganglion cells [8]. LF is thought to be a marker of the physiological and preterm aging of cells.

AGEs are a complex group of heterogeneous compounds formed when reducing sugars react with amino acids in proteins. AGEs can be present in food, and can be formed in the human body. AGEs are found in higher concentrations in older adults. There is increased evidence supporting role of AGEs in the development of chronic degenerative diseases, such as cardiovascular disease, Alzheimer disease, and with complications of diabetes mellitus. Several studies in animal models, and in humans, show that the restriction of dietary AGEs has positive effects on wound healing, insulin resistance, and cardiovascular disease. The restriction of AGEs in food has been reported to increase life span in animal models [9].

In an interesting study, researchers fed a diet derived AGE and LF to *Drosophila*. It was observed that chronic ingestion of these substances accelerates the functional decline in *Drosophila*—quite similar to that which occurs during normal aging. This observation confirms the role of AGE- and LF-like unwanted products in cellular damage and in the aging process [10]. As a diet consisting of toxic substances like AGE and LF can expedite the aging process, it is also possible to slow down the process through therapeutic diet and lifestyle modifications.

We find striking similarities between the concept of *Ama* from Ayurveda, and AGEs and LF. As described in the "Primer," *Ama* is a sticky, toxic waste product, which is a result of undigested food. It clogs *Srotasa*, and plays a vital role in pathogenesis. We hypothesize that the Ayurveda treatment for digesting and eliminating *Ama* will reduce AGE and LF. It will be interesting to study whether such nonenzymatic, nondegradable, complex substances can be degraded and removed by *Panchakarma* and *Rasayana* therapy.

Senescence

Senescence is the state or process of aging. Organismal senescence is characterized by declining ability to respond to stress, increasing homeostatic imbalance, and an increased risk of disease. Currently, such degenerative changes are considered to be irreversible, and finally to lead to death. Some researchers consider aging a disease, and think it is treatable and preventable. Numerous species are known to show very low signs of aging. The best known examples include trees like the bristlecone pine; fish like the sturgeon and the rockfish; and invertebrates like the quahog, sea anemone, and lobster.

Cellular senescence has been attributed to the shortening of telomeres with each cell cycle. Telomeres are the ends of eukaryotic chromosomes. When telomeres become too short, the cells die. The length of telomeres is a reliable predictor of the aging process. Researchers have shown that genetically manipulating mice to shut down their telomerase enzyme resulted in the mice displaying many disorders and premature aging. When the enzyme telomerase was reactivated, it repaired the damaged tissues, and reversed the signs of aging [11]. Telomere length is maintained by the telomerase enzyme in immortal cells like germ cells, keratinocyte stem cells, and even in cancer cells. Mortal cell lines can be immortalized by the activation of their telomerase gene. In most of the cancerous cells, the telomerase gene undergoes mutation, which leads to uncontrolled growth.

In an interesting experiment, two scientists, Drs Hayflick and Moorhead, from the Wistar Institute in Philadelphia, observed that it was possible to grow and maintain normal diploid fibroblasts for several months. However, they noted a curious phenomenon that the cells could not be subcultivated more than about 50 times. They concluded that some intrinsic factor/s now known as *Hayflick factors* accumulate in these cells, until the cells are *senesced*. They proposed that this cellular phenomenon could be relevant for organismal aging. It is now clear that the senescent response can be triggered by a wide variety of cellular stresses, including the loss of telomeres, or DNA damage. Hayflick discovered the phenomenon known as senescence in 1961. He discovered that isolated cells also show a limited ability to divide in culture, which is known as the Hayflick limit.

Interestingly, HeLa cells are the exception to the Heyflick limit. In October 1951, George Gey, an expert in tissue culture at John Hopkins, showed that it is possible to grow human cells continuously in the laboratory. He named these cells *HeLa cells*. This name was given in memory of Henrietta Lacks, a patient of cervical cancer, whose biopsy sample led to these immortal cells. HeLa cells have the capacity to continuously grow without any control. The HeLa cells showed scientists a way to understand the barriers which separate normal cells from their cancer cells. These same barriers now appear to be intimately connected to the process of aging. Scientists hope that the reasons which make the cancer cells immortal, may also show a way to extend life span. Subsequently, scientists have found many similarities between the biological processes of cancer and aging [12].

Although, aging remains the major cause of mortality, scientists feel that anti-aging and life-extension research has been greatly underfunded. Scientists have been able to extend the life span of mice to about 2.5 times greater than its normal span, and to about 10 times its normal span in yeast and nematodes. This has given hope for delaying senescence and reversing aging, or at least significantly delaying aging in humans.

Rejuvenation

Progressive damage to macromolecules, cells, tissues, and organs is responsible for aging. Rejuvenation processes attempt to prevent, or repair such damage.

Actually, rejuvenation means delaying or reversing aging and is distinctly different from mere extension of the life span. Rejuvenation requires the repair of the aging-related damage, and also the replacement of damaged tissue with new tissue. Rejuvenation may result in life extension, but not vice versa. The rejuvenation process is very close to the Ayurveda concept of *Rasayana*, which actually incorporates both rejuvenation and regeneration.

It is known that levels of many important hormones, such as human growth hormone, testosterone, thyroid, and estrogen/progesterone; as well as erythropoietin, insulin, dehydroepiandrosterone (DHEA), and melatonin decline with age. Scientists hypothesized that replacing these hormones might be beneficial in preventing, repairing, and restoring aging-related damage to many body tissues, organs, and functions. In history, a few such experiments have been successful in partially rejuvenating laboratory animals.

In the 1920s, a French surgeon, Serge Voronoff, tried grafting monkey testicle tissue onto the testicles of men. He considered this to be an option for rejuvenation therapy. Later, he was not able to scientifically support his claims. Stem cell regenerative medicine involves implantation of stem cells from culture into an existing tissue structure. Currently, scientists are also working on experiments involving genetic repair using a retrovirus to insert a new gene at specific positions on chromosomes. It is essential to get stem cells from the bone marrow for conducting these experiments.

A pioneering work in regenerative therapy has been done by a leading English biomedical gerontologist, Aubrey de Grey. Dr de Grey has proposed Strategies for Engineered Negligible Senescence (SENS). Dr Grey cofounded the SENS Foundation to expedite progress in regenerative medicine. He is also editor-in-chief of the peer-reviewed journal, *Rejuvenation Research*. According to Dr Grey, cell or tissue loss can be repaired, or even reversed by various means. For example, suitable exercise can repair or regain muscle tissue mass. Many other tissues may require specific growth factors, or stem cells to regrow and repair. It is possible to eliminate aged cells with the help of the immune system, or they can be destroyed by gene therapy. According to Dr Grey, even macromolecular damage, or cross-linking can be reversed by specific drugs or enzymes. Vaccination can be used to remove extracellular waste products such as amyloid proteins responsible for dementia and Alzheimer disease.

Regeneration

Regeneration generally means the regrowth; regenerative medicine deals with the science of tissue regeneration or tissue engineering [13]. We, as humans, have a very limited capacity to regenerate, except in some organs, such as the liver. Human skin is constantly being renewed and repaired. Interestingly, many animals can regenerate complex body parts, even after significant damage, or actual loss of a significant amount of tissue. For instance, the flatworm can regenerate the head or tail from either section. Fish can regenerate parts of the brain, eye, kidney, heart, and fins. Frogs can regenerate limbs, tail, and brain and eye tissue only in the early developmental stage, as tadpoles. Salamanders can regenerate tissues of the limbs, heart, tail, brain, and eyes throughout life as well as organs like the kidney, brain, and spinal cord [14].

A small animal species, known as the *Hydra*, has an extensive capacity to regenerate by decoupling the aging process. Scientists feel that *Hydra* are able to escape aging as a result of high levels of cell proliferation; and that their regenerative ability is probably due to effective stem cell maintenance and telomere dynamics, which prevent damage, and protect against disease and pathogens [15]. The underlying molecular mechanisms responsible for the unlimited life span of the *Hydra* have yet to be explored. Scientists have identified a factor known as FoxO, putatively responsible for continuous self-renewal. It was observed that overexpression of FoxO factor increased stem cell proliferation. Downregulation of FoxO resulted in a drastic reduction of growth rate [16].

Some animals, like the *Hydra*, can regenerate their cells and tissues. It is not clearly known when, in the process of evolution, we lost these capabilities. It is hypothesized that blastema, a kind of stem cell which accumulates at an injury site, may be involved in the regenerative process. We still do not know if blastema is a kind of multipotent stem cell. Specific tissues like muscle, nerve, and skin seem to have specific stem cells. Therefore, in attempt to clone and engineer human tissues, the biology of adult tissue regeneration remains very challenging. In an adult human, the wound repair process commonly leads to a scar, with a mass of fibrotic tissue. However, in a process resembling regeneration, injured tissues can be recreated, avoiding scar formation. Clearly, such regenerating ability is present in humans during the very early stage of development. However, subsequently it is lost, as age advances. Some organisms seem to retain the ability to regenerate tissue throughout adult life. If we understand the underlying biological mechanisms, so that we can regain the lost capabilities of regeneration, it would indeed be a major breakthrough in the history of biomedical sciences [17]. We feel that tissue-specific treatments and medicines used in Ayurveda are worth exploring. These *Rasayana* drugs will be discussed in later sections.

Peter Reddien, a biologist from Massachusetts Institute of Technology, has shown that even a single pluripotent stem cell of the flatworm can regenerate a whole animal. However, many animals cannot use pluripotent cells for regeneration. Many animal tissues like muscle, nerve, or skin have their own set of stem cells. Normally, a muscle stem cell is not able to differentiate as a skin cell, and vice versa. Such tissue-specific, multipotent stem cells are probably present in humans. The reason why some stem cells are pluripotent and some are multipotent is not clearly known.

Differentiated cells can multiply to replace lost tissue, such as in the case of the zebrafish, where cardiomyocytes divide to replenish lost cardiac tissue. Such regenerative phenomenon has been observed in newborn-mouse hearts, however, as the mice grow and mature, these properties are lost. Regenerative research attempts to find ways to regain the capacity of differentiated cells to divide and produce new tissue in humans. Scientists are also trying to learn how injury can lead to stem cell regeneration of the missing part, instead of leading to the formation of scar tissue. Salamanders and frogs use tissue stem cells to regenerate a whole limb, whereas humans form scars. Based on current research, scientist feel that cell–cell signaling plays an important role in these processes, which apparently in mammals, is lost.

According to Dr Irving Weissman, a biologist from Stanford University School of Medicine, tissue-specific adult stem cells are responsible for the ability of mammals to regrow the tips of fingers or toes, which have been lost during injury or surgery. The mice model is genetically well-documented, and better suited to study the process of limb regeneration. Dr Weissman's group has shown that such damage is repaired by specialized adult stem cells, which otherwise would have remained stable as a specific tissue type. These cells initiate action only when the first sign of damage occurs, and start working independently to regenerate bone, skin, tendon, blood vessels, and nerves. The division of functions among these stem cells is restricted to the original tissue type [18].

In contrast, the blastema theory suggests that a new pluripotent cell type can be formed from specialized cells. These pluripotent cells can change their nature, and can regenerate into all the tissue types of the limb.

A German group, headed by stem cell scientist, Dr Elly Tanaka, observed similar regenerative nature of pluripotent cells in salamanders, but it was not known whether this is possible in mammals. This finding changes the current understanding of limb regeneration as originating from pluripotent blastema cells, to being the result of the action of tissue-specific stem and progenitor cells.

A house surgeon at Montreal General Hospital showed that the regrowth of bone, nail, and skin was possible even after amputation of the distal phalanx. This is one of the earliest examples of regeneration of a digit-tip in an adult human [19]. Earlier, it was reported that children up to the age of 10 or so can regrow the tip of the digit within a month or so after injury. Regenerative therapies have reported regaining the lost fingertip. There are also a few reports claiming that human toes, human ribs, human liver, and even human kidneys can be regenerated. It is known that the liver has the unique ability to regenerate from even a quarter portion, mainly because the hepatocytes have the required potency to induce proliferation of hepatocytes [20]. The regenerative capacity of the human kidney however, is very limited. The regeneration of the acute tubular component after injury is known. Scientists have also claimed that the regeneration of the glomerulus can occur. Recently, the role of kidney stem cells in tubular regeneration was shown. Studies on the capacity of bone marrow stem cells to differentiate into renal cells are in progress [21].

Several animals can regenerate heart damage, but in mammals, heart muscle cells, known as cardiomyocytes cannot proliferate. However, studies on rats with cardiac injury have shown that treatment with heparin-binding fibroblast growth factor 1, and mitogen activated protein kinase inhibitors can regenerate heart and improve cardiac function [22].

Admittedly, studies on the regeneration of mammalian cells, tissues, and organs have had very limited success, and seem too far from producing actual benefits in near term. However, we propose that traditional knowledge may help to expedite the much-desired breakthroughs. In this context, it will be interesting to study more about the concept of Ayurveda.

THE AYURVEDIC CONCEPT OF AGING

As we discussed earlier, aging is not a disease, but a normal stage of human development. The term used for "body" in Sanskrit is *shareera*, meaning an entity which decays and withers. Thus, body aging is considered a natural physiological process starting right from birth. According to Ayurveda, childhood, youth, and old age are dominated by *Kapha*, *Pitta*, and *Vata* respectively. Thus *Vata Dosha* predominates during old age. Among the three *Prakriti* types, *Kapha Prakriti* persons are likely to have longer life spans, while *Vata Prakriti* persons may have shorter life spans.

The Tokyo Centenarian Study has reported an association between personality types and longevity [23]. Seventy cognitively intact Japanese centenarians aged 100–106 years and 1812 people aged 60–84 years were included in this study. The results show that compared to controls, there was a higher level of openness, conscientiousness, and extraversion among the centenarians. Interestingly, according to Ayurveda, these traits belong to *Kapha Prakriti*.

The Concept of *Rasayana*

Rasayana is one of the eight major branches of Ayurveda. *Rasayana* comprise lifestyle, diet, and botanical, mineral, or animal products, which have specific properties to enhance growth, retard aging, stimulate immunity, and induce tissue regeneration. The word *Rasayana* means "the pathway for the essence of nutrition (*Rasa*) toward optimal nourishment of *Dhatu* elements (*ayana*)." Charaka defines *Rasayana* as the means of procuring the best qualities of different tissues. *Rasayana* therapy improves longevity, memory, intelligence, health, youth, complexion, voice, and motor and sensory strength. The Ayurvedic surgeon, Sushruta, mentions that *Rasayana* are capable of pacifying all sufferings.

Rasayana are classified on the basis of actions, processes, or materials. For instance, herbal and animal materials, and minerals are a part of medicinal *Rasayana*, while *Achara Rasayana* comprise ethical behavior, and guidance for healthy lifestyle.

Sharangdhara Samhita describes which attributes of life are degraded as age progresses (Figure 10.2). *Charaka Samhita* has a section, *Vayasthapana*, meaning "longevity promoters," which describes many drugs for longevity. A few examples of medicinal plants in the *Rasayana* category include Guduchi (*Tinospora cordifolia*), Haritaki (*Terminalia chebula*), Amalaki (*Embelica officinale*), Mukta (pearl), Shweta (*Clitoria ternatea*), Jeevanti (*Leptadenia reticulata*), Shatavari (*Asparagus racemosus*), Mandukaparni (*Centella asiatica*), Shaliparni (*Desmodium gangeticum*), and Punarnava (*Boerhavia diffusa*). Many of these plants are used in longevity-promoting products such as Chyavanprash and Brahma *Rasayana* (BR). The following section gives importance of *Rasayana* in healthy aging, rejuvenation, and regeneration.

A number of formulations from botanical, animal, and mineral substances are included as *Rasayana* medicines. It was believed that the *Rasayana* drugs

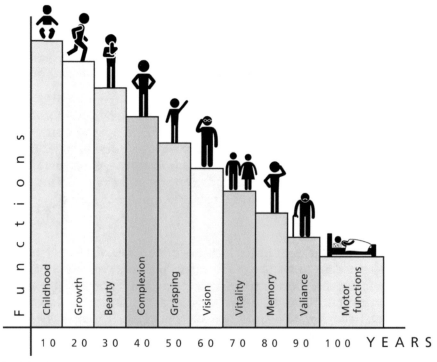

FIGURE 10.2

Ayurveda description of degrading capacities due to aging.

produced pharmacological effects through inherent qualities. Rejuvenation therapy is more than just the administration of medicinal formulations. The full benefits of therapy cannot be achieved unless the body and mind are pure, and the individual follows a code of virtuous conduct.

Rasayana therapy is a specialized practice involving rejuvenation recipes and dietary regimens; and specific conduct, behavior, and drugs for promoting health. A study of *Rasayana* therapy is difficult under experimental conditions. A large number of formulations with the potential to increase life span, enhance memory, sharpen intellect, increase resistance to disease, brighten complexion, promote digestion, speed up tissue repair, strengthen the body, and minimize the ill effects of old age are available—all of which need to be researched. *Rasayana* drugs have rejuvenation, antiaging, and immunomodulatory properties.

Aging and Immunomodulation

Several studies on *Rasayana* have made important contributions to ethnopharmacology, especially as it relates to inflammation and immunopharmacology. Some interesting research findings are available, and indicate the usefulness of *Rasayana* drugs on obesity [24], anxiety [25], arthritis [26,27], inflammation [28,29], immune stimulation [30], and in natural product drug discovery [31]. These studies clearly show the importance of traditional knowledge systems and ethnopharmacology in bioprospecting safe, and effective medicines and treatments [32].

Reviews of the current literature available on *Rasayanas* indicate that immunomodulation is the most studied property/activity as it relates to *Rasayana* [33]. Researchers have studied a few selected *Rasayana* plants, including *Withania somnifera* (Ashwagandha), *A. racemosus* (Shatavari), *T. cordifolia* (Guduchi), *Embelica officinale* (Amalaki), and *Semecarpus anacardium* (Bhallataka); and have reported immunomodulatory activity for various standardized extracts, and formulations prepared from these plants. Researchers also evaluated their potential as antistress [34], anxiolytic [25], adaptogenic [35], and immuno- [36] and myeloprotectants [37]. A study reported that ashwagandha is a better and safer drug than ginseng [38]. Work on antiaging activities of Ayurvedic medicines in topical application forms has also been encouraging [39,40]. Such evidence-based investigations are important to properly position Ayurvedic herbal medicine in the competitive international market.

The author's group has studied the pharmacodynamics of ashwagandha, shatavari, and guduchi in experimentally-induced tumors and infection in mouse models for immunomodulation and Th1–Th2 balance [41]. Studies on in vivo cytokine modulation, using flow cytometry, showed that a 100 mg/kg dose resulted in a significant Th1 response (IL-2, IFN-g), in comparison to levamisole and cyclosporin. In immune-suppressed animals, ashwagandha exhibited significant dose-dependent potentiation of cellular and humoral immune response, comparable to levamisole, and faster recovery of CD4+ T cells percentages compared to the control, and cyclosporin [42]. The study indicated immunostasis activity, and suggests its use where Th1–Th2 modulation is required. This activity has shown significant benefits as immunoadjuvants, when studied on mortality and morbidity associated with DPT, and the potentiating, protective effects of vaccine.

Newer vaccines such as subunit and DNA vaccines are weakly immunogenic and require adjuvants. We hypothesized that *Rasayana* may offer better and safer immunodrugs that can be used as adjuvants in vaccines and cancer treatments [43]. Researchers from our group used a modified Kendrick test that involved the challenge of live pertussis cells intracerebrally, where a significant increase in antibody titer, reduced mortality, and improvement in overall health was observed [44]. This observation has immense importance in the vaccine industry, to obtain more efficient and sustained immunostimulation resulting in increased yield of immune sera, and immunobiologicals [44]. These studies indicate applications of *Rasayana* as potential immunoadjuvants, that also offer direct therapeutic benefits resulting in lower morbidity and mortality [45]. A project to develop a vaccine adjuvant was successfully completed in collaboration with Savitribai Phule Pune University and the Serum Institute of India. This effort led to four Indian and one United States patent in the area of vaccine adjuvant [46–48]. Using an Ayurveda-inspired reverse pharmacology approach, bioactive fractions have been developed as potential vaccine adjuvants.

Most cancer chemotherapeutic agents are immune-suppressants and cytotoxic. Researchers have used cyclophosphamide-induced immunosuppression to screen plant-derived drugs for anticancer and cyto-protective potential, and to demonstrate myelo- and immunoprotective activity in ascitic sarcoma-bearing animals. Researchers carried out activity-related extractions to identify the best performing candidate drugs. This work resulted in a United States patent in the area of cancer adjuvants. This product will have importance in cancer therapeutics, especially to counter untoward effects of chemotherapy, without compromising the anticancer activity of chemotherapy [49]. These attempts to determine the scientific evidence-base for herbal medicines are important and exemplary.

Immune response requires the timely interplay of multiple cell types, proteins, tissues, and organs within specific microenvironments to maintain immune homeostasis. Through a series of steps called the immune response, the immune system attacks organisms and substances that invade body and cause disease. The repertoire of immune defenses used by the body in defending against infections and other internal changes such as cancer, involves humoral, as well as cellular immunity, and the regulators of the immune system, such as cytokines. The most likely hypothesis explaining the apparently diverse actions of *Rasayana* was that the plants were modulating as endogenous systems of the body, setting into motion a cascade of events leading to multiple effects in the immune system.

Immunomodulating agents that are free from side effects, and those that can be administered for long duration—if possible throughout life—to obtain a continuous immune activation, are highly desirable for prevention of diseases. The significance of *Rasayana* as immunomodulating agents compared to other conventional immunomodulators, is that they activate immune function without altering the other basic parameters of the body.

Humoral immunity, in terms of antibody production and cell-mediated immunity, in terms of delayed type hypersensitivity, has been stimulated by various *Rasayana* herbs and formulations. Percentage neutrophil adhesion, phagocytic activity, and Th1–Th2 balance is used to measure the innate immune system activation by the *Rasayana*, which also have the potential to be used as vaccine adjuvants. *A. racemosus* and *W. somnifera* enhanced protective immune response by improving serological and hematological parameters, following DPT vaccination.

Free Radical Scavenging

Free radicals are charged, metabolic by-products, which can attack cells, disrupting cellular membrane and mitochondria, and react with and damage nucleic acids, proteins, and enzymes present in the body. Free radicals play an important role in aging, as shown in several animal studies. Oxidative stress occurs as a result of high metabolic rates, increased oxygen tension, and the presence of redox-active xenobiotics that compromise normal cellular antioxidants. Superoxide is often referred to as the "primary ROS," as most of the other ROS and RNS arise from it. Apart from aging, free radicals are also known to be involved in more than a 100 diseases. The disease preventive action of *Rasayana* has been researched by studying the inhibition of free radical stress. Indian Pharmacologists Govindarajan et al. have described the role of Ayurvedic *Rasayana* herbs in disease management through antioxidant approach [50].

Rasayana plants like *W. somnifera* and *E. officinale* have been studied for their free-radical scavenging properties in cold stress-induced rodent models. Enzymes like superoxide dismutase (SOD), glutathione peroxidase, and glutathione reductase and catalase have been used to assess the antioxidant potential of *Rasayana* drugs. Mangathayaru et al. have reported the doxorubicin-induced DNA damage protective activity of *Cynodon dactylon* (Durva) in spectral studies in vitro [51]. These studies indicate that *Rasayana* drugs hold potential in decelerating the aging process and imparting youthfulness or longevity to the individual.

Neuro-cognitive Improvement

Diseases like Parkinson's disease, Alzheimer disease, and Huntington's chorea occur as a result of neurodegenerative processes. Ayurveda texts describe a set of rejuvenative measures to impart biological sustenance to the bodily tissues. Some of the *Rasayana* are organ and tissue-specific. Those specific to brain tissue are called *Medhya Rasayana*, or brain tonics. Such *Rasayana* retard brain aging, and help in the regeneration of neural tissues—besides producing antistress, adaptogenic, and memory enhancing effect. The popular *Medhya Rasayana* are Shankhapuspi (*Convolvulus pluricaulis*), Ashwagandha (*W. somnifera*), Brahmi (*Bacopa monnieri*), and Mandukaparni (*C. asiatica*).

The use of *W. somnifera* in various CNS disorders, particularly its indication in epilepsy, stress, and in neurodegenerative diseases such as Parkinson's and Alzheimer disorders, tardive dyskinesia, cerebral ischemia, and even in the management of drug addiction has been reviewed by Kulkarni and Dhir [52]. The investigations by Bhattacharya et al. support the use of *W. somnifera* as a mood stabilizer in clinical conditions of anxiety and depression [53]. *A. racemosus* significantly reversed scopolamine and sodium nitrite-induced increase in transfer latency, indicating antiamnesic activity. Further, *A. racemosus* inhibited acetylcholinesterase enzyme in specific brain regions (prefrontal cortex, hippocampus, and hypothalamus), in a dose-dependent manner [54].

BR, which contains *Clitoria ternatea*, *Acorus calamus*, and *E. officinalis*, has been described in Ayurveda as improving intelligence, memory power, and immunity [27]. When BR was administered in the elderly it has shown a memory-restorative effect useful in the treatment of dementia by the decreasing levels of whole-brain acetylcholinesterase activity in mice models.

Radioprotective Effect

In spite of adverse effects, ionizing radiation—alone or in combination with other therapies—is one of the most commonly used treatments for cancer. The early or acute effects of irradiation result from the death of a large number of cells in tissues, with a rapid turnover rate. These include effects on the epidermal layer or skin, gastrointestinal epithelium, and in the hematopoietic system. Injury resulting from the irradiation of biological tissue is a consequence of the transfer of radiation energy to critical macromolecules or, indirectly, through the action of free radicals. Antioxidant enzymes act as the first-line defense against free radicals. The most important enzymes include SOD, catalase, and glutathione peroxidase (GPx).

The radioprotective agents are chemicals that reduce the biological effects of radiation by the scavenging of free radicals or by repairing radiation injury. Radiation exposure-induced increase in serum and liver lipid peroxides was significantly reduced in mice models by treatment with BR, while an increase in SOD, catalase, and GSH levels was observed. The results indicate that BR could ameliorate the oxidative damage produced in the body by radiation, and may be useful as an adjuvant during radiation therapy.

The administration of Triphala—a combination of *E. officinalis*, *T. bellerica*, and *T. chebula*—resulted in an increase in the radiation tolerance. Triphala provided protection against both gastrointestinal and hemopoetic death through free radical scavenging activity [55].

Aphrodisiac Activity

Throughout the ages, men and women have incessantly pursued all means to enhance, maintain, and bring back their sexual ability, or to stimulate the sexual desire for the opposite sex. One of the most common methods has been the use of aphrodisiacs. Herbal medicines are a major source of aphrodisiacs, and have been used worldwide for thousands of years by different cultures and civilizations to enhance sexual performance. Natural herbs that are popularly considered to have properties to enhance potency and sexual functioning, or to improve sexual performance, currently make up the large segment of current herbal market. On a broader scale, herbal aphrodisiacs may also include products that have adaptogenic, tonic, revitalizing, and rejuvenating properties.

A special class of *Rasayana* drugs known as *vrishya* (vitality promoter) and *vajikarana* (libido enhancer) are aphrodisiacs drugs, which improve sexual potency. They are specially recommended to people suffering from sexual insufficiency and people in advancing age; those losing interest in sexual act, or failing in sexual performance. Numerous plants are indicated as having been used as sexual stimulants. *Anacyclus pyrethrum*, *A. racemosus*, *Chlorophytum borivilianum*, and *Curculigo orchioides* are some of the examples described in modern scientific literature.

In a study conducted by Sharma et al. petroleum ether extract of *A. pyrethrum* roots showed equal aphrodisiac activity with testosterone. The herb was found to be effective in rats tested after a lapse of 7 and 15 days of discontinuation of treatment [56]. Thus, the suggestion is that the drug has a prolonged effect and capacitates the treated rats for improved sexual potential. Aqueous extracts of *A. racemosus*, *C. borivilianum* roots, and rhizomes of *C. orchioides* had pronounced anabolic effect, and showed significant increase in the sexual behavior of animals as reflected by a reduction of mount latency, ejaculation latency, postejaculatory latency, intromission latency, and an increase of mount frequency. The observed effects appear to be attributable to the testosterone-like effects of the extracts. Nitric oxide-based intervention may also be involved as observable from the improved penile erection [57]. These reports authenticate the claim for the usefulness of these herbs as aphrodisiacs, and provide a scientific basis for their purported traditional usage as *Vajikarana*.

Antiulcer Activity

Peptic ulcer disease (PUD), encompassing gastric and duodenal ulcer, is the most prevalent intestinal disorder. The pathophysiology of PUD involves an imbalance between offensive (acid, pepsin, and *H. pylori*) and defensive factors (mucin, prostaglandin, bicarbonate, nitric oxide, and growth factors). There are two main approaches for treating peptic ulcer. The first deals with reducing the production of gastric acid and the second with reinforcing gastric mucosal protection. Current drug therapies for PUD include proton pump inhibitors, histamine receptor blockers, drugs affecting the mucosal barrier, and prostaglandin analogues. However, the clinical evaluation of these drugs showed development of tolerance, incidence of relapses, and side effects that make their efficacy arguable. This has been the rationale for the development of new antiulcer drugs, which includes herbal drugs. *A. racemosus*, *B. monniera*, *C. asiatica*, and *C. pluricaulis* are some of the *Rasayana* drugs which have been characterized for their gastroduodenal ulcer protective activity.

A. racemosus showed significant protection against acute gastric ulcers induced by cold restraint stress (CRS), pyloric ligation, aspirin plus pyloric ligation, and duodenal ulcers induced by cysteamine, and acetic acid in animal models [30]. However, *A. racemosus* was ineffective against aspirin- and ethanol-induced gastric ulcers. *B. monniera*, *C. pluricaulis*, and *C. asiatica* showed dose-dependent antiulcerogenic effects on various gastric ulcer animal models induced by ethanol, aspirin, CRS, and pylorus ligation [31]. Treatment with *C. pluricaulis* and *C. asiatica* showed no effect on acid–pepsin secretion or increased mucin secretion, while it decreased cell shedding with no effect on cell proliferation, while *B. monniera* showed significant antioxidant effect per se, in stressed animals. All four *Rasayana* herbs studied for their protective effect on gastric ulceration have shown that they act by strengthening the mucosal defensive factors, mucin secretion, life span of mucosal cells, and glycoproteins. They did not show any significant effect on the mucosal offensive factors such as acid–pepsin.

Anticancer Activity

Cancer is the abnormal growth of cells in the human body, the overgrowth of which can ultimately lead to death. Many genetic changes, behavioral defects, and infections are linked to the development of cancer. Cancer cells usually invade and destroy normal cells. Every year, millions of people are diagnosed with cancer, leading to death. Serious side effects, and the toxicity of chemotherapy and radiation therapy to normal cells or tissues, drive many cancer patients to seek alternative and/or complementary methods of treatment.

In 1997, Menon et al. studied the effect of five *Rasayana* formulations, BR, *Aswagandha Rasayana* (AR), *Narasimha Rasayana*, *Amrithaprasam* (AP), chyavanaprash (CP) and extract of *E. officinalis* for their antimetastatic activity, using B16F-10 melanoma cells in C57BL/6 mice. On oral administration, BR and AR significantly reduced the lung nodule formation, reduced hydroxyproline and serum sialic acid indicating the inhibition of metastasis. Further, BR and AR also increased the life span of the animal. Other *Rasayana* did not show any significant activity. BR and AR stimulated antibody-dependent, complement-mediated, tumor cell lysis, and natural killer cell activity, while AR was also found to activate macrophages [58].

The anticancer properties of *Rasayana* has been attributed to their immuno-stimulating nature. Administration of *Rasayana* was found to significantly enhance the proliferation of spleen and bone marrow cells in mice models—especially in the presence of mitogen. Esterase activity was found to be enhanced in bone marrow cells, indicating increased maturation of cells of lymphoid lineage. *Rasayana* also enhanced humoral immune response, as seen from the increased number of antibody-forming cells and circulating antibody titer.

BR treatment to Copenhagen rats injected with MAT-LyLu cells resulted in a decreased palpable tumor incidence, a delay in the tumor occurrence, lower mean tumor volumes—by as much as 14% to 35%—and significant reduction in tumor weight and lung metastasis, in comparison to untreated controls. BR treatment showed a significant reduction in Factor VIII expression, compared to the control, indicating reduced angiogenesis. BR treated tumor specimens showed a decrease in the proangiogenic factors. Methanolic extract of BR was found to inhibit the proliferation, tube formation, cell migration, and attachment of HUVEC on matrigel in a dose-dependent manner. *Rasayana* formulations are inexpensive preparations that have little or no adverse side effects and hold the potential to be a lead chemopreventive agent in cancer management [59].

Treatment of ascitic sarcoma-bearing mice with a formulation of total extracts of *W. somnifera* and *T. cordifolia*, and alkaloid-free, polar fraction of *W. somnifera* resulted in immunoprotection, and protection from Cyclo-phosphamide-induced myeloprotection, by significantly increasing white cell counts and hemolytic antibody titers. Treatment with these candidate drugs will be important in the development of a supportive treatment to be used with cancer chemotherapy [37].

Human beings are desirous of longevity, youthfulness, and health. Realization of these goals is possible by promoting rejuvenation and the healing and regeneration of living tissues in the body. *Rasayana* is not just a drug therapy, but it is a specialized procedure practiced in the form of rejuvenative recipes, dietary regimens, and special health-promoting conduct and behavior. Scientific studies have proven the efficacious role of *Rasayana* drugs in preventive medicine and in the management of chronic degenerative diseases. Unlike modern drugs, they may not possess sharply defined pharmaceutical activities if used in holistic traditional forms; hence they may be treated as soft and safe medications that can be used as dietary supplements taken daily to support healthy living.

Regeneration Activity

Several *Rasayana* medicinal plants have shown tissue regenerative and protective activities. A short review of few important examples may be useful at this stage. *E. officinale* known as *Amalaki*, is one of the main ingredients of chyavanprash, and has been shown to possess chondroprotective activity, through inhibition of hyaluronidase and collagenase-type 2 activities in vitro. Its combination with *Shorea robusta* resin and Ayurvedic zinc have shown activities in wound healing, fractures, anemia, corneal ulcers, brain, and deoxyribonucleic acid (DNA) damage in experimental models. *Amalaki Rasayana* has effectively demonstrated reduction in DNA damage in brain cells, demonstrating its genomic stability in neurons and astrocytes. The same formulation demonstrated an increase in median life span and starvation resistance in a *Drosophila melanogaster* mode. A recent study on the traditional formulation, known as *Dhanvantar Kashaya* (a decoction of herbs having regeneration property), has demonstrated beneficial activity on Wharton jelly mesenchymal stem cells. The decoction increased the proliferation rate, decreased the turnover time, and delayed senescence. Curcumin had been shown to stimulate developmental and adult hippocampal neurogenesis, and enhance neural plasticity and repair.

RASAYANA AND STEM CELLS

Cell differentiation is an essential process for the development, growth, reproduction, and longevity of all multicellular organisms. Regulation of differentiation has remained a matter of serious investigation during past four decades. Recent discoveries consisting of novel experiments are changing many established concepts. Experts agree there is a need to rethink our suppositions about stem and differentiated cells [60].

Against this background, it will be interesting to understand Ayurvedic concepts of development, rejuvenation, regeneration, and longevity. Now it is known that even adult stem cells have the capacity to divide and the potential for differentiation. Currently, intentionally directed differentiation of human pluripotent stem cells into organ and tissues remains a major challenge. We feel that a better understanding of the concept of *Dhatu* (tissue) specific *Rasayana* from Ayurveda may help to address this challenge.

There is a need to undertake research on the effect of *Dhatu*-specific *Rasayana* on adult or embryonic stem cells or stem cell lines. A systems approach needs to be followed by stem cell biologists and Ayurveda experts to understand complex molecular pathways through which *Rasayana* acts, and the way in which it contributes to steering cellular differentiation pathways in a predictable manner.

As discussed earlier, *Rasayana* are primarily useful for nourishing various tissues in the body. According to Ayurveda there are seven *Dhatu* (tissues). The nourishment to *Dhatu* is expected to be provided in their sequence of development. Thus, the tissue sequence as per Ayurveda is *Rasa* (~protoplasm), *Rakta* (~blood), *Mamsa* (~muscle), *Meda* (~lipids), *Asthi* (~bones), *Majja* (~nerves and bone marrow), and *Shukra* (~sperm). Respective *Dhatu* are thought to be dominant in the respective tissues.

It was hypothesized that specific *Dhatu* may constitute stem cells in a particular tissue, which can develop into the next *Dhatu* in sequence. To support our hypothesis, we rely on a few known observations. Firstly, the adult bone marrow stromal cells can differentiate into neural cells in vitro by treating them in the presence of epidermal growth factor or brain-derived neurotrophic factor (BDNF). Second, human adipose tissue obtained from liposuction procedures has been used to isolate a fibroblast-like cell population, called processed lipoaspirate (LPA) cells. In vitro studies with LPA cells demonstrated differentiation into adipogenic, chrondogenic, myogenic, and osteogenic cells. We feel that the Ayurvedic concept of *Dhatu* needs to be explored from this perspective. *Rasayana* can change the quality and quantity of *Dhatu*, and are recommended for nourishing specific *Dhatu*. Thus, if nervous tissue is depleted, *Medhya Rasayana*, which is targeted to nervous tissue can be used. When heart tissue is depleted, heart specific *Hridya Rasayana* can be used (Figure 10.3).

According to Ayurveda concepts, the *Dhatu* concept seems to be the foundation of *Rasayana* therapy, tissue regeneration, and *Kayakalpa*. We hypothesize that treatment with *Dhatu*-specific *Rasayana* may show proliferative effects on stem cells, causing them to differentiate in a particular direction. Thus, *Hridya Rasayana* will expedite the differentiation of stem cells into cardiac cells; *Medhya Rasayana* will expedite the proliferation and differentiation of stem cells into nerve cells; and *Balya Rasayana* will expedite the proliferation and differentiation of stem cells into muscle cells.

Some preliminary experiments on *Medhya Rasayana* have supported our hypothesis. It was observed that when stem cells were treated with *Medhya Rasayana*, an expression of nestin was evident. It is known that nestin is an early marker of neuronal stem cells differentiation. The expression of nestin indicates that *Medhya Rasayana* has triggered neuronal stem cell differentiation. We hope that *Dhatu*-specific *Rasayana* can play a major role in steering cellular differentiation pathways in a predictable manner. This may have a large impact on translational research, regenerative biology, and stem cell therapy [61].

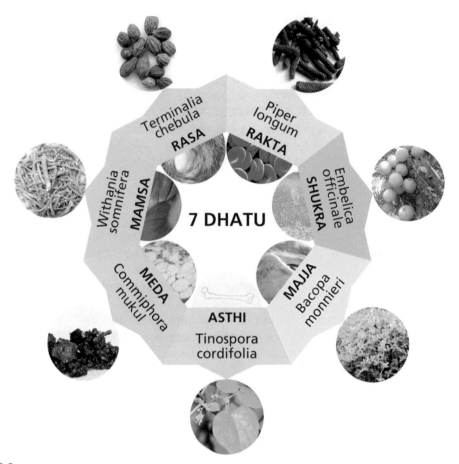

FIGURE 10.3

Rasayana formulations have *Dhatu*-specific effects. The figure shows botanical drugs acting on seven *Dhatu*. This is one of the major criteria for selection of individualized *Rasayana*.

REJUVENATION AND *KAYAKALPA*

Kaya is "body," and *Kalpa* means "rejuvenation regime." *Kayakalpa* literally means "a regime for total rejuvenation of the body." The term also suggests ways and means of prolonging life. It is actually considered as rebirth of the body, where a person undergoing the process is supposed to gain new tissues, improved organs, and a recharged mind. *Kayakalpa* is thought of as a good procedure for health, vitality, longevity, and higher consciousness. Ayurveda has described a detailed process of *Rasayana* therapy known as *Kuti Praveshik Rasayana* (an indoor therapy for longevity), which is a controlled intervention. Here, a volunteer is kept in total isolation in a scientifically-designed cottage with three concentric courts. The person then follows a strict regime as described by

Ayurveda. This includes fasting, consumption of specific *Rasayana*, and meditation. Several historical notes are available where *Kayakalpa* was performed on individuals. Some report success, while some have not. A book by T.S. Ananthmurthy, *Biography of Shriman Tapasviji Maharaj*, mentions that this great soul, who underwent *Kayakalpa Rasayana*, lived for 185 years. He also administered *Kayakalpa* to Pandit Madan Mohan Malveeya, founder and vice chancellor of Banaras Hindu University, although this was not very successful.

Indications, inclusion criteria, medical ethics, and objectives of the treatment are important factors to be considered for *Kayakalpa* treatment. Total rejuvenation of body with new teeth, darker hair, supple musculature, improved eyesight, skin complexion, and regaining the strength of youth are documented in the biography of Tapasviji Maharaj. Amalaki (*E. officinale*) and Haritaki (*T. chebula*) are considered to be promising drugs for rejuvenation. Gujarat Ayurveda University has created a typical hut as mentioned in the ancient texts of Ayurveda. *Kayakalpa* involves various stages, such as purifying the body by fasting and some *Panchakarma* processes. This is followed by nourishment—giving specific medicinal *Rasayana* preparations. During the process, lifestyle is also strictly controlled, and involves use of Yoga and meditation.

Kayakalpa consists of a strict regime where an individual undergoing *Kayakalpa* is kept in a *Kuti* (cottage) or a specially designed cell. As a preparatory process, an individual undergoes certain body purification protocols, including *Panchakarma* or five cleansing processes: medical emesis, purgation, oleation, steam bath, sauna, and massage. The individual undergoing *Kayakalpa* is made to fast, followed by administration of a *Rasayana*-rich, special diet specifically designed—depending on *Prakriti* and other parameters (Figure 10.4).

At one time *Kayakalpa* was seen as a fantasy or myth, and scientists were skeptical about the regeneration capability of human cells, tissues, and organs. However, given new research on aging, regeneration, and stem cells, these possibilities seem to be practical and possible. For example, the Avatar project describes steps to move toward human immortality with a robotic interface. (http://2045.com/)

EXTENSION OF LIFE SPAN

Extension of life span has always been alluring to humanity. And why just to humanity? Actually, according to evolutionary biologists, all living species have only two options—either be immortal or reproduce. Mortality is the law of nature—applicable to animal and vegetative life forms. However, people aspire to live longer, or even might want to become immortal. In Indian mythology there are examples of seven immortal men known as *Chiranjeevi*. Scientists have also been intrigued by this idea. Several studies on small animals such as

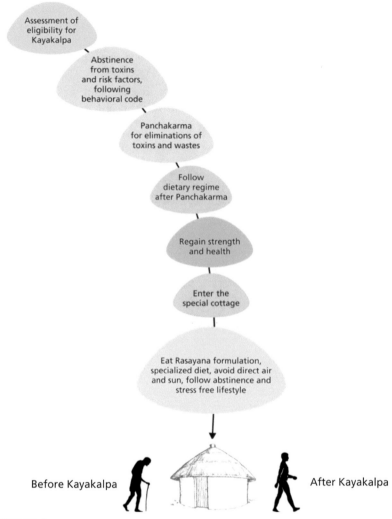

FIGURE 10.4

The figure explains process of *Kayakalpa*. The volunteer has to be free from diseases and should undergo sessions for removal of toxins and accumulated wastes. The person is expected to stay in a specially designed cottage having walls built as three concentric circles. This arrangement protects the person from direct sunlight and air. The most important therapy in the cottage is consumption of *Rasayana*, controlled diet, and lifestyle that includes meditation and refraining from physical and mental stress. The *Rasayana* is administered in escalating dose that may replace usual diet. This therapy is to be followed for up to 3 months. This process is under continuous observation of experienced Vaidya.

C. elegans and insects such as drosophila, have shown that it is possible to extend life span.

The Methuselah Fly presents a ground-breaking project on the biology of aging. In this project, the mutant line of *D. melanogaster* known as methuselah (mth), was shown to display about a 35% increase in average life span. The mth flies also showed increased resistance to stress due to starvation, high temperature, and free radicals [62]. In another study, scientists also showed that overexpression of a single gene of superoxide dismutase, can extend the life span up to 40% longer than usual. The life span extension observed in these flies is possibly due to enhanced reactive oxygen metabolism [63]. These studies support the importance of antioxidant enzymes like SOD and catalyze in aging, as well as the theory of oxidative stress as the main pathway for the aging process.

Many antioxidants, vitamins, fruits, vegetables, and herbal drugs have been studied for their roles in the extension of life span. This includes cranberry [64], apple [65], blueberry, cinnamon, green and black tea, pomegranate, sesame, curcumin, morin, pycnogenol, quercetin, and taxifolin. Critical analysis does not support the idea that isolated phytonutrient antioxidants and anti-inflammatories are potential longevity therapeutics. However, consumption of whole fruits and vegetables seems to be associated with enhanced health and life span.

Rasayana medicines like *E. officinale*, and *W. sominfera* [66] have shown some promise. Preliminary experiments on life span extension by appropriately modified *Rasayana* in animal models, have indicated an approximately 50% extension in the normal life span [67]. Studies on two Ayurvedic formulations have revealed formulation-specific effects on several parameters of the fly's life, in general agreement with their recommended human usages in Ayurvedic practices [68]. However, most of these studies are on small animals, involving very preliminary experimental methodology, and are insufficient to support a definitive conclusion for their putative effects in humans.

Calorie reduction (CR) is considered to be responsible for slowing cellular growth, and extending the time between cell divisions. Some advantages of CR have been discussed in Chapter 7. Here, a few specific advantages focused on aging and longevity are discussed in the context of longevity. CR is known to extend longevity, probably through sirtuins, which is a family of nicotinamide adenine dinucleotide-dependent enzymes [69]. A few studies on CR have shown an increase in mitochondrial activity in organisms, which is associated with salutary effects [70]. Sirtuins are also shown to regulate energy metabolism and response to caloric restriction in mice [71]. CR works on many diverse species, including mice, yeast, and *Drosophila*, and also appears to increase life span in primates, as was shown by a study of the United States' National Institute of Health. A few studies have also suggested that, in combination, CR and rapamycin may have a synergistic effect in extending longevity, that is not to be had by a treatment with either, alone [72].

Drug companies are currently searching for ways to mimic the life span-extending effects of caloric restriction without having to severely reduce food consumption. An important and promising compound seems to be rapamycin, which is thought to play a major role in life span extension and cancer prevention. Mammalian target of rapamycin (mTOR), also known as serine or threonine kinase, has been shown to play an important role in this regulation. Inhibition of the mTOR signaling pathway by genetic or pharmacological intervention has been shown to extend life span in invertebrates, including yeast, nematodes, and fruit flies, but its role in mammalian species is not known. American cell biologists Harrison et al. reported that rapamycin, an inhibitor of the mTOR pathway, can extend the life span of mice by about 14%. Scientists think that rapamycin may extend life span by postponing death from cancer and also by retarding mechanisms of aging [73].

Several drugs and biological and nutraceutical compounds claim retardation of aging in multiple species, including mammals. Inhibition of the mTOR pathway is also known to enhance the efficiency of reprogramming in induced pluripotent stem cells (iPSCs). The effects of longevity-promoting compounds on iPSC induction was studied, including resveratrol and fisetin as sirtuin activators; spermidine as an autophagy inducer; phosphoinositide 3-kinase as a PI3K inhibitor; and curcumin as an antioxidant. All these chemicals have been reported to promote somatic cell reprogramming [74]. Researchers have also identified a relationship between the aging of human muscle cells and nicotinamide phosphoribosyltransferase also known as Nampt or visfatin. A few studies have indicated that Nampt is a longevity protein that can give stress-resistant life to human cells [75].

YOGA AND LONGEVITY

The Yoga interventions practices are studied for their effects on metabolic rate, sympathetic activation, neuro-hormonal modifications, and aging mechanisms. A lower basal metabolic rate (BMR) is associated with delayed aging, which is achieved by Yoga practices. When observed after 6 months of Yoga practice, the BMR of the Yoga practitioners was found to be significantly lower than that of the non-yoga control group [76]. Another study from the same group of researchers demonstrated that Yoga training stabilized autonomic nervous system response, various stressors, and lowers metabolic rate [77].

Amounts of growth hormone (GH) and DHEA-sulfate in the body are reduced with advancing age. Yoga practitioners showed a significant rise in GH and DHEA-sulfate, even after just 12 weeks of Yoga practice, when compared with wait-listed controls. This research indicates possible mechanisms of the antiaging effects of Yoga. The rise in these hormones may also have a possibility of "reversing" aging, which is currently perceived as being in the realm of fantasy [78].

Aging is known to be associated with a decline in cardiovascular, autonomic function, and BDNF. A study exploring the role of yoga in improving age-related degenerative changes in cardiometabolic risk profile, autonomic function, stress, and BDNF levels, was done on healthy, active males who practiced Yoga 1 h daily, for 3 months. The study revealed that various crucial parameters, such as heart rate, blood pressure, cardiac output, myocardial oxygen consumption, and total cholesterol were significantly decreased following yogic practice. Levels of heart rate variability, total power and skin conductance were increased significantly following yogic practice. The study also reported decreased catecholamine, cortisol, and ACTH; and an increase in serotonin, dopamine, and BDNF following yogic practice. This study strongly supports the notion that yogic practices help in the prevention of age-related degeneration by changing cardiometabolic risk factors, autonomic function, and BDNF [79].

The importance of lifestyle is known to play a major role in the process of aging. While Ayurveda offers medicine, and a therapeutic-based approach to healthy aging, Yoga offers several body-mind related interventions for healthy aging. Regular practice of Yoga *asanas*, (physical posture-based stretching and relaxation exercise) greatly helps in physical fitness. The breathing, meditation, and relaxation techniques help in improving mental health. Regular practice of Yoga has shown to be beneficial in maintenance of health for all ages—especially for the elderly.

Some research highlighting the importance of lifestyle in healthy aging will be helpful in understanding the value of adapting a holistic lifestyle. For example, in the small island country Okinawa, situated between China and Japan, longevity is a feature of the community. People (90–100 years of age), whose work involves physical activity, work 7–8 h days. The literature is replete with references to the Okinawan's health. The contributing factors for Okinawan longevity can be summarized as follows: They live in harmony in *joint families*; and there is intergenerational togetherness in the family. They eat together, and as a rule, they only consume up to 80% of the amount dictated by their appetites. They do not consume processed food or red meat. They consume plenty of fish, fresh vegetables, and fruits. They exercise regularly, and even in their nineties, they work on their farms. They are relatively stress-free in their approach to life. The social milieu is conducive to tranquility, and peace of mind, as is exemplified by their eating together at table, and by their healthy socializing. All this is in total consistency with what Ayurveda prescribes as *Achara Rasayana*. In the truest sense, this is a behavioral medicine prescription for a healthy lifestyle. For that matter, any ancient tradition advocates similar doctrines. These basic rules have rewarded the Okinawan people with enviable health and wellness [80].

In another scientific study conducted by the famous cardiologist, Dean Ornish and Nobel Laureate, Elizabeth Blackburn, experimental evidence was provided to support the role of lifestyle modification in arresting the aging process [81]. Telomere shortness in human beings is used as marker of aging, disease, and premature morbidity. Telomerase enzyme is known to prevent shortening of telomeres at the end of linear chromosomes, which require the telomeres for their stability. This study observed that comprehensive lifestyle intervention including diet, activity, stress management, and social support were found to be associated with an increase in relative telomere length.

Evidently, Yoga practices combined with lifestyle changes advocated in Ayurveda, and with appropriate use of *Rasayana*, may offer the means for optimum longevity, and freedom from disabilities and chronic diseases in the elderly. In short, Ayurveda recommends wholesome food and a nutritious, healthy diet that is suitable for the particular individual. These recommendations consider individual nature or *Prakriti* and season. There should be movement, activity, and productivity. Exercise and the regular practice of Yoga and meditation restore the balance of the mind. Ayurveda also emphasizes the importance of diagnosing and treating disease in the early stages. In addition to treatments, Ayurveda stresses at having a personal philosophy of life and suggests that greed and the pursuit of material gain should be minimized, at least during old age.

According to Indian philosophy, life is divided in to four phases in a 100 year life span. The first 25 years of an individual's life is for gaining knowledge, and the capacity-building. The next 25 years are about family responsibilities. The next 25 years are about contributing to social and environmental responsibilities and concerns. The next 25 years make up the final phase of life, which is about individual liberation—the individual is expected to get connected to universal consciousness and to a commitment to ethics, values, and spirituality.

The elderly population should not be considered a burden, which is evident in most insurance policies, since they charge more when you are old. In contrast, the elderly should be considered as a valuable source of wisdom. At the same time, the elderly population must try to use their rich experience of life for the benefit of society. To protect the health and preserve the productivity of the elderly population is one of the greatest challenges facing humanity today.

POTENTIAL AREAS FOR FUTURE RESEARCH

Besides being mentioned in classical *Ayurveda* texts, and in the claims made by traditional physicians, it is noted that evidence from scientific studies is now emerging to demonstrate the benefits of *Rasayana* drugs. However, the definite lacuna in all the research publications reviewed in this article is any investigations of Ayurveda. This has led to superficial correlations, where the plant drugs have remained mere candidates for testing against selected pharmaco-logical reactions. Delving deeper into the Ayurvedic concept behind *Rasayana* therapy can help identify better candidates and models for study. For example, it is indicated that *Rasayana* drugs act through nutrition dynamics by improv-ing the quality of plasma (*Rasa* enhancing), by normalizing the digestion and metabolism (*Agni*), and by improved tissue perfusion of nutrients (*Srotasa*) at molecular level. None of the literature cited in this review has considered this logic of *Rasayana* action in their studies. Future research involving these episte-mological considerations may give deeper scientific insights into mechanisms of *Rasayana* actions.

The Evidence-base for the Efficacy of *Rasayana* Medications is threefold:

First, from literary and conceptual evidence; second, from experience-based evidence, and the long tradition of use; and third, from new, scientific clinical, and experimental studies. Currently, the latter are not adequate, but it is worth-while to point out all the evidence to provide leads for further studies. Drug discovery strategies based on natural products and traditional medicines are reemerging as attractive options. A reverse pharmacology approach, inspired by traditional medicine and Ayurveda, can offer a smart strategy for finding new drug candidates, and in doing so, facilitate the discovery process and the development of rational, synergistic, botanical formulations. The optimistic view toward enhancing longevity and advancing knowledge of the biology of aging can impart new insights for *Kayakalpa* therapy. This can be a high impact endeavor aimed at increasing our knowledge about healthy aging.

Ayurveda and Yoga are about integrating body, mind, and spirit. Both of these knowledge systems have a Vedic origin. The Vedic aspirations are to have a healthy life span of 100 years. Vedic literature proclaims "Let's witness the supreme spirit for one hundred autumns, let's live for one hundred autumns, and preserve our vital senses and organs for one hundred autumns." It further states that all senses should be functional and an individual should never lose self-respect and self-esteem until they draw their last breath. In a true sense, this is healthy aging. This aspiration can become reality through having a dis-ciplined lifestyle, tranquility of thought, and thinking beyond oneself, in one's general attitude toward self, and society. This can be described as spirituality, where a person thinks of vital philosophical questions such as "Who am I? Why am I here? What is my responsibility toward life? What is my relationship with the animal and vegetable kingdom, and the environment?"

In short, longevity or immortality cannot be attained through taking mirac-ulous drugs like rapamycin or any elixir. Longevity without a clear goal and purpose for life will be the greatest curse of mankind. The WHO has given a message by endorsing the importance of "adding life to years" rather than merely "adding years to life." In Indian mythology there is an example of

Ashvatthama—one of the seven immortal souls who is edgily wandering all over the world with bleeding wounds in search of healing, death, or liberation. We hope the new generation of god-men scientists, who are trying to clone human beings as a commercial venture, may learn from science fictions like Avatar—if not from the mythological *Ashvatthama*—to use their technological skills to improve the quality of life, rather than for its artificial extension. Death is a natural process. We need to prepare ourselves to face death as a natural, inevitable transformation. Yoga and Ayurveda can help us to live a healthy life, with purpose, and fearlessly accept death, with dignity. Amen!

REFERENCES

[1] Sebastiani P, Bae H, Sun FX, Andersen SL, Daw EW, Malovini A, et al. Meta-analysis of genetic variants associated with human exceptional longevity. Aging (Albany NY) 2013;5(9):653–61.

[2] Fonseca LM, Testoni I. The emergence of thanatology and current practice in death education. OMEGA–(Westport). 2011;64(2):157–69.

[3] Harman D. The aging process. Proc Natl Acad Sci USA 1981;78(11):7124–8.

[4] López-Otín C, Blasco MA, Partridge L, Serrano M, Kroemer G. The hallmarks of aging. Cell 2013;153(6):1194–217.

[5] Wickens AP. Ageing and the free radical theory. Respir Physiol 2001;128(3):379–91.

[6] Speakman JR, Selman C. The free-radical damage theory: accumulating evidence against a simple link of oxidative stress to ageing and lifespan. BioEssays 2011;33(4):255–9.

[7] Nowotny K, Jung T, Grune T, Höhn A. Accumulation of modified proteins and aggregate formation in aging. Exp Gerontol 2014;57:122–31.

[8] Jung T, Bader N, Grune T. Lipofuscin: formation, distribution, and metabolic consequences. Ann NY Acad Sci 2007;1119:97–111.

[9] Luevano-Contreras C, Chapman-Novakofski K. Dietary advanced glycation end products and aging. Nutrients 2010;2(12):1247–65.

[10] Tsakiri EN, Iliaki KK, Höhn A, Grimm S, Papassideri IS, Grune T, et al. Diet-derived advanced glycation end products or lipofuscin disrupts proteostasis and reduces life span in *Drosophila melanogaster*. Free Radic Biol Med 2013;65:1155–63.

[11] Jaskelioff M, Muller FL, Paik J-H, Thomas E, Jiang S, Adams AC, et al. Telomerase reactivation reverses tissue degeneration in aged telomerase-deficient mice. Nature 2011;469(7328):102–6.

[12] Finkel T, Serrano M, Blasco MA. The common biology of cancer and ageing. Nature 2007;448(7155):767–74.

[13] http://www.eurostemcell.org.

[14] Brockes JP, Kumar A. Appendage regeneration in adult vertebrates and implications for regenerative medicine. Science 2005;310(5756):1919–23.

[15] Philipp EER, Abele D. Masters of longevity: lessons from long-lived bivalves – a mini-review. Gerontology 2010;56(1):55–65.

[16] Boehm AM, Khalturin K, Erxleben FA, Hemmrich G, Klostermeier UC, Lopez-Quintero JA, et al. FoxO is a critical regulator of stem cell maintenance in immortal Hydra. Ann Neurosci 2013;20(1):17. http://dx.doi.org/10.5214/ans.0972.7531.200107.

[17] Gurtner GC, Werner S, Barrandon Y, Longaker MT. Wound repair and regeneration. Nature 2008;453(7193):314–21.

[18] Baker M. Irving Weissman: creating a standard for stem cell therapies and culturing the unorthodox. Nat Rep Stem Cells 2009. http://dx.doi.org/10.1038/stemcells.2009.107.

[19] Wicker J, Kamler K. Current concepts in limb regeneration: a hand surgeon's perspective. Ann NY Acad Sci 2009;1172:95–109.

[20] Michalopoulos GK. Liver regeneration. Science 1997;276(5309):60–6.

[21] Singaravelu K, Padanilam BJ. In vitro differentiation of MSC into cells with a renal tubular epithelial-like phenotype. Ren Fail 2009;31(6):492–502.

[22] Engel FB, Hsieh PCH, Lee RT, Keating MT. FGF1/p38 MAP kinase inhibitor therapy induces cardiomyocyte mitosis, reduces scarring, and rescues function after myocardial infarction. Proc Natl Acad Sci USA 2006;103(42):15546–51.

[23] Masui Y, Gondo Y, Inagaki H, Hirose N. Do personality characteristics predict longevity? Findings from the Tokyo centenarian study. Age (Dordr) 2006;28(4):353–61.

[24] Paranjpe P, Patki P, Patwardhan B. Ayurvedic treatment of obesity: a randomised double-blind, placebo-controlled clinical trial. J Ethnopharmacol 1990;29(1):1–11.

[25] Jadhav RB, Patwardhan B. Anti-anxiety activity of *Celastrus paniculatus*. Indian J Nat Prod 2003;19(3):16–9.

[26] Kulkarni RR, Jog V, Gandage S, Patki P, Patwardhan B. Efficacy of ayurvedic formulation in rheumatoid arthritis: a randomized, double blind, placebo controlled crossover study. Indian J Pharmacol 1992;24(2):98–101.

[27] Kulkarni RR, Patki PS, Jog VP, Gandage SG, Patwardhan B. Treatment of osteoarthritis with a herbomineral formulation: a double-blind, placebo-controlled, cross-over study. J Ethnopharmacol 1991;33(1–2):91–5.

[28] Saraf M, Patwardhan B. Pharmacological studies on *S. brevistigma* part I: antiallergic activity. Indian Drugs 1988;26(2):49–54.

[29] Saraf M, Patwardhan B. Pharmacological studies on *S. brevistigma* part II: brochodialator activity. Indian Drugs 1988;26(2):54–7.

[30] Patwardhan B, Kalbag D, Patki PS, Nagsampagi B. Search of immunomodulatory agents: a review. Indian Drugs 1990;28(2):56–63.

[31] Patwardhan B. Ayurveda and future drug development. Int J Altern Complement Med 1992;10(12):9–11.

[32] Patwardhan B. Ethnopharmacology and drug discovery. J Ethnopharmacol 2005;100(1–2):50–2.

[33] Balasubramani SP, Venkatasubramanian P, Kukkupuni SK, Patwardhan B. Plant-based Rasayana drugs from Ayurveda. Chin J Integr Med 2011;17(2):88–94.

[34] Patil M, Patki P, Kamath HV, Patwardhan B. Antistress activity of *Tinospora cordifolia* (wild) miers. Indian Drugs 1997;34(4):211–5.

[35] Ziauddin M, Phansalkar N, Patki P, Diwanay S, Patwardhan B. Studies on the immunomodulatory effects of ashwagandha. J Ethnopharmacol 1996;50(2):69–76.

[36] Agarwal R, Diwanay S, Patki P, Patwardhan B. Studies on immunomodulatory activity of *Withania somnifera* (ashwagandha) extracts in experimental immune inflammation. J Ethnopharmacol 1999;67(1):27–35.

[37] Diwanay S, Chitre D, Patwardhan B. Immunoprotection by botanical drugs in cancer chemotherapy. J Ethnopharmacol 2004;90(1):49–55.

[38] Grandhi A, Mujumdar AM, Patwardhan B. A comparative pharmacological investigation of Ashwagandha and Ginseng. J Ethnopharmacol 1994;44(3):131–5.

[39] Datta HS, Mitra SK, Paramesh R, Patwardhan B. Theories and management of aging: modern and ayurveda perspectives. Evid Based Complement Altern Med 2011;2011:528527. http://dx.doi.org/10.1093/ecam/nep005.

[40] Datta HS, Mitra SK, Patwardhan B. Wound healing activity of topical application forms based on ayurveda. Evid Based Complement Altern Med 2011;2011:134378. http://dx.doi.org/10.1093/ecam/nep015.

[41] Gautam M, Saha S, Bani S, Kaul A, Mishra S, Patil D, et al. Immunomodulatory activity of *Asparagus racemosus* on systemic Th1/Th2 immunity: implications for immunoadjuvant potential. J Ethnopharmacol 2009;121(2):241–7.

[42] Bani S, Gautam M, Sheikh FA, Khan B, Satti NK, Suri KA, et al. Selective Th1 up-regulating activity of *Withania somnifera* aqueous extract in an experimental system using flow cytometry. J Ethnopharmacol 2006;107(1):107–15.

[43] Patwardhan B, Gautam M. Botanical immunodrugs: scope and opportunities. Drug Discov Today 2005;10(7):495–502.

[44] Gautam M, Diwanay SS, Gairola S, Shinde YS, Jadhav SS, Patwardhan BK. Immune response modulation to DPT vaccine by aqueous extract of *Withania somnifera* in experimental system. Int Immunopharmacol 2004;4(6):841–9.

[45] Gautam M, Gairola S, Jadhav S, Patwardhan B. Ethnopharmacology in vaccine adjuvant discovery. Vaccine 2008;26(41):5239–40.

[46] Jadhav SS, Patwardhan B, Gautam M. Adjuvant composition for vaccine. United States Patent, US 8501 186 B2, August 6, 2013.

[47] Patwardhan B, Gautam M. Aqueous extracts of plant products. India; 1247/Mum/2003, 2006. (Indian Patent).

[48] Patwardhan B, Gautam M. Process for making immunological adjuvants. India; 1253/Mum/2003, 2006. (Indian Patent).

[49] Diwanay S, Gautam M, Patwardhan B. Cytoprotection and immunomodulation in cancer therapy. Curr Med Chem Anticancer Agents 2004;4(6):479–90.

[50] Govindarajan R, Vijayakumar M, Pushpangadan P. Antioxidant approach to disease management and the role of "Rasayana" herbs of Ayurveda. J Ethnopharmacol 2005;99(2):165–78.

[51] Mangathayaru K, Umadevi M, Reddy CU. Evaluation of the immunomodulatory and DNA protective activities of the shoots of Cynodon dactylon. J Ethnopharmacol 2009;123(1):181–4.

[52] Kulkarni SK, Dhir A. *Withania somnifera*: an Indian ginseng. Prog Neuropsychopharmacol Biol Psychiatry 2008;32(5):1093–105.

[53] Bhattacharya SK, Bhattacharya A, Sairam K, Ghosal S. Anxiolytic-antidepressant activity of *Withania somnifera* glycowithanolides: an experimental study. Phytomedicine 2000;7(6):463–9.

[54] Ojha R, Sahu AN, Muruganandam AV, Singh GK, Krishnamurthy S. *Asparagus racemosus* enhances memory and protects against amnesia in rodent models. Brain Cogn 2010;74(1):1–9.

[55] Jagetia GC, Malagi KJ, Baliga MS, Venkatesh P, Veruva RR. Triphala, an ayurvedic rasayana drug, protects mice against radiation-induced lethality by free-radical scavenging. J Altern Complement Med 2004;10(6):971–8.

[56] Sharma V, Thakur M, Chauhan NS, Dixit VK. Effects of petroleum ether extract of *Anacyclus pyrethrum* DC. on sexual behavior in male rats. J Chin Integr Med 2010;8(8):767–73.

[57] Thakur M, Chauhan NS, Bhargava S, Dixit VK. A comparative study on aphrodisiac activity of some ayurvedic herbs in male albino rats. Arch Sex Behav 2009;38(6):1009–15.

[58] Menon LG, Kuttan R, Kuttan G. Effect of rasayanas in the inhibition of lung metastasis induced by B16F-10 melanoma cells. J Exp Clin Cancer Res 1997;16(4):365–8.

[59] Gaddipati JP, Rajeshkumar NV, Thangapazham RL, Sharma A, Warren J, Mog SR, et al. Protective effect of a polyherbal preparation, Brahma rasayana against tumor growth and lung metastasis in rat prostate model system. J Exp Ther Oncol 2005;4(3):203–12.

[60] Sánchez Alvarado A, Yamanaka S. Rethinking differentiation: stem cells, regeneration, and plasticity. Cell 2014;157(1):110–9.

[61] Joshi K, Bhonde R. Insights from Ayurveda for translational stem cell research. Ayurveda Integr Med 2014;5(1):4–10.

[62] Lin YJ, Seroude L, Benzer S. Extended life-span and stress resistance in the *Drosophila* mutant methuselah. Science 1998;282(5390):943–6.

[63] Parkes TL, Elia AJ, Dickinson D, Hilliker AJ, Phillips JP, Boulianne GL. Extension of *Drosophila* lifespan by overexpression of human SOD1 in motorneurons. Nat Genet 1998;19(2):171–4.

[64] Sun Y, Yolitz J, Alberico T, Sun X, Zou S. Lifespan extension by cranberry supplementation partially requires SOD2 and is life stage independent. Exp Gerontol 2014;50:57–63.

[65] Vayndorf EM, Lee SS, Liu RH. Whole apple extracts increase lifespan, healthspan and resistance to stress in *Caenorhabditis elegans*. J Funct Foods 2013;5(3):1235–43.

[66] Kumar R, Gupta K, Saharia K, Pradhan D, Subramaniam JR. *Withania somnifera* root extract extends lifespan of *Caenorhabditis elegans*. Ann Neurosci 2013;20(1):13–6.

[67] Priyadarshini S, Ashadevi JS, Nagarjun V, Prasanna KS. Increase in *Drosophila melanogaster* longevity due to rasayana diet: preliminary results. J Ayurveda Integr Med 2010;1(2):114–9.

[68] Dwivedi V, Anandan EM, Mony RS, Muraleedharan TS, Valiathan MS, Mutsuddi M, et al. In vivo effects of traditional Ayurvedic formulations in *Drosophila melanogaster* model relate with therapeutic applications. PLoS One 2012;7(5):e37113. http://dx.doi.org/10.1371/journal.pone.0037113.

[69] Libert S, Guarente L. Metabolic and neuropsychiatric effects of calorie restriction and sirtuins. Annu Rev Physiol 2013;75:669–84.

[70] Guarente L. Mitochondria-a nexus for aging, calorie restriction, and sirtuins? Cell 2008;132(2):171–6.

[71] Boily G, Seifert EL, Bevilacqua L, He XH, Sabourin G, Estey C, et al. SirT1 regulates energy metabolism and response to caloric restriction in mice. PLoS One 2008;3(3):e1759. http://dx.doi.org/10.1371/journal.pone.0001759.

[72] Fok WC, Bokov A, Gelfond J, Yu Z, Zhang Y, Doderer M, et al. Combined treatment of rapamycin and dietary restriction has a larger effect on the transcriptome and metabolome of liver. Aging Cell 2014;13(2):311–9.

[73] Harrison DE, Strong R, Sharp ZD, Nelson JF, Astle CM, Flurkey K, et al. Rapamycin fed late in life extends lifespan in genetically heterogeneous mice. Nature 2009;460(7253):392–5.

[74] Chen T, Shen L, Yu J, Wan H, Guo A, Chen J, et al. Rapamycin and other longevity-promoting compounds enhance the generation of mouse induced pluripotent stem cells. Aging Cell 2011;10(5):908–11.

[75] Van Der Veer E, Ho C, O'Neil C, Barbosa N, Scott R, Cregan SP, et al. Extension of human cell lifespan by nicotinamide phosphoribosyltransferase. J Biol Chem 2007;282(15):10841–5.

[76] Chaya MS, Kurpad AV, Nagendra HR, Nagarathna R. The effect of long term combined yoga practice on the basal metabolic rate of healthy adults. BMC Complement Altern Med 2006;6:28. http://dx.doi.org/10.1186/1472-6882-6-28.

[77] Chaya MS, Nagendra HR. Long-term effect of yogic practices on diurnal metabolic rates of healthy subjects. Int J Yoga 2008;1(1):27–32.

[78] Chatterjee S, Mondal S. Effect of regular yogic training on growth hormone and dehydroepiandrosterone sulfate as an endocrine marker of aging. Evid Based Complement Altern Med 2014;2014:240581. http://dx.doi.org/10.1155/2014/240581.

[79] Pal R, Singh SN, Chatterjee A, Saha M. Age-related changes in cardiovascular system, autonomic functions, and levels of BDNF of healthy active males: role of yogic practice. Age (Dordr). 2014;36(4):9683.

[80] George K. Okinawa: the history of an island people. Tokyo: Tuttle Publishing; 2013.

[81] Ornish D, Lin J, Chan JM, Epel E, Kemp C, Weidner G, et al. Effect of comprehensive lifestyle changes on telomerase activity and telomere length in men with biopsy-proven low-risk prostate cancer: 5-year follow-up of a descriptive pilot study. Lancet Oncol 2013;14(11):1112–20.

CHAPTER 11

Personalized Approaches for Health

Every individual is different from another; hence, each should be considered as a unique entity. As many variations as there are in the universe—all are seen in a human being.

Charaka Samhita, 400 BC

Diversity is inherent in beauty of nature. Thousands of species from microscopic viruses, bacteria, and fungi, to herbs, shrubs, trees, insects, and vertebrates, live interdependently, harmoniously, and symbiotically, and struggle to exist. All evolve, and many become extinct as a natural process of evolution. This rich biodiversity makes our planet beautiful. But this diversity also exists among similar species as well. Human beings are no exception. We are the same, but at the same time, we are different.

Everyone has a unique identity. Interestingly, this diversity is not found only in physical appearance, but is also present at the psychological level. Sometime we call this the personality. According to some estimates, over 5000 major ethnic groups are spread over all the continents. There is a huge diversity of people, cultures, cults, languages, and religions. We have different shades of skin colors, different postures, different physical stature, appearances, and responses. At the same time, we are all a part of nature. We are governed by the laws of nature. We are made up of the same building blocks of nucleic acids and share many genes with other species. We have birth, growth, and death phases like every living creature. We are creatures—just of different types. We have tried to make sense of these differences using various criteria, such as geography, religion, and profession; and we have created castes, creed, races, and societies. This diversity has many connotations in the social, psychological, and political dimensions. Understanding the basis of this diversity is very important when we address issues related to health and disease.

In this chapter, we review various theories, approaches, and scientific advances in human typology, diversities, and classifications along with their limitations—mainly in relation to health and disease.

In 1892, Sir William Osler, who is regarded as the father of modern medicine, stated "If it were not for the great variability among individuals,

293

medicine might as well be a science, and not an art." This was the general view in the field of medicine until the twentieth century. What the physicians did while making necessary clinical judgments about diagnosis and prescribing medicines, was often referred to as an art, mainly because of the lack of objective data available in making decisions for the care of individual patients. However, with the advent of the Human Genome Project, scientists were able to identify inherited differences between individuals. This understanding can greatly help to predict each patient's response to a particular medicine. As rightly stated by the physician and geneticist, Alan Roses, "if Sir William Osler was alive today, he would be re-considering his view of medicine as an art, not a science" [1]. While medical practice may continue to be an art, today's, per se, has not remained merely intuitive, or judgmental, but has become more scientific and evidence based. However, variations in each person make the practice of medicine challenging. Efforts to study these variations have led to interesting discoveries and theories of human classification. Since an understanding of individual variation and classification of humans is crucial to bring more precision to diagnostics and therapeutics, it is important to know a short history of the attempts to study human variations and classifications.

CLASSIFYING HUMANS

Philosophically, humans have long been considered animals. Plato referred to humans as featherless, biped animals; while Aristotle defined the human being as a rational, or political animal. Ancient Hebrews described humans as living souls who breathe. Living things were believed to produce their own kind, and humans were believed to comprise a single kind. Christian thought considered animals soulless creatures, even until the 1990s when none other than Pope John Paul II proclaimed that animals also have souls, like humans [2]. During the Middle Ages, Europeans believed that humanity was divided into three races, one for each of the sons of Noah. The indigenous or aboriginal people were thought to be soulless animals by leaders in the time of Western Colonialism.

In the past, another system of classification, which was based on the person's type of work or profession, was very controversial in India. This belief led to the caste system, which severely affected social equality, and justice. A more systematic classification was proposed in 1758 by Swedish naturalist and physician, Carl Linnaeus, who defined humans as *Homo sapiens*. Initially, races were considered as a human subspecies. However, modern genetic research has clearly shown that such inherited differences do not match common racial divisions.

History of Classification

In the Western world, Hippocrates proposed four temperaments based on dominance of four humors: black bile, yellow bile, phlegm, and blood [3]. Galen, in his dissertation, *De temperamentis*, elaborated upon these constitution types further, describing melancholic, choleric, phlegmatic, and sanguine-based types as being determined by the dominance of the respective humors. The melancholic type people are introverted, creative, thoughtful, and sensitive. They are deeply involved in any task and have a predisposition to depression. Hence, one of the types of depression is known as melancholia. Choleric types are extroverted, aggressive, energetic, and passionate. They are leaders and dictate terms to others. Phlegmatic types are relaxed, calm, sluggish, rational, observant, and affectionate. Sanguine types are social, impulsive, talkative, and overly confident.

The traditional chinese medicine (TCM) theory is based on five elements, and the energies (*qui*), yin and yang, which flow through the body channels, and meridians. TCM classifies humanity into nine constitutional types: balanced constitution, qui-deficiency constitution, yang-deficiency constitution, yin-deficiency constitution, phlegm-dampness constitution, heat-dampness constitution, blood stasis constitution, quiz-stagnation constitution, and inherited special constitution [4,5]. These constitutional types are diagnosed based on anatomical, physiological, and psychological characteristics. Each constitutional type is prone to certain diseases. For example, individuals with phlegm-dampness constitution are thought to be susceptible for hyperlipidemia and diabetes [6].

Sasang constitutional medicine is a part of Korean traditional medicine. This system assesses anatomical characteristics, temperament, and other symptoms of an individual, and assigns them into any of the four constitutional types: tae-eum, so-yang, so-eum, and tae-yang. Recently, a tool for quantitative assessment of Sasang constitutional type has been developed. This tool integrates questionnaire information with face, body shape, and voice analysis [7]. Many foods and herbs are thought to be specific to the respective type of constitution, and may cause adverse effects in persons of another type. This shows how traditional concepts from different parts of the world had similarities in human classification approaches. Galen's approach and the Chinese, Korean, and Hippocratic approaches seem very similar to the classification based on the concept of *Prakriti* in Ayurveda. Interestingly, taeeumin, soyangin, and soeumin are reported to be similar to *Kapha* (K), *Pitta* (P), and *Vata* (V) *Prakriti* types, respectively [8]. Details about the concept *Prakriti* are discussed in a separate section later.

As modern science progressed, several scientists attempted to understand the nature of individual variations, and possible ways and means to classify, stratify, or type them into appropriate clusters.

Taxonomic Classification

Carl Linnaeus is known as the father of modern taxonomy or science of classification. He proposed that organisms can be classified according to the phylogenetic and evolutionary process. He developed a system for classifying plants and animals, based on a hierarchy of categories, ranging from kingdom to species. This considers three main domains. The human is classified under eukaryote domain, which is multicellular organisms having a membrane-bound nucleus. Human belongs to the Animalia kingdom that comprises all animals. Humans are further classified in the phylum, chordata, which includes the vertebrate animals. The Mammalia class of this phylum comprises warm-blooded vertebrates, in which mammary glands are present in females. Humans are part of the order primates, under the family Hominidae, which is a cluster of chimpanzees, gorillas, and orangutans. Among these, humans have the characteristic tendency of performing complex tasks such as the use of tools. The word *Homo sapiens* means "wise man" in Latin.

Linnaeus suggested human classification in his book, *Systema Naturae*. He proposed four main groups (races) within the genus, *Homo*, based on characteristics like color, and continent: Americanus, Asiaticus, Africanus, and Europeaus. Americanus were supposed to be people with redskin, black hair, and sparse beards. They were thought to be stubborn and prone to anger. They were free and governed by traditions. In Linnaeus' classification, the Asiaticus were thought to have yellowish skin, black hair, and brown eyes. They were supposed to be severe, conceited, and governed by opinion. Africanus were thought to have dark skin, curly hair, and possess cunning, passive, and inattentive attributes. They were thought to be ruled by impulse. Linnaeus described Europeaus as adaptable, changeable, clever, inventive, and law abiding. Linnaeus also suggested other kinds of humans: wild men; dwarfs; troglodytes, or cave dwellers; and lazy Patagonians, or hunter-gatherers. Efforts to categorize humans into large, and distinct populations, or groups based on anatomical, cultural, ethnic, genetic, geographical, historical, linguistic, religious, and social affiliation are still in progress.

Typology in Anthropology

In anthropology, several terms are used to classify, or characterize human populations: species, race, caste, creed, and traits. Each term has a specific meaning, though sometimes they may be used interchangeably. Generally, species are a taxonomic group whose members can interbreed. Race is taxonomic group, which is a subdivision of a species. The race is thought to be a consequence of geographical isolation within a species. Caste relates to the social status, or position conferred by a system based on class, or stratified according to ritual purity. Creed is any system of principles, or beliefs, and trait is a distinguishing feature of the personal nature. Socioeconomic class relates more to the category where people with the same social, economic, or educational status are included. These are a few representative and well-known attempts to categorize human populations using certain approaches or parameters.

Human species can be characterized by physical traits that are readily observable from a distance, such as head shape, skin color, hair form, body build, and stature. In the twentieth century, anthropologists used a typological model to classify people from different ethnic regions into three main races: the Negroid race, the Caucasoid race, and the Mongoloid race. This racial classification system was proposed by American anthropologist, Carleton Coon, in 1962. The anthropology-based, typological model assumes that humans can be assigned to a particular race according to physical traits [9]. However, this approach is currently disregarded.

Today, it is believed that morphological traits may be due to geographical, environmental, and climatic variations in specific regions, resulting in genetic and epigenetic changes, which may lead to the development of physical traits. It is becoming increasingly clear that just the presence of a genetic variation need not lead to development of a particular trait. Several environmental factors are thought to be involved in the development of any trait. The presence of certain genetic factors can enhance or repress other genetic factors. Genes can be turned on and off. Many factors control genes from being turned "on" or "off." It is also known that proteins encoded by genes can be modified, which can affect the genes' normal cellular functions. A branch of science dealing with these issues is now known as epigenetics, which actually means "beyond or above genome."

Genes are responsible for a trait. Therefore, the genotype is responsible for the trait. Observable expression of these genes is known as the phenotype. It is known that an organism's physical properties are a consequence of the inheritance of genes. Therefore, understanding the genotype–phenotype distinction is very important.

Behavioral or physical traits may vary based on several factors, environments, and circumstances. Genetic factors may also influence certain environmental factors leading to a particular trait. For example, a person may be genetically known to have a risk for developing cancer from smoking, however, if the person does not smoke, cancer may not develop. Therefore, it is now believed that classifying human populations only on a genetic basis may not be a correct method. This new understanding of how epigenetics operates may explain the many hurdles faced along the way to the development of personalized medicine. It had been hoped that personalized medicine would be one of the fruits of the Human Genome Project.

Personality Types

Human personality can also be studied using trait theory involving habitual patterns of behavior, thought, and emotions. The measurement of traits is important in this approach. It is believed that traits differ across individuals but are relatively stable. As we know, the nature of some people may be extroverted, whereas others may be introverted, which can influence their behavior. Most of the trait models, including those of ancient Greek philosophy, believe that extraversion and introversion are the main basis of human personality. This has led psychologists and clinical researchers to find out more about the mind–body relationships. The mental state is known to affect physical health, but the reasons behind this are not known.

Many psychologists classify human population broadly into two categories known as Type A and Type B, as two most common personality types. Type A people are relatively hostile, impatient, and competitive. Type B people are supposed to be more calm and cool. These patterns of behavior can either raise or lower chances of developing specific diseases, such as proneness to coronary heart disease in Type A behavior [10]. In the 1950s, eminent cardiologists Meyer Friedman and Ray Rosenman showed that certain populations with Type A personality behavior have a potential risk for heart disease. They also showed that Type A behavior doubles the risk of erectile dysfunction in healthy individuals [11,12].

This theory continues to remain controversial in the scientific and medical communities. Ayurveda and Yoga also use personality types as psychological *Prakriti* to understand patients and their behavior, as it relates to health and disease. The proponents of personality theory did not consider *Prakriti* concept involving *Satva, Rajas,* and *Tamas* attributes. The *Satva* type has distinct consciousness, intelligence, sensitivity, and memory. *Rajas* type has more energy, initiative, enthusiasm, and physical activity. *Tamas* type has more anger and lethargy as key attributes. This classification of personality attributes is helpful in knowing the nature and response of any individual in healthy, stressful, and pathological conditions. The personality types suggest general attitudes, likings, hobbies, activities, and other psychological attributes. Interestingly, Ayurveda also classifies diet into three similar categories: *Satvik, Rajasik,* and *Tamasik*. This was discussed in Chapter 7 (Food and Diet). Whether influence of epigenetic and dietetic factors can transform one type into others will be an interesting area of research in nutrigenomics.

Traits and Human Psychology

American psychologist, Dr Gordon Allport, is a pioneer in the study of traits. He referred to traits as dispositions, and believed that certain principal, or cardinal traits are central to an individual personality. Common traits can be recognized and may vary between cultures. Dr Allport also thought that it is possible to recognize any individual by ascertaining their cardinal traits. Later, an expert in psychological phenomena, Dr Hans Eysenck from Kings College, London, proposed that personality can be classified into three major traits. Many other researchers feel that factors like humor, economic status, and appearance are also needed to describe the human personality. A prominent trait found in nearly all models is neuroticism or emotional instability [13,14]. Many psychologists currently believe that five main factors, including openness, conscientiousness, extraversion, neuroticism, and the degree of emotional stability are adequate. Identifying the traits and structure of human personality still remains one of the most fundamental goals in modern psychology [15].

Behaviors are complex traits involving multiple genes that are affected by a variety of other factors. English scientist, Sir Francis Galton, studied human behavior systematically. Human behavioral genetics, a relatively new field, seeks to understand both the genetic and environmental contributions to individual variations in human behavior. No single gene is responsible for a particular behavior. For any typical physical appearance, disorder, or behavior, genes are not in isolation.

The Somatotypes

In the 1940s, American psychologist, William Herbert Sheldon, developed an interesting theory by associating body types with human temperament types. Sheldon proposed that the human physique can be classified according to the relative contribution of three fundamental elements. He called them somatotypes, after the three germ layers of embryonic development: the endoderm, which develops into the digestive tract; the mesoderm, which develops into muscle, heart, and blood vessels; and the ectoderm, which forms the skin and nervous system [16].

Ectomorphic type is characterized by long, thin muscles and limbs, and low fat storage—usually referred to as slim. Ectomorphs are not predisposed to store fat or build muscle. The mesomorphic type is characterized by medium bones, solid torso, low fat levels, and wide shoulders, with a narrow waist—usually referred to as muscular. Mesomorphs are predisposed to build muscle, but not store fat. The endomorphic type is characterized by increased fat storage, a wide waist, and large bone structure, usually referred to as fat. Endomorphs are predisposed to store fat.

Sheldon's method of somatotyping was modified by American anthropologists Barbara Heath and Lindsay Carter, based on several tools for precise quantification of shape and composition of human body. The somatotyping is done with anthropometric and photographic methods. The scores of relative endomorphy, mesomorphy, and ectomorphy are calculated based on standard tables and preset equations [17].

There is evidence that different physiques carry cultural stereotypes. For example, endomorphs are likely to be perceived as slow, sloppy, and lazy. Mesomorphs, in contrast, are typically popular, and hardworking, whereas ectomorphs are often viewed as intelligent, but fearful.

The body type descriptions could be modulated by body composition. Certain diets, exercises, and training techniques may have a role in modulating body compositions. During starvation, an endomorph may resemble an ectomorph, while an athletic mesomorph may look like an endomorph as a result of loss of muscle, and adipose mass, or simply due to the aging process. However, certain characteristics of the somatotype cannot be changed. For example, the bone structure is a fixed characteristic, except for a few changes due to the reduction in the distance between joints due to aging or physical deformities. Even cultural conditions may lead to a tendency to change temperaments. Because of many such limitations, the constitutional psychology approach was not accepted by the scientific world, and is not in use much today. During Sheldon's time, Western scientists generally may have been aware about the various Eastern traditional ways in which people are classified for the purpose of treatments. If Sheldon would have visited India, or had studied concept of *Prakriti* in Ayurveda, perhaps constitutional psychology would have gone on a different route. An eminent physician–scientist from India, R. D. Lele has found a correlation of somatotypes with *Prakriti*, where he links mesomorphs to *Kapha*, endomorphs to *Pitta*, and ectomorphs to *Vata* types [18]. Sheldon's classification might serve as a basis for *Prakriti* assessment.

Thus, while several approaches were proposed in the past, a reasonably sound and scientifically acceptable way to classify human population is not available. Admittedly, new hopes have emerged because of the recent understanding of epigenetics, and advances in genomics, and other omics technologies.

THE SCIENCE OF GENOMICS

Advances in genomics have produced a deeper understanding of genetic variations among human populations. The large data from the Human Genome Project have opened new avenues for scientific research. A major challenge for today's modern medicine is to understand the relationships, and interactions between genetic variations, and environmental triggers of disease. To address this challenge, the emerging science of epigenetics draws from genomics, proteomics, bioinformatics, metabolomics, and metabonomics in providing a new, logical framework to elucidate disease etiology, and connections between seemingly disparate disease states. Advances in genomics have also raised many ethical, and political questions about social, and cultural notions of race, ethnicity, and human genetic variation.

Scientists know that all human beings are genetically the same. Through scientific analysis, the Human Genome Project established the fact that all humans are 99.9% same. The genetic difference between individual humans is only about 0.1%. It is now known that the human genome is not very different from that of the chimpanzee or bonobo genome—with 98.8% similarity. The DNA difference of humans from gorillas, another of the African apes, is only about 1.6%. Most importantly, chimpanzees, bonobos, and humans all show almost the same amount of difference from gorillas. The DNA evidence indicates that the human and ape, or for that matter, any animals, share a number of genetic similarities. Scientists have shown that human DNA is part of the evolutionary tree embedded within the great apes [19]. What Erasmus Darwin once said has been proven to be true: "There is a common filament or thread of life that runs through all the living beings." The whole DNA molecule constitutes the genome, which is present in the nucleus of every cell. It consists of genes with the molecular codes to regulate the output of genes, and for the output of proteins, which are the building blocks of our body. Genetic codes in DNA direct the development of cells, triggers and controls the growth of tissues, and operates the physiology of blood, bone, and organs—including brains. The structure and function of DNA, and the basic metabolic processes are absolutely similar in all animals, including humans. We can no longer accept the religious or traditional notions that human beings are, as entities, unique in nature, and disregard animals as *soulless* creatures. The science of genomics has now established the phylogenetic relationships between human and animals, and the soul (if we accept its presence at all), is not the monopoly of human beings.

Modern humans have been living, traveling, migrating, and dispersing over the past one hundred thousand years—facing large variations in geo-climatic conditions. This probably has resulted in several patterns of phenotypic variation including the color of skin, eyes, hair, physical stature, and behaviors. These visible variations have exerted powerful influences on the lives of individuals and the experiences of groups.

These differences in appearance may have contributed to the development of ideas about *race* and *ethnicity*. People formerly believed that these inherited differences may distinguish humans. The racial, ethnic, and ancestral categories in genetics research can imply that group differences arise directly through differing allele frequencies, with little influence from socially mediated mechanisms. At the same time, careful investigations of the biological, environmental, social, and psychological attributes associated with these categories will be an essential component of cross-disciplinary research [20].

While humans have great similarities with animals, and they may be genetically 99.9% same, the remaining 0.1% makes every individual different. These differences are mainly because of single nucleotide polymorphism or SNPs. These differences also lead to genotypes and phenotypes. The genotype–phenotype distinction is important in genetics. Genotype gives complete hereditary information of an organism, while phenotype provides its actual observed properties, such as physical appearance, or behavior. This distinction is very important to understand human classification.

English physiologist, Archibald Garrod, first proposed that genetic variants might modulate variability in drug actions. Enzyme polymorphism was fairly well known, as was the understanding that enzymatic defects may lead to some *inborn errors of metabolism*. Sharp-eyed observations during World War II noted the high incidence of paralysis after the use of succinylcholine, and a higher incidence of hemolytic anemia resulting from antimalarial drugs therapy, especially among blacks. This was the beginning of the understanding that the same category of drugs may not have similar effects or side effects in different human populations. This was the beginning of pharmacogenomics.

Emergence of Pharmacogenomics

People from different regions look different, think and behave differently, eat different foods, and live in different environments. It is known that their genetic makeup is also different. Still, until recently, it was believed that all people would respond similarly to any drug or treatment. All these years, physicians were aware of varied responses to the same drug in different patients, however, they did not understand the basis of this phenomenon. The term pharmacogenomics has been used more recently to endorse the idea that variable drug response may reflect sets of variants within an individual or across a population. Pharmacogenomics can be defined as the technology that analyzes how the genetic makeup of an individual affects his or her response to drugs. As the word suggests, it combines the knowledge of pharmacology and of genomics. It deals with the influence of genetic variation, on drug responses in patients by correlating gene expression, or SNPs with a drug's efficacy or toxicity.

Pharmacogenomics aims to develop a rational means to optimize drug therapy, with respect to patients' genotype, to ensure maximum efficacy, with minimal adverse effects. Such approaches promise the advent of *personalized medicine*, in which drugs and drug combinations are optimized for each individual's unique genetic makeup. Pharmacogenomics is slowly and steadily making inroads in therapeutics toward personalized medicine. Already, about 10% of FDA-approved drugs in the United States contain pharmacogenomic information [21]. Personalized medicine requires accurate diagnostic tests for the patients who are expected to benefit from targeted therapies. Knowledge of specific molecular diagnostics to predict a better response to the medication helps the clinicians to better target patients' treatment. Several such compilations of drugs and companion diagnostics have been approved by the FDA for clinical practice; however, sufficient postmarketing surveillance data that can be used for producing better outcomes are still emerging [22].

Pharmacogenomics studies the effects of genes on the human response to drugs, with the goal of minimizing drug reactions, which can be life-threatening, and maximizing the desired therapeutic effect, which can cure terrible diseases. According to a pharmacologist from the Mayo Clinic, Richard Weinshilboum, modern drugs are very powerful agents that can do great good, but also great harm. Several known genes are responsible for variances in drug metabolism and response. The most common are the cytochrome P450 (CYP) genes, which encode enzymes that influence the metabolism of more than 80% of current prescription drugs. Codeine, clopidogrel, tamoxifen, and warfarin are a few examples of medications which are processed through CYP. Patient genotypes are usually categorized into predicted phenotypes as an extensive metabolizer phenotype. An extensive metabolizer is considered normal, while other CYP metabolism phenotypes include intermediate, ultrarapid, and poor. Each phenotype is based on the allelic variation within the individual genotype.

Strong scientific evidence suggests that patients affected by genetic polymorphisms may experience severe or even lethal toxicities if these differences are not well understood. This is particularly true in oncology, where many anticancer chemotherapeutic drugs may be effective only in patients with specific genetic subtypes, while others may actually show toxicities without any therapeutic benefits. Pharmacogenomics has applications in illnesses like cancer, cardiovascular disorders, depression, bipolar disorder, attention deficit disorders, HIV, tuberculosis, asthma, and diabetes. In cancer treatment, pharmacogenomics tests are used to identify which patients are most likely to respond to certain cancer drugs. In behavioral health, pharmacogenomic tests provide tools for physicians, and caregivers to better manage medication selection, and side effect amelioration. Pharmacogenomics is also known as companion diagnostics, meaning that the tests are many times bundled with drugs.

A study by renowned neurobiologist, Dr Candace Pert, also supported the "one shoe does not fit all" maxim, and indicates a need to consider individual variation, stratification, and systematic subgrouping. While at NIH, Dr Pert invented a new HIV entry inhibitor known as Peptide T. In a large, multicentric, placebo-controlled, clinical trial, Peptide T was not found to be significantly different from a placebo. However, Peptide T was associated with improved performance with respect to memory and learning in the subgroup of patients with more severe cognitive impairment [23]. Several such examples are now known in many diseases where the rate of efficacy of a standard drug may be as low as 25% in conditions like cancer; while this rate may be about 50–60% in rheumatoid arthritis, diabetes, schizophrenia, and depression. In these patients, pharmacogenomics enables predictions of responders, and nonresponders based on the individual, genetic makeup—leading to the personalized medicine.

The Ayurvedic concept that "every individual is different" is very similar to pharmacogenomics. Ayurverdic practitioners diagnose and treat each person separately, no matter that they might report similar symptoms. This is because the individual's *Prakriti* type or genetic fingerprint, or genotype is uniquely different, and requires a personalized regimen consisting of drug, diet, and life-style modifications. If these principles had been known to geneticists and pharmacologists earlier, today's medicine would have been much more advanced, with improved, and targeted efficacy, and better safety. If someone like Dr Weinshilboum, who is considered the father of pharmacogenomics, would have known about the concept of *Prakriti*, the discipline of pharmacogenomics could have arrived much earlier.

Interestingly, the concept of *Prakriti* is actually about healthy individuals. Of course, it can also help in diagnosis and therapeutics. However, genomics may not be able to able to give us predications about healthy individuals. Genomics can predict disease predispositions based on the presence or absence of certain genes, and by predicting which genes are likely to be turned on, or off, based on environmental, or other compounding factors. Genomics may give us detailed descriptions of the genes responsible for various diseases, but not much information about health. Even knowing the whole sequence of the human genome can hardly tell us about the status and future of individual health. Going beyond genomics is necessary in order to understand how the environment and behaviors can be responsible for inheritable changes, when the genome remains unchanged. This science of epigenetics is seen as a future hope, providing answers to many puzzles. We feel that a detailed understanding of Ayurvedic concepts like *Prakriti* may actually facilitate this process.

Metabonomics

The human body contains thousands of small molecules such as hormones and other molecules present in blood, urine, and biological samples. These are metabolites, and their whole composite, is the metabolome. The study of metabolites, and the metabolome, involving measurement, characterization, and analysis, is called metabolomics. Metabolites are the small, molecular intermediates, and products of metabolism. Primary metabolites are involved in the normal life cycle, while secondary metabolites have important ecological functions. The Human Metabolome Database contains over 41,500 entries, including water-soluble, and lipid-soluble metabolites, 1600 drugs, 3100 common toxins, and 28,000 food components, and additives. The database also includes over 5680 genes and protein sequences, and 440 human metabolic and disease pathways (www.hmdb.ca).

Common metabolic pathways like glycolysis or tri-carboxylic acid cycle, related to glucose breakdown are known. However, the study of thousands of metabolites, and their interdependent involvements in large networks of reactions, is quite complex. Metabolomics is the study of all this. Therefore metabolic phenotyping remains very important in the understanding of health and disease. Moreover, trillions of microbial cells present in our body have their own metabolome, which is constantly interacting with the human metabolome. This is known as the metabonome, and studies related to such supercomplex interactions are known as metabonomics. The importance and relevance of understanding such complex interactions has already been discussed in the chapter on systems biology.

The current objective of modern medicine is to develop personalized therapies tailored to an individual's pathophysiology. Patients' stratification based on genetic and/or phenotypic biology is usually viewed in terms of maximizing drug safety and efficacy. Every individual is uniquely different, and that is why everyone responds differently—even to similar therapeutic interventions. The generalized approach in therapeutics is one of the probable reasons for increased adverse drug reactions. This is a serious cause for concern, especially idiosyncratic toxicity, which is revealed when large populations are exposed to new therapeutics. These variations in response to therapies are mainly due to differences in the metabolic phenotypes [24]. Ayurveda gives a lot of importance to the health of the whole digestive track, gut, or *Maha Srotasa*. More details of these terms are given in the "Primer" on Ayurveda as an annexure.

The relationships between individual genomic and phenotypic variations in response to drug treatment are still poorly understood. Genetic information alone may not be sufficient to understand these differences. Factors such as gender, age, diet, gut microbiota, physical activity, latent disease, type of medication, hormones, and stress levels can impact the prevalence, and risk of disease. Most personalized approaches to drug treatment so far have been based on measuring genotype variation, and polymorphisms in drug-metabolizing enzymes (DME), such as CYP. However, mere pharmacogenomic predictions of drug metabolism and toxicity are not sufficient. Individual variations are also dependent on symbiotic relationships with a consortium of gut microbes and their individual variations.

Large numbers of microorganisms are present in the human body. Over 100 trillion bacteria, and other microorganisms, almost 10 times more than that of human cells are present. They have symbiotic relation with our physiological systems. They are mainly present in gut which actually functions like any *organ*. The gut microbiota or intestinal flora has many functions. The composition and nature of gut microbes is known to modify human phenotypes, influence health, disease etiology, and also affect drug metabolism, toxicity, and efficacy [25]. This is known as pharmacobonomics. Many experts are of the view that consideration of pharmacogenomics, and pharmacometabonomics may be required for personalized health care, and individualized drug therapy.

Emergence of Epigenetics

The unique identity of a genome is based on the DNA sequence, however, the terminal phenotypes may still vary due to developmental, nutritional, and environmental reasons. The gene expressions may be altered without any variation in the DNA sequence. The emerging science of epigenetics describes heritable alterations of gene expression that do not involve DNA sequence variations. Mechanisms responsible for epigenetic changes include DNA methylation and histone modification. The methylation process may knock down certain genes; this is known as *gene silencing*. These processes are influenced by experiences, food habits, and environmental exposures, and can change the way a genome functions without changing the DNA itself. Moreover, these changes could be passed on to the next generation. Now it is known that the epigenetic status may also influence drug response, and drugs may modulate the epigenetic status. Therefore, no longer are the discussions of personalized medicine restricted to genomics and metabolomics but are increasingly influenced by epigenetics.

PERSONALIZED MEDICINE AND AYURVEDA

The human personality is resultant of several attributes like physical, psychological, behavioral, emotional, and mental conditions. The four temperaments described by Hippocrates are purely psychological attributes. The Ayurveda concept of *Prakriti* is more comprehensive. It includes anatomical, structural features; and physiological, functional, and genetic features, together with behavioral, and psychological features. The Ayurvedic concept of *Prakriti* cuts across most of the existing classifications based on race, caste, creed, color, and physical, or mental frames. Somatotype may be considered as phenotype, which can change, whereas *Prakriti* is persistent, individual nature—or a trait which remains unchanged.

Specific permutations and combinations of all variables considered by Ayurveda practically extend to infinity. Therefore Ayurveda states that "every individual is different." Ayurvedic knowledge helps in defining characteristics of every individual as a unique entity in relation to the external environment. This forms the basis of personalized decisions about health and medicine. Such comprehensive understanding involving body–mind–spirit, in consonance with nature makes Ayurveda a holistic, yet personalized medicine.

Prakriti in Health and Disease

Ayurveda has a concept of three *Dosha*. There are three distinct types of *Dosha* known as *Kapha*, *Pitta*, and *Vata* each having distinct physical, physiological, and psychological characteristics (Figure 11.1). Every person is supposed to have various proportions of all the three *Dosha*. According to

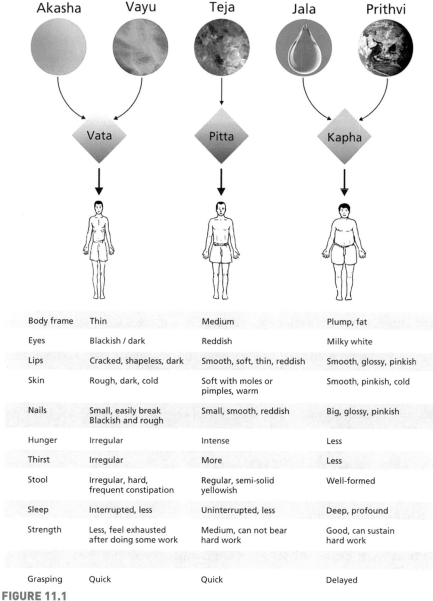

Body frame	Thin	Medium	Plump, fat
Eyes	Blackish / dark	Reddish	Milky white
Lips	Cracked, shapeless, dark	Smooth, soft, thin, reddish	Smooth, glossy, pinkish
Skin	Rough, dark, cold	Soft with moles or pimples, warm	Smooth, pinkish, cold
Nails	Small, easily break Blackish and rough	Small, smooth, reddish	Big, glossy, pinkish
Hunger	Irregular	Intense	Less
Thirst	Irregular	More	Less
Stool	Irregular, hard, frequent constipation	Regular, semi-solid yellowish	Well-formed
Sleep	Interrupted, less	Uninterrupted, less	Deep, profound
Strength	Less, feel exhausted after doing some work	Medium, can not bear hard work	Good, can sustain hard work
Grasping	Quick	Quick	Delayed

FIGURE 11.1

Unique characteristics of *Vata*, *Pitta*, and *Kapha Prakriti*.

Ayurveda, the nature of every individual is determined by the proportional dominance of these *Dosha*. The resultant expression of three *Dosha* in every person is known as *Prakriti*. The *Prakriti* types—*Kapha Prakriti*, *Pitta Prakriti*, and *Vata Prakriti*—classify the human population in three broad categories.

According to Ayurveda, any imbalance and vitiation of *Dosha* is considered a main reason of ill health. Generally, normal *Prakriti* type may be considered as healthy genotype, which has a key role in management of health, and prevention and treatment of disease. The therapeutic strategies are based on *Prakriti*-specific interventions, through lifestyle and diet modifications, and drugs and detox procedures aimed at regaining the balance, or homeostasis. The therapeutic interventions may bring the state back to its *trait*, or phenotype, but do not alter the inherent nature, or constitution. Therefore, *Prakriti* has an important role in the selection of drugs, doses, and vehicle. For example, *Piper longum* may be the drug of choice for asthma treatment for *Kapha Prakriti* patients, but it is contraindicated for *Pitta Prakriti* patients. Health promotion regimens suggested by Ayurveda are easy ways of avoiding *Dosha* imbalance, and in maintaining them at a normal or healthy state.

Almost every individual has specific proportions of all the three V, P, K *Dosha Prakriti* types. They are as specific, and individualized as the DNA sequence-based genetic makeup. *Prakriti*-specific treatment, including medicine, diet, and lifestyle is a distinctive feature of Ayurveda.

A few studies have demonstrated an association between *Prakriti* subgroups and etiology in autoimmune conditions. In a control study comparing the rheumatoid arthritis cohort with controls, genes related to inflammatory pathways were found dominant in *Vata* subgroup, and those for oxidative stress were significantly associated with the *Pitta*, and *Kapha* subgroups [26].

The association of *Prakriti* with cardiovascular risk factors, inflammatory markers, and insulin resistance in patients with coronary artery disease has been studied. This study concluded that proneness of *Vata–Kapha* type to have high triglycerides—VLDL, and LDL, and lower HDL. Inflammatory markers (IL6, TNF alpha, hsCRP, and HOMA IR) were high in this group [27]. *Prakriti*-based response in platelet aggregation (MPA) was observed in normal, healthy participants, and it was highest in *Vata* and *Pitta Prakriti* [28].

An observational study reports on the constitutional typing of individuals with known idiopathic Parkinson's disease (PD) [29]. This study shows *Vata Prakriti* patients are at higher risk for PD when compared with non-Parkinson's controls. That Parkinson's disease is a degenerative disorder of the central nervous system can be considered as a manifestation of *Vata* vitiation. *Prakriti* is an important determinant of prognosis. More systematically designed studies may provide data on *Prakriti* determinants like diet, season, nutritional status, and *Dosha* status of infants.

The role of *Prakriti* may be useful as a guiding principle for healthy progeny, and also to prevent many diseases, which may have fetal or preconception origin. David Barker's hypothesis of fetal origins of adult disease suggests that events during early development have a profound impact on risk of development of adult disease in future [30]. For example, low birth weight as a surrogate marker of poor fetal growth, and nutrition is linked to coronary artery disease, hypertension, obesity, and insulin resistance during adulthood [31].

ASSESSMENT OF *PRAKRITI*

The precise identification of *Prakriti* in current clinical practice is quite challenging because of the subjective and qualitative nature of evaluation. Current questionnaires and methods of *Prakriti* assessment require validated tools, which are accurate and reliable [32]. Complex genomic studies also require advanced, and high-throughput technologies, and computational bioinformatics tools to make some sense of the vast amounts of genomic data. Along these same lines, the large knowledge base of Ayurveda involving over 30,000 verses from classic texts, providing specific logic and algorithmic messages has been developed into software (decision support system). For this purpose, a collaborative project between the Government of India's Center for Development of Advanced Computing (C-DAC), and the author's institute, University of Pune, to develop AyuSoft as a decision support system was embarked upon. AyuSoft converts the logic of classical Ayurvedic texts into authentic, intelligent, and interactive knowledge repositories, with the help of complex analytical tools. The AyuSoft database includes more than five hundred thousand records, and captures information from nine classic texts. AyuSoft is a suite of software with knowledge, and data mining tools, and a query-based decision support system for research, and clinical practice. It offers a *Prakriti* assessment tool, through an interactive application. The data mining tool enables precise information searches using Boolean operators. Information related to diseases, causative factors, symptoms, treatment guidelines, drugs, dietary recipes, lifestyle changes, and treatment procedures can be searched through complex queries employing any number of combinations of search strings. A search engine was also developed as a part of AyuSoft.

A comprehensive database of human constitution, disease, detailed symptoms, treatment logic, and drug activity libraries may facilitate individualized, standardized, and uniform treatment based on principles of Ayurveda. *Prakriti*-based preventive medicine may have the potential to strengthen personalized health management.

GENETIC BASIS OF *PRAKRITI*

Even a broad level understanding of the *Prakriti* concept indicates that it may have a genetic connotation. While the data on Human Genome Project were being deciphered, genomic-based drug discovery, and personalized medicine was emerging on the horizon, a team of scientists from Pune, India, hypothesized that *Prakriti* has a genetic connotation that can provide a tool for classifying human populations based on broad phenotype clusters. As a proof of this concept, a first study evaluated 76 subjects both for their *Prakriti* and HLA DRB1 types. This study revealed that the HLA DRB1 allele distribution was significantly different among V, P, and K types. The study concluded that Ayurveda-based phenomes may provide a model to study multigenic traits, possibly offering a new approach for correlating genotypes with phenotypes in human classification.

In another set of experiments, the same team of scientists from the University of Pune hypothesized that *Prakriti* might relate to drug metabolism and genetic polymorphism of DME. Interindividual variability in drug response can be attributed to polymorphism in genes encoding different DME, drug transporters, and enzymes involved in DNA biosynthesis, and repair. Gene polymorphism precipitates in different phenotypic subpopulations of drug metabolizers. Poor metabolizers (PM) have high plasma concentration of the drug for longer periods, and so retain drugs in the body for longer times. Intermediate metabolizers retain drugs in the body for normal time periods. Extensive metabolizers (EM) retain drugs in the body for the least amount of time and plasma concentrations are high for shorter periods.

Metabolic variability in different *Prakriti* types was studied using the DME CYP2C19/CYP2C9 gene polymorphism model. The distribution of CYP2C19 and CYP2C9 genotypes was investigated in 132 healthy individuals of different *Prakriti* classes. The results obtained suggest a possible association of CYP2C19 gene polymorphism with *Prakriti* phenotypes. EM genotypes were predominant in *Pitta* type, while PM genotypes were highest in *Kapha*. It is interesting to note that the genotype specific to the extensive metabolizer group was present only in *Pitta*, while the genotype typical of the poor metabolizer group was observed in *Kapha*, as expected. Similarly, in the case of gene polymorphism, we observed that the occurrence of EM genotypes was significantly higher in *Pitta Prakriti*. Thus, this study demonstrated a probable genomic basis for metabolic differences attributable to *Prakriti*, possibly providing a new approach to pharmacogenomics [33].

Another set of studies from India could establish a link between variations in EGLN1, and high-altitude adaptation, as well as susceptibility to high-altitude pulmonary edema. This study takes the lead from gene expression and genetic differences in normal individuals identified from three contrasting constitution types described in Ayurveda [34]. Of course, these studies involved only a few genes, and for more definitive relevance, genome-wide expression, and biochemical differences studies are required. A study undertaken by scientists from India showed that, overexpression of genes related to immune response was predominant in *Pitta*, whereas expression of genes related to cellular processes was more apparent in *Vata*, and upregulation of genes involved in cellular biosynthesis was predominant in *Kapha*. A number of hub genes like TLR-4, FAS, and HLA-DQB1 linked to complex diseases, and a few genes like DPYD, ABCC1, and FTL, associated with the outcome of cancer treatment, were present in data sets of genes differentially expressed among *Prakriti* types. Despite a few limitations, this study provided evidence that differences exist across extreme constitutional types, at the gene expression level.

Currently, under the scheme, Science Initiatives in Ayurveda, the Government of India's Department of Science and Technology, has initiated an ambitious project on genomic variation analysis, and gene expression profiling, using principles of Ayurveda. The study demonstrated immune-phenotype patterns in three distinct *Prakriti*. The expression of lymphocyte subset CD markers (CD14, CD25, and CD56) was different in the three *Prakriti* types. The data suggest increased level of CD25 and CD56 in *Kapha* dominant volunteers—indicating better immune response [35]. This feature of *Kapha Prakriti* is also described in Ayurveda texts. This initiative supports robust research with scientific precision, to further elucidate the fundamental concepts of Ayurveda.

Of course, these are preliminary studies, pointing the way as we begin to understand individual variations, and the genetic basis of *Prakriti*.

AYUGENOMICS

From perspective of modern biology, this concept could be well explained if we look at disease in terms of altered expression of the genes, and protein interactions influenced by external factors, such as nutrition, environment, hormones, and stress. Even though the genome is unaltered throughout the life, gene expression is influenced by external factors. Comparative genomics could differentiate between disease prone and disease resistant populations, populations which are drug responsive, and those that are nonresponders [36]. SNP mapping, RFLP, and micro-, and minisat mapping associated with the three *Prakriti* types will enable an understanding of their genomic basis. The modulation of genes or the differential expression leading to complex phenotypic differences as manifestations of K, P, and V, could be tracked by technologies such as microarrays.

AyuGenomics is an effort to develop new strategies of drug discovery, and personalized medicine by integrating Ayurveda with the modern technologies of genomics, proteomics, and metabolomics. Understanding the possible genetic and epigenetic relationship between *Prakriti* and the genome is important. Functionally, this will involve the creation of three organized databases. The first relates to the human constitution as genotypes; the second relates to disease constitution as phenotype; and the third relates to drug constitution or *Prakriti*. Following the logic of Ayurveda, these databases are capable of intelligently communicating with each other, which may result in a personalized prescription. In this systems approach, we visualize optimal use of traditional knowledge, and modern technologies to aid the practitioner's decision-making, and in the creation of personalized prescriptions. A separate technology similar to any vending machine may be developed for dispensing these prescriptions.

In short, Ayurvedic therapeutics is a complex process involving several variables. *Prakriti* is the essence of variables that help practitioners to design appropriate treatments, suited for an individual in a particular stage, environment, and lifestyle. *Prakriti* is a healthy genotype, and *Vikriti* is a disease phenotype. *Prakriti* denotes healthy physiology, and *Vikriti* denotes disease pathology. The transition of *Vikriti* to restore *Prakriti* can be attempted by interventions in diet, and lifestyle modifications, and, if necessary, therapeutic procedures and medicines. This is akin to a paint shop where you can design an original shade by systematically mixing several colors, in several proportions. In Ayurveda, the original shade, or *Prakriti*, can be restored by balancing the vitiated shade or *Vikriti*, through appropriate interventions. This is akin to a bar code that gives a unique identity to any product, or a DNA sequence that gives a unique identity to an individual.

The promise of personalized medicine through the genomic route may be a drawn-out, complicated, and expensive route. We feel that experiential rigor, and advances of omics technologies, integrated with the knowledge and wisdom of Ayurveda may provide a better, faster, and more affordable route as we move toward personalized medicine.

APPLICATIONS FOR THE FUTURE

Ayurveda has indicated the importance of human stratification; what has been done so far is apply the technology to verify basic concepts. In the aforementioned studies related to AyuGenomics, the traits, and respective proneness to disease are made obvious. AyuGenomics studies demonstrated expected upregulations, downregulations, and selective expressions of specific genes. These studies have provided confidence to probe further. Now *Prakriti* research need not be limited to the study of a few genes. Whole genome sequencing and next-generation sequencing may provide insights to Ayurveda concepts (Figure 11.2).

Epigenetic Applications

Epigenetic regulations are heritable changes in gene expression that occur in the absence of alterations in DNA sequences. Various epigenetic mechanisms include histone modifications and DNA methylations. Such epigenetic

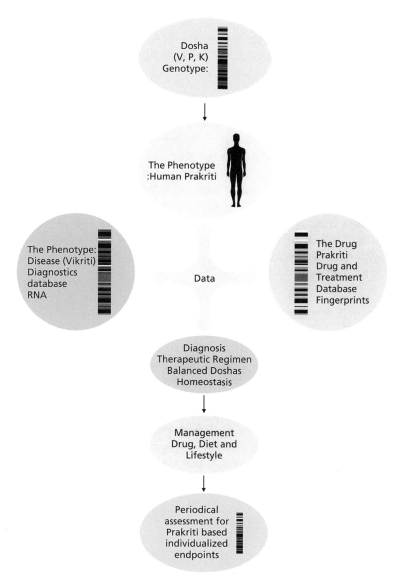

FIGURE 11.2
Integrative approach for personalized medicine.

modifications, despite being heritable, and stably maintained, are also potentially reversible, and there is scope for the development of epigenetic therapies.

"*Prakriti* Programming" of Progeny

David Barker first proposed the concept of fetal origins of adult disease, which suggests that events during early development have a profound impact on risk for development of adult disease in future. For example, low birth weight as a surrogate marker of poor fetal growth, and nutrition is linked to coronary artery disease, hypertension, obesity, and insulin resistance during adulthood. Ayurveda considers method of procreation as a planned and systematic process that aims at balancing or alleviating vitiated *Dosha* of the parents with the help of *Panchakarma*, and modulating certain *Dosha* in the progeny with certain drugs, and diet and lifestyle modifications. The Ayurveda treatment for healthy pregnancy is popularly known as *Garbhasamskar* in India. However, there is a need of systematic studies on possible manipulation of the preconception, or intrauterine environment to influence the gene expression, and methylation. How might parental influences stably alter the *Prakriti* phenotype? Could such effects involve the *programming* of gene expression toward better health? How might such stable influences occur? Answers to these questions may reveal new knowledge about programming the healthy progeny through additional care, with diet and selected therapies, which are time-tested, and safe.

Prakriti and Predictive Medicine

A routine, normal health checkup, according to modern medicine, involves mostly pathological tests, and actually is not an assessment of *health*. *Prakriti* evaluation may be useful as an easy health assessment tool and can provide valuable guidance in disease prevention, prognosis, and treatment. The assessment of the health of an individual requires the consideration of the *propensity* of that individual to various health and disease states. This specifies a baseline of health and different normal ranges for the respective individuals. In a clinical trial on osteoarthritis patients, the authors observed different body mass indexes specific to *Kapha*, *Pitta*, and *Vata Prakriti*. In another study, researchers found different levels of thyroid hormones in three distinct *Prakriti* subtypes. Longitudinal cohort studies following distinct *Prakriti* types may provide more information about disease proclivity.

Prakriti–environment interaction affects *Dosha* status of an individual. The focus of Ayurveda management is the relative balance of *Dosha*, considering the *Prakriti* of the respective individual. Understanding *Prakriti* helps to control or prevent *Dosha* vitiation. Theoretically, if we can design a balanced state of *Dosha* or a *Sama Prakriti* of the fetus, then we can achieve good immune status, and longer life. At least by considering the family history, and possible disease risks in the progeny, modulation of *Prakriti* of a fetus can play an important role in the prevention of respective diseases. Ayurveda considers method of procreation as a planned and systematic process that aims at balancing or alleviating vitiated *Dosha* of parents, with the help of *Panchakarma*, and modulating certain *Dosha* in progeny with drugs, diet, and lifestyle modifications.

Understanding Physiological Variations

Each *Prakriti* type has a specific predisposition for diseases, due to the inherent dominance of specific *Dosha*. The dominant *Dosha* expresses certain characteristics in a typical way; some of those could be regarded as pathological. The stratification of clinical and biological variables based on *Prakriti* may explain the underlying physiological variation. Exploration of physiological responses (with a focus on biochemistry and immune functions) in various *Prakriti* types is needed to understand mechanisms of disease proneness in various *Prakriti*. The variation may also redefine concepts of *normality* that exist today.

Leads to Modern Diagnostics

Genome-wide association studies (GWAS) focus on assessing genetic variants in different individuals. A group of patients is compared with healthy individuals to assess the relative presence of genetic variants in both the groups. These studies are crucial in diagnostic research, as they find associations of particular genotypes with diseases. GWAS studies face the problems of false positive or false negative findings due to participants' heterogeneity [37]. Several efforts are underway to study the inherent variation among people [38,39]. *Prakriti*, is an important variable which considers several anatomical, physiological, and psychological variables on a logical foundation, and may help in stratifying individuals. This classification may help in understanding variations in GWAS. This could be the contribution of *Prakriti* to modern diagnostics.

Prakriti, *Dosha*, and Chronobiology

The study of chronobiology in the context of *Prakriti* will be very interesting. Each *Dosha* has specific diurnal variations and those are dominant in the respective *Prakriti*. The variations in physiological functions pertaining to *Prakriti* are an important aspect for study. We may find a relationship of *Prakriti* to neurotransmitters, metabolism, and sleep patterns.

Research on *Prakriti* assessment methods is needed. Validated assessment tools with better accuracy and reliability are needed. The current questionnaires and tools are important in Indian (Asian) contexts. We need various versions of these tools, specific to various populations, with relevant interpretation of Ayurveda terms, and application of specific *Prakriti*-determining variables.

Prakriti-based stratification will help to predict and reduce adverse events of drugs with narrow therapeutic ranges. The clinical trials, especially on Ayurveda intervention, can use *Prakriti* as a determinant to reduce adverse drug events by administering *Prakriti*-specific *anupana* (vehicle) and dose adjustment.

Together, modern medicine and biomedical sciences are moving toward personalized medicine. However, next-generation sequencing and the robust use of bioinformatics are not enough to understand individual variations among human beings. Newer efforts like predictive, preventive, promotive, participatory, and personalized medicine are good, but their translation into clinical decision-making still seems like a distant dream. High-throughput technologies can turn up huge amounts of data, where meaningful and exciting biological understanding can be lost.

At present, the speed of technology is much faster than the scientists' ability to make sense from the constantly emerging gigantic data. We fear that laudable initiatives like P3, P4, or P5 medicine are likely to get buried under the overload of data. We feel that integration of the genomic approach with traditional knowledge systems like Ayurveda and Yoga may facilitate personalized health, and not just medicine. The concepts like *Dosha, Prakriti, Guna, Srotas, Agni*, from Ayurveda can be studied with help of omics, and computational technologies in order to understand the science behind the ancient wisdom. With this approach, we feel that personalized health and medicine can be a reality.

REFERENCES

[1] Roses AD. Pharmacogenetics and the practice of medicine. Nature 2000;405(6788):857–65.

[2] Preece R, Fraser D. The status of animals in biblical and Christian thought: a study in colliding values. Soc Animals 2000;8(3):245–63.

[3] www.nlm.nih.gov/exhibition/shakespeare/fourhumors.html.

[4] Wong W, Lam CL, Wong VT, Yang ZM, Ziea ET, Kwan AK. Validation of the constitution in Chinese medicine questionnaire: does the traditional Chinese medicine concept of body constitution exist? Evid Based Complement Altern Med 2013;2013:481491. http://dx.doi.org/10.1155/2013/481491.

[5] Yao SL, Zhang ZZ, Yang XS, Xu X, Cao J, Xie GY et al. Analysis of composite traditional Chinese medicine constitution: an investigation of 974 volunteers. J Chin Integr Med 2012;10(5):508–15.

[6] Wang J, Wang Q, Li L, Li Y, Zhang H, Zheng L et al. Phlegm-dampness constitution: genomics, susceptibility, adjustment and treatment with traditional Chinese medicine. Am J Chin Med 2013;41(2):253–62.

[7] Do JH, Jang E, Ku B, Jang JS, Kim H, Kim JY. Development of an integrated Sasang constitution diagnosis method using face, body shape, voice, and questionnaire information. BMC Complement Altern Med 2012;12:85. http://dx.doi.org/10.1186/1472-6882-12-85.

[8] Kim D. A comparative study of Korean oriental medicine & Indian traditional medicine. J Korean Orient Med 2005;26:201–16.

[9] Coon C. The origin of races. 1st ed. New York: Alfed A. Knopf, Inc; 1962.

[10] Sirri L, Fava GA, Guidi J, Porcelli P, Rafanelli C, Bellomo A, et al. Type A behaviour: a reappraisal of its characteristics in cardiovascular disease. Int J Clin Pract 2012;66(9):854–61.

[11] Friedman M, Rosenman R. Association of specific overt behavior pattern with blood and cardiovascular findings. JAMA 1959;169(12):1286–96.

[12] Ragland DR, Brand RJ. Type A behavior and mortality from coronary heart disease. N Engl J Med 1988;318(2):65–9.

[13] Eysenck HJ. Biological basis of personality. Nature 1963;199:1031–4.

[14] Eysenck HJ. Dimensions of personality: 16, 5 or 3?—Criteria for a taxonomic paradigm. Pers Individ Differ 1991;12(8):773–90.

[15] Poropat AE. A meta-analysis of the five-factor model of personality and academic performance. Psychol Bull 2009;135(2):322–38.

[16] Sheldon HW. Atlas of man. New York: Gramercy Publishing Company; 1954.

[17] Carter LJE. The Heath-Carter anthropometric somatotype instruction manual; 2003.

[18] Lele RD. Ayurveda and modern medicine. Mumbai, India: Bharatiya Vidya Bhavan; 2001.

[19] Bruce Elizabeth AF. Humans and apes are genetically very similar. Nature 1978;276(5685):264–5.

[20] Race, Ethnicity and GWG. The use of racial, ethnic, and ancestral categories in human genetics research. Am J Hum Genet 2005;77(4):519–32.

[21] Frueh FW, Amur S, Mummaneni P, Epstein RS, Aubert RE, DeLuca TM, et al. Pharmacogenomic biomarker information in drug labels approved by the United States food and drug administration: prevalence of related drug use. Pharmacotherapy 2008;28(8):992–8.

[22] Hamburg MA, Collins FS. The path to personalized medicine. N Engl J Med 2010;363(4): 301–4.

[23] Heseltine PN, Goodkin K, Atkinson JH, Vitiello B, Rochon J, Heaton RK, et al. Randomized double-blind placebo-controlled trial of peptide T for HIV-associated cognitive impairment. Arch Neurol 1998;55(1):41–51.

[24] Holmes E, Wilson ID, Nicholson JK. Metabolic phenotyping in health and disease. Cell 2008;134(5):714–7.

[25] Li M, Wang B, Zhang M, Rantalainen M, Wang S, Zhou H, et al. Symbiotic gut microbes modulate human metabolic phenotypes. Proc Natl Acad Sci USA 2008;105(6):2117–22.

[26] Juyal RC, Negi S, Wakhode P, Bhat S, Bhat B, Thelma BK. Potential of Ayurgenomics approach in complex trait research: leads from a pilot study on rheumatoid arthritis. PLoS One 2012;7(9):e45752. http://dx.doi.org/10.1371/journal.pone.0045752.

[27] Mahalle N, Pendse N, Kulkarni M, Naik S. Association of constitutional type of Ayurveda with cardiovascular risk factors, inflammatory markers and insulin resistance. J Ayurveda Integr Med 2012;3(3):150–7.

[28] Bhalerao S, Deshpande T, Thatte U. Prakriti (Ayurvedic concept of constitution) and variations in platelet aggregation. BMC Complement Altern Med 2012;12:248. http://dx.doi.org/10.1186/1472-6882-12-248.

[29] Manyam BV, Kumar A. Ayurvedic constitution (prakruti) identifies risk factor of developing Parkinson's disease. J Altern Complement Med 2013;19(7):644–9.

[30] Barker D, Eriksson J, Forsen T, Osmond C. Fetal origins of adult disease: strength of effects and biological basis. Int J Epidemiol 2002;31(6):1235–9.

[31] Calkins K, Devaskar SU. Fetal origins of adult disease. Curr Problems Pediatr Adolesc Health Care 2011;41(6):158–76.

[32] Sanjeev R. Development and validation of a Prototype Prakriti Analysis Tool (PPAT): inferences from a pilot study. Ayu 2012;33(2)209–18.

[33] Joshi K, Ghodke Y, Patwardhan B. Traditional medicine to modern pharmacogenomics: ayurveda Prakriti type and CYP2C19 gene polymorphism associated with the metabolic variability. Evid Based Complement Altern Med 2011;2011:249528. http://dx.doi.org/10.1093/ecam/nep206.

[34] Aggarwal S, Negi S, Jha P, Singh PK, Stobdan T, Pasha MA, et al. EGLN1 involvement in high-altitude adaptation revealed through genetic analysis of extreme constitution types defined in Ayurveda. Proc Natl Acad Sci USA 2010;107(44):18961–6.

[35] Rotti H, Guruprasad KP, Nayak J, Kabekkodu SP, Kukreja HSK, et al. Immunophenotyping of normal individuals classified on the basis of human dosha prakriti. J Ayurveda Integr Med 2014;5(1):43–9.

[36] Wilson JF, Weale ME, Smith AC, Gratrix F, Fletcher B, Thomas MG, et al. Population genetic structure of variable drug response. Nat Genet 2001;29(3):265–9.

[37] Begum F, Ghosh D, Tseng GC, Feingold E. Comprehensive literature review and statistical considerations for GWAS meta-analysis. Nucleic Acids Res 2012;40(9):3777–84.

[38] Melén E, Granell R, Kogevinas M, Strachan D, Gonzalez JR, Wjst M, et al. Genome-wide association study of body mass index in 23 000 individuals with and without asthma. Clin Exp Allergy 2013;43(4):463–74.

[39] Zhang G, Karns R, Sun G, Indugula SR, Cheng H, Havas-Augustin D, et al. Finding missing heritability in less significant loci and allelic heterogeneity: genetic variation in human height. PLoS One 2012;7(12):e51211. http://dx.doi.org/10.1371/journal.pone.0051211.

CHAPTER 12

Integrative Approaches for the Future

Our great mother, Eywa, does not take sides; she only protects the balance of life.

James Cameron, Avatar, 2009

FOUNDATION FOR INTEGRATION

In earlier chapters, we discussed the fact that our current understanding of health care is rooted more in terms of absence of disease. Today's health care consists of treatment through medical care. With the integration of modern science and technology, the practice of modern medicine became specialized, mechanical, compartmentalized, and commercialized. Today's medicine and health care have been shaped by powerful advances in science and technology, the impact of the industrial era, and the resultant changes in the socioeconomic fabric of global society, and increasing commercialization of the service professions. In the process, the practice of modern medicine has become too symptom-oriented, and algorithmic; now the focus on pathology and on diagnostic reports and technology overshadows the clinical acumen of doctors, and affects their potential bedside interface or approach toward healing.

Western modern medicine has contributed to a substantial reduction of mortality from infectious disease; this has led to a significant reduction of infant and child mortality, and morbidity, and to resultant longevity. Advances in modern surgery have saved countless lives that were lost due to surgical emergencies. However, the limitations of these advances have clearly failed to make an impact on chronic conditions, and lifestyle disorders, which are predominate worldwide—both in industrialized and developing societies. The rising cost of medical care has imposed further limitations on the impact of modern medicine on global health.

It is time for the focus of medicine to shift from illness to wellness, from treatment to prevention and early diagnosis, and from a generalized approach to personalized medicine. Treating Illness is passive, and individual-based, while Wellness requires participatory approach (Figure 12.1). Ayurveda and Yoga highlight the importance of participating together in the pursuit of health and

319

FIGURE 12.1

Ayurveda and Yoga approach teach importance of working together, humility, and respect for nature. Ayurveda and Yoga underline that it is in our hand to keep the tree of life green by paying heed to inner and outer harmony with nature or let the imbalance contribute to its withering away.

wellness—with humility and respect for the environment—and in defining our own role and responsibility in the process. This situation gives the *raison d'être* for replacing the reductionist approach with the integrative approach.

Today, pharmaceutical and insurance companies are defining health and disease, and driving medical practice in clinics and hospitals. These actors, with their technological orientation, and their profit-based motive for medical practice, have distorted the patient–doctor relationship into becoming one of "conveyer belt" commerce.

There is a massive erosion of trust, ethics, and values that once elevated the past practice of medicine into a truly noble vocation. A major tectonic disruption is needed in the philosophy and technology of modern medicine in order for it to evolve as the medicine of tomorrow. We hope that integrative approaches can catalyze the overdue paradigm shift that medicine needs.

BROADER VISION OF INTEGRATION

The value of integrative approaches is becoming increasingly clear in many superspecialty areas such as oncology, cardiology, neurology, dermatology, psychiatry, and geriatrics. This book is about *Integrative Approaches for Health*, and not just combining medicine, treatments, and therapies. Our model necessitates a broader vision, and an open mind characterized by integrity, objectivity, and the ability to comprehend wholeness.

It is high time to discard the silos mentality. Modern medicine alone cannot fulfill our current requirements. Biomedical professionals should not monopolize medical care or health care. The limitations of modern medicine—especially in managing chronic, behavioral, and lifestyle diseases—are becoming clearer. The present approach of evidence-based medicine must not become too rigid, and restrictively protocol-based. Modern medicine cannot lose sight of the person behind the patient.

At the same time, Ayurveda and Yoga professionals cannot continue to position themselves as ancient traditions anymore. They cannot exist merely on pride, and past glory; nor can they remain dogmatically ritualistic. They must learn the methodology of science, and the understanding of evidence. They must come out of relying on unproven anecdotal evidence, and embrace scientific research attitude. They must be open to questioning, and experimentation. While retaining their pride in their heritage, they must move with the times to contribute to the further development and growth of their ancient wisdom.

It is heartening to note that the concept of integrative medicine (IM) is now getting wider recognition from the credible scientific bodies like the Institute of Medicine (IOM) of the National Academies in the United States. A preface in the landmark IOM report on IM is indicative of the future trend. In this report, Dr Ralph Snyderman, Chair, Planning Committee for the Summit on Integrative Medicine and the Health of the Public states: "Of course, no single approach could be identified as the solution, but it was broadly agreed that health, and health care must be centered on the needs of the individual throughout his or her life, supporting the individual's capability to improve health and well-being, to predict and prevent chronic disease, and to treat it effectively and coherently when it occurs. Approaches to care must be evidence-based, yet caring and compassionate. Fortunately, many such integrative approaches already exist on which demonstration projects might be built to identify and validate the best integrative solutions to the various health-care delivery needs" [1].

Another important report from the Bravewell Collaborative states that "Integrative medicine is an approach to care that can be easily incorporated by all medical specialties, and professional disciplines, and by all health-care systems. Its use will not only improve health care for patients, it can also enhance the cost effectiveness of health-care delivery for providers and payors."

According to Dr Lawrence Green from the University of California, San Francisco, "Research in integrative medicine can shift the spotlight from mediating variables that focus on the mechanisms of change, to the moderating variables that focus on the characteristics of individual people, and the context in which they live." Dr Don Berwick of the Institute for Healthcare Improvement in the United States aptly summarized the essence of integrative spirit in his talk at the IOM conference when he stated, "The sources of suffering are in separateness, and the remedy is in remembering we are in this together."

Undoubtedly, there is a growing consensus among top medical professionals and establishments that the integrative approach is simple, scientific, affordable, and whole-person, patient-centric; and can transform the current, sick health care system into the vibrant, evidence-based, holistic, and humane health care of the future [2].

PRINCIPLES FOR INTEGRATION

True integration is not to be mistaken for a mere cross-practice, or pluralistic system of medicine. The integrative approach is required at many levels.

1. Integration of Epistemology: Philosophy and Science
2. Integration of Philosophy: Western and Eastern
3. Integration of Logic: Reductive and Holistic
4. Integration of Science: Basic and Applied
5. Integration of Systems: Macrocosm and Microcosm
6. Integration of Theory: Linear and Nonlinear
7. Integration Concepts: Health-Disease-Illness-Wellness
8. Integration of Entities: Body-Brain-Mind-Spirit
9. Integration of Biosocial Organization: Personal-Social-Ecological-Spiritual
10. Integration of Determinants: Nutrition, Environment, Lifestyle, and Genetics
11. Integration of Interventions: Drug-Diet-Lifestyle-Behavior
12. Integration of Evidence: Experimental and Experiential
13. Integration of Strategy: Protection, Promotion, Prediction, Prevention, Participation, and Personalized (P6)

We hope that the overarching P6 strategy can emerge as a rainbow of IM—bringing more hope for the people, and a brighter future for universal health care (Figure 12.2).

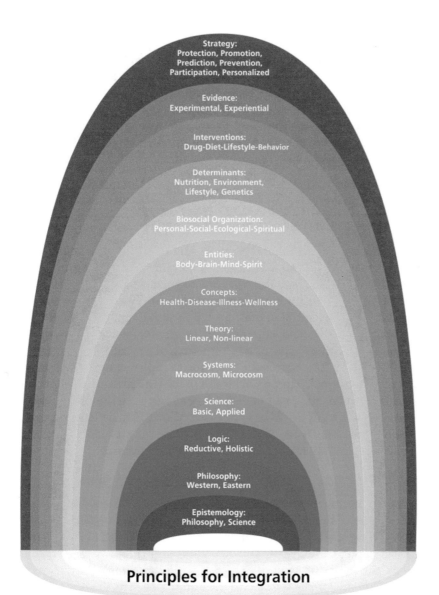

Principles for Integration

FIGURE 12.2

The figure suggests principles of Integrative Medicine evolving from philosophy to strategies for holistic approach. Tomorrow's medicine will be focused on health protection, promotion and personalized, preventive, participatory and personalized approached of medicine.

ATTEMPTS AT INTEGRATION

The Global Initiatives

Many developed and developing countries have started strategically moving toward integrative approaches for medicine. This trend is in-line with popular demand by the patient community, and will prevent much of the excess use of potent, modern drugs, which cause iatrogenic illnesses. The WHO has also recommended the judicious use of T&CM practices. Many countries such as Australia, Canada, China, Germany, Ghana, Hungary, India, Indonesia, Italy, Japan, Korea, Malaysia, Nigeria, Norway, South Africa, Uganda, and the United States have attempted various types of integration.

In some countries, integration is legal and formalized, in many others, the process is still evolving. For instance, Ghanaian health authorities took a major step to resolve a seemingly contentious issue in the fight against HIV/AIDS in the country by encouraging T&CM practitioners [3]. The Uganda Law Reform Commission has a law for the recognition, and the protection, and practice of T&CM. This enables national institutions, and international organizations to promote, and integrate herbal medicine into their development plans. Traditional and Modern Health Practitioners Together against Aids and Other Diseases (THETA), is an Ugandan organization in which traditional, and modern health practitioners work together [4]. The integration of T&CM into the Nigerian health care delivery system has been studied for legal implications, and complications [5]. The National Center for Complementary and Integrative Health of the National Institutes of Health in the United States presents a best-case scenario where an open-minded, science-, and evidence-driven approach is adopted for the study of T&CM approaches. Globally, many researchers, institutions, initiatives, and scientific journals are promoting integrative approaches for health, through clinical services, and biomedical research.

The Indian Scenario

In India, pioneering institutes like the Banaras Hindu University have adopted the integration of modern medicine superspecialties, and traditional practices such as Ayurveda and Yoga. A few universities and centers of excellence at Jamnagar, Jaipur, Coimbatore, Kottakal, Pune, and Udupi are endeavoring to maintain the purity of Ayurveda practice, while remaining open to modern research. Similarly, many specialized institutions, such as the Bihar School of Yoga in Munger, Kaivalyadhama Yoga Institute, Pune, the Swami Vivekananda Yoga Anusandhan Samstha (SVYASA), and the National Institute of Mental Health and Neurosciences (NIMHANS) in Bangalore have been involved in the development of unique, integrative models of Yoga, and biomedical research.

Several eminent physician-scientists from India are doing exemplary work in IM—just to mention a few Dr B.M. Hegde of Manipal University; Dr R.D. Lele of Jaslok Hospital; Dr Rama and Ashok Vaidya of Kasturba Health Society; Dr G.N. Qazi from Hamdard University; Dr Arvind Chopra of the Center for Rheumatic Diseases; Dr Saravu Narahari of the Institute of Applied Dermatology; and Dr Seetharaman of Vivekananda Memorial Hospital in Karnataka. From the Ayurveda and Yoga community, Dr H.R. Nagendra of SVYASA; Dr Krishna Kumar, and Dr Ram Manohar of Arya Vaidya Pharmacy; Dr Gangadharan of M.S. Ramaiah Indic Center for Ayurveda and Integrative Medicine; Dr Issac Mathai of the Soukya International Holistic Health Center; Dr R.H. Singh, and Dr G.P. Dubey of Varanasi; Dr M.S. Baghel of Jamnagar; and Dr Abhimanyu Kumar of Delhi, among others, are promoting the cause of IM. Distinguished scientists and National Research Professors Dr R.A. Mashelkar and Dr M.S. Valiathan have been promoting high-end research projects in the area of traditional medicine and Ayurvedic biology. While modesty forbids mentioning our role in this process, we, the authors, also have made a contribution in advocating the integrative approach to health.

A renowned cardiac surgeon from India, Dr Naresh Trehan and Ayurveda physician, Dr Geethakrishnan, have taken the lead in the systematic integration of modern and ancient treatments at the state of the art Medanta Hospitals near New Delhi; they are developing a sustainable prototype for IM [6]. As a result of two decades of hard work by Darshan Shankar and his team, in January 2014, the government of Karnataka conferred university status upon the Foundation for Revitalisation of Local Health Tradition (FRLHT), as The Institute of Transdisciplinary Health Sciences and Technology (IHST). Their health center functions as the Institute of Ayurveda and Integrative Medicine (I-AIM). The establishment, in 2010, of the peer-reviewed *Journal of Ayurveda and Integrative Medicine* (www.jaim.in), is another step in the direction of modern and ancient treatments—but there is still a long way to go.

Recently, India has taken steps toward mainstreaming Ayurveda, Yoga, Unani, Siddha, and Homeopathy (AYUSH) systems in the public health system. To the end of achieving universal health care in India by 2020, the creation of the Integrated National Health System has been proposed [7]. This will involve mainstreaming AYUSH in public health, especially at village levels for primary care.

The United States and Australia
Historically, the integrative approach to health and medicine were part of traditional systems, especially in China and India. In contrast, IM is a relatively recent phenomenon in the West. The Western adoption of IM is mainly in response to the global health care crises. The West, including the United States, hopes to develop affordable health care systems [8]. Several efforts have emerged in the past few years to strategize, adopt, and propagate IM practices. In July 1999, a historic meeting happened at the Fetzer Institute in Kalamazoo, Michigan where representatives from eight medical institutions, agreed to establish The Consortium

on Integrative Medicine. This consortium includes Duke University, Harvard University, Stanford University, the University of California, San Francisco, the University of Arizona, the University of Maryland, the University of Massachusetts, and the University of Minnesota. This consortium, along with the International Society for Complementary Medicine Research (ISCMR), organized an international research congress on integrative health and medicine. The ninth congress of the ISCMR was held during May 13–17, 2014, in Miami, Florida. Proceedings and deliberations in these conferences have greatly helped IM networks to grow during the past few years.

In the United States, an important initiative in this sector, known as the Bravewell Collaborative, believes that IM puts the patient at the center of care, and addresses the full range of physical, emotional, mental, social, spiritual, and environmental influences that affect a person's health. A personalized strategy of IM considers the patient's unique conditions, needs, and circumstances. It uses the most appropriate interventions, with help of scientific disciplines, to heal illness and disease, and help people regain, and maintain optimal health.

A survey by the Bravewell Collaborative concludes: "Integrative medicine is now an established part of health care in the United States." It noted that IM is practiced in many hospitals, and the practice is informed by a common knowledge base. The centers where IM is practiced follow a set of core values, such as patient-centered care, and collaborative and cooperative approaches between patients and practitioners, and among practitioners. While, in practice, the Collaborative's optimistic conclusions appear to be ahead of their time, a good beginning seems to have been made in a few leading centers of medicine. However, we have far to go before we succeed in altering our current, medical practice philosophy, in favor of the new orientation, which considers the mind, body, and spiritual domains—thus ushering in the new, transformative era of tomorrow's medicine.

The integrative approach is supported in Australia through governmental and professional organizations such as the Australian Medical Association, the Royal Australian College of General Practitioners, and the Australasian Integrative Medicine Association [9]. There is organizational support for economic and social concerns in other regions of the world, including the European Union, and the Americas [10]. Generally, the Western IM approach involves an interplay of various systems of medicine and therapies, including modern medicine, and complementary, alternative, and traditional medicine.

Integration works best when it is based on self-regulation, in relation to standards of practice and training. It needs a central or regional system for drug control and evaluation, and maintenance of good manufacturing practice, and a comprehensive program of research. When conventional medicine dominates complementary medicine, there is a risk of losing essential features of complementary medicine. Professional conflicts can arise. It is advisable that patient-friendly policy should ensure that complementary medicine remains affordable and available to people. Substantial investment is necessary for the development of effective services of complementary medicine. Underinvestment contemplates a risk of perpetuating poor standards of practice, services, and products [11].

As a scientific platform, the focus of IM should be on the promotion of innovative, efficient, evidence-based, affordable treatments, and also on prevention, wellness, and patient-centric care. Dr Andrew Weil of the University of Arizona, is a pioneering evangelist of the IM movement in the United States. In 1994, he established the first Integrative Medicine Center. According to Dr Weil, "IM is healing-oriented medicine that takes account of the whole person, including all aspects of lifestyle. It emphasizes the therapeutic relationship between practitioner and patient, is informed by evidence, and makes use of all appropriate therapies." Therefore, it is quite different than alternative therapies, which are sometimes used *instead of* conventional, modern medicine interventions, often without evidence of effectiveness [12].

The concept of self-healing is often ignored in modern medicine. We agree with Dr Weil that IM is the future of medicine, because it best serves people in the most cost-effective manner [13]. However, we feel that the integrative approach should not be restricted only to medicine, treatment, and therapeutics. We feel that integrative approaches should cut across all areas concerning health and medicine. There should be integrative approaches to biology, pathophysiology, nutrition, growth, aging, and the whole life cycle in order to maintain health and wellness.

In a recent survey of 29 IM centers in the United States, The Bravewell Collaborative found significant success using integrative practices to treat chronic pain; and found positive results in treating gastrointestinal conditions, depression and anxiety, cancer, and chronic stress. According to Dr Donald Abrams from the University of California, San Francisco, chronic health issues cost the United States economy more than $1 trillion a year; it is essential to find the most effective ways to treat, and prevent the most prevalent conditions. All participating centers were affiliated with hospitals, health systems, and/or medical, and nursing schools. Patient services include adult care, geriatric care, adolescent care, obstetrics/gynecology care, pediatric care, and end-of-life care. The integrative interventions include nutrition supplements, Yoga, meditation, traditional Chinese medicine/acupuncture, massage, and pharmaceuticals. Findings from the report—which evaluated trends in prevention and wellness, patient outcomes, emerging norms of care, and reimbursement—strongly suggest that the practice of IM holds promise for increasing the effectiveness of care, and improving people's health [14].

The University of Arizona Integrative Health Center, in affiliation with the District Medical Group has developed a hybrid financing approach that provides health insurance reimbursement for T&CM treatments. It is anticipated that detailed evaluations of such clinics will provide future practice models for complementary and IM. Another effort by Duke Integrative Medicine has demonstrated improvements in measures of diabetes, diabetes risk, and weight management; and in the risk for cardiovascular disease, and stroke. It suggests that supporting and enabling individuals to make major lifestyle changes for the improvement of their health might have the potential to reduce the rates, and morbidity of chronic disease, and to impact myriad aspects of health care. Several other centers have reported interesting, and encouraging success stories in their endorsements of the usefulness of IM. The Chopra Center, founded by Dr. Deepak Chopra and David Simon provides an integrative approach to total wellbeing through self-awareness, and the practice of yoga, meditation and Ayurveda.

The European Response

An innovative approach has been proposed by The European Association for Predictive, Preventive and Personalized Medicine (EPMA). The EPMA rightly believes that the current medical and health care system is mired in late and delayed diagnoses, which delay therapeutic interventions. By the time a diagnosis is made, the disease is already established. Obviously, treating such established disease is a major problem, and is considered a major challenge of the twenty-first century. To disrupt this vicious cycle, the EPMA has proposed a new paradigm consisting of three important sequential steps: predictive, preventive, and personalized medicine (PPPM). The EPMA is active in more than 40 countries worldwide, and promotes communication among professionals, including medical doctors, biotechnologists, computer scientists, health care providers, policy-makers, and educators.

This three-dimensional approach has as its first step, predictive medicine. This involves using sophisticated diagnostics for prognosis; early diagnosis is made, before the disease manifests. Predictive diagnostics is considered as the basis for concomitant, targeted prevention of a potential disease condition. The third step is personalized medicine, which includes patient profiling for individualized treatment algorithms, and medical approaches tailored to the patient. The EPMA position paper describes PPPM as the core of the European Union's Horizon 2020 strategy [15]. Excellence in science, industrial partnerships, and societal challenges is the triad of priorities under this ambitious, innovative, and forward-looking program. It is hoped that this far-reaching approach will be coupled with a strategic plan for translating it into practice within the next few years.

MODELS OF INTEGRATION

We wish to reiterate that there might not be a single, uniform model of integration. In countries such as China and India, T&CM practices are active and alive [16]. Here, the integrative approach can be furthered by strengthening T&CM systems, and by using modern medicine to offer additional benefits like surgery, and emergency and specialized services. The Chinese experience of integration began in 1955, when it was proposed that Chinese and Western medicine be combined to boost the health care of the Chinese populace. The Ministry of Public Health has medical professionals from both Chinese, and modern medicine. Today, medical students in China take mandatory courses in both Western and traditional medicine. They actively implement their cross-cultural knowledge in hospitals, and teaching clinics. As a result, Chinese physicians are familiar with the strong and weak points of both medical systems, and can choose the right combination to maximize positive effects.

Admittedly, China has moved ahead with several partnerships, and international collaborations—especially with United States universities, while Indian efforts in building such a mutually beneficial exchange are still in their infancy [17]. India and other countries might learn from the Chinese model of IM, where T&CM and modern medicine synergistically coexist, while maintaining the integrity of individual systems [18]. Based on studying integration in China, a recent report prepared under aegis of the Indian government's AYUSH department, indicates a need for the integration of Ayurveda into the health care system [19].

Historically, Ayurveda has been progressive, dynamic, and inclusive, which propagates an integrative approach.

A quote from Charaka is appropriate: "The science of life shall never attain finality. Therefore, humility and relentless industry should characterize your endeavor and approach to knowledge. The entire world consists of teachers for the wise. Knowledge conducive to health, longevity, fame, and excellence, coming even from an unfamiliar source, should be received, assimilated and utilized with earnestness."

The idea of integrative health pivoting on Ayurveda and Yoga is a nascent and evolving concept. It is quite different from the notion of integrated medicine. The term *integrated* implies past, something that has already been agreed on and frozen. The term *integrative* implies continuity—something that is evolving. The integrative approach has a creative and exploratory intent, wherein science seeks tenable relationships with tradition, without sacrificing its own integrity. Such an integrative approach implies a serious effort to establish foundational, theoretical, experimental, and functional relationships between Ayurveda, Yoga, and biomedical, and health sciences. Indicators of its success will be an enhanced quality of health care to the community at the functional level, and improved cross-cultural understanding at the foundational and theoretical level.

The whole and its parts are related, but the key point to be understood is that the relationship is not one-to-one, because the whole is not equal to the parts, nor does the sum of parts add up to make the whole. Therefore, one should not seek equivalence in developing the relationship between Ayurveda, Yoga, and biomedical sciences; in doing so, one will either reduce the whole to a part, or assume that the part represents the whole, and thus develop a distorted understanding [20]. Yoga philosophy emphasizes three core concepts: actions of body, responses of mind, and power of speech. Here, speech is not to be taken merely as words or talk, but includes every expression. Thus, the extent of integration in Ayurveda and Yoga is much deeper.

In presenting our advocacy, we wish to reiterate that IM is not a hodgepodge of modern medicine, and T&CM. In our opinion, the real spirit of the integrative approach envisages the coherence of philosophical, conceptual, and scientific and professional domains related to health and medicine. We feel the rigor, methodologies, and technologies of scientific biomedical research can be applied to understand, and validate concepts, procedures, medicines, and therapies from T&CM systems such as Ayurveda and Yoga. We feel that such attempts, and cross-pollination between Eastern and Western philosophies, ancient and modern concepts, and holistic and reductive sciences will give rise to new integrative approaches. The new integrative approaches may not be limited to the practice of medicine in clinics or in hospitals, rather, these approaches should be ingrained as health and wellness, sociocultural practices. There may not emerge a universal approach, but there could be several models of integration based on regional or individual requirements. A few suggestions regarding integrative approaches and models are offered throughout this book.

Collaboration and integration between biomedical sciences, Ayurveda and Yoga can be very fruitful. There are certain, incredible details of the parts, uncovered by science that can enrich the understanding of the whole; similarly, there are new perceptions and insights that are revealed in a holistic view that can fundamentally alter the partial view. This is indeed a complex and challenging task, compounded because of its transdisciplinary nature; almost like *riding a tiger* [21]. We urge more attention and investments for biomedical research in basic concepts of health from Ayurveda and Yoga.

RESPECTING MUTUAL STRENGTHS

It is not prudent to think that T&CM is the answer to all the unmet challenges in modern medicine. Many proponents of T&CM take a position where modern medicine is disparaged because of the prevalence of the side effects of biomedical treatments; T&CM is glorified as safer, and more people-friendly. In fact, T&CM might not have *much* role in many of the acute conditions and infections, where modern medicine provides potent remedies. At the same time, because of the existing and emerging limitations of modern medicine, especially in the treatment of chronic, degenerative, and psychosomatic diseases, T&CM may have a larger role to play. For instance, the effective use of preventive and immunity-building measures through *Swasthavritta* and *Panchakarma* can reduce the incidence of many acute conditions. However, there is a need to do a critical reassessment of the strengths and weaknesses of each of these approaches; this will lead to correct strategies for future medical therapy, and health care. While we should respect mutual strengths, it is equally important to develop a mutual appreciation of diverse approaches, views, and philosophies from all cultures and traditions.

The true spirit of integration is in avoiding the temptation of taking any sides—be it modern medicine, Ayurveda, Yoga, or any other system. True integration is about a scientific and unbiased attempt to strike a mutual, trust-based balance between the various systems in the best interest of the people. As does Eywa in the film, *Avatar*, it is important to protect the balance of health.

This has been a long journey as we arrive at the end of this book. We pause and emphatically affirm that it is high time for modern medicine, and T&CM, including Ayurveda and Yoga, to seamlessly integrate, and emerge as the mainstream medicine of tomorrow. Future doctors need to become more human, humble, and humane in their approach to patients in distress. They need to give priority to patient care, rather than give in to a hubristic intoxication with any particular system—be it modern medicine, or T&CM. The compassion and care shown by a good doctor is said to be more powerful than any existing medicines, or therapies. The necessary curriculum reforms in medical education are necessary to open minds of budding doctors and to shape their thoughts for embracing integrative approaches.

The sooner a roadmap emerges to hasten the process of integration, the better it will be for healing today's sick planet. Besides curing illness, the quintessential mission of medicine is to prevent disease, promote health, restore vigor, and help ensure the highest quality of life, at every stage of life. This is a prodigious dream, and an overarching aspiration of humanity; it is not going to be realized easily or spontaneously. The entire global medical community—practitioners, scientists, and specialists—should meet this grand challenge, and embrace integrative approaches to health care, thus furthering the goal of health for all.

This is the promise implicit in the Vedic exhortation:

"*Sarvetra sukhinah santu, sarve santu niramayah, Sarve bhadrami pashyaantu makashchit dukhamaapnuyaat.*"

"May everyone be happy and healthy, May everyone be blissful, and may there be no trace of sufferings and sorrow."

REFERENCES

[1] Schultz AM, Chao SM, Mcginnis JM. Integrative medicine and the health of the public: a summary of the February 2009 summit; 2009.

[2] The Bravewell Collaborative Report. Integrative medicine wellbeing. 2010. p. 46.

[3] Afele M. Ghana seeks integration of traditional medicine practice. January 18, 2001. Available from: http://www.modernghana.com/news/11948/1/ghana-seeks-integration-of-traditional-medicine-pr.html.

[4] The World Bank. Traditional medicine practice in contemporary Uganda.

[5] Ajai O. The integration of traditional medicine into the Nigerian health care delivery system: legal implications and complications. Med Law 1990;9(1):685–99.

[6] Pillai GKG, Sharma P. Finding a sustainable prototype for integrative medicine. J Ayurveda Integr Med 2014;5(3):134–8.

[7] Reddy KS, Patel V, Jha P, Paul VK, Kumar AKS, Dandona L. Towards achievement of universal health care in India by 2020: a call to action. Lancet 2011;377(9767):760–8.

[8] Weil A. Integrative medicine: a vital part of the new health care system, testimony before the committee on health, education, labor, & pensions. US. Senate 2009. Available from: http://www.help.senate.gov/imo/media/doc/Weil.pdf.

[9] Cohen MM. CAM practitioners and "regular" doctors: is integration possible? Med J Aust 2004;180(12):645–6.

[10] MacLennan AH, Wilson DH, Taylor AW. The escalating cost and prevalence of alternative medicine. Prev Med Balt 2002;35(2):166–73.

[11] Bodeker G. Lessons on integration from the developing world's experience. BMJ 2001;322(7279):164–7.

[12] Abrams DI, Weil AT. What's the alternative? N Engl J Med 2012;366(23):2232.

[13] Weil A. The state of the integrative medicine in the US and Western World. Chin J Integr Med 2011;17(1):6–10.

[14] Integrative Medicine in America. How integrative medicine is being practiced in clinical centers across the United States. The Bravewell Collaborative; 2012.

[15] Golubnitschaja O, Watson ID, Topic E, Sandberg S, Ferrari MCV. Position paper of the EPMA and EFLM: a global vision of the consolidated promotion of an integrative medical approach to advance health care. EPMA J 2013;4(1):12. http://dx.doi.org/10.1186/1878-5085-4-12.

[16] Patwardhan B, Warude D, Pushpangadan P, Bhatt N. Ayurveda and traditional Chinese medicine: a comparative overview. Evid Based Complement Altern Med 2005;2(4):465–73.

[17] Haramati A. New Indo-US partnership in Ayurveda. J Ayurveda Integr Med 2010;1(2):89–90.

[18] Robinson N. Integrative medicine – traditional Chinese medicine, a model? Chin J Integr Med 2011;17(1):21–5.

[19] Chandra S. Status of Indian medicine and folk healing. Part I and part II. 2013. Available from: http://issuu.com/knowledgeforall/docs/ayush_report_partii/3.

[20] Shankar D. Conceptual framework for new models of integrative medicine. J Ayurveda Integr Med 2010;1(1):3–5.

[21] Patwardhan B. Ayurveda and integrative medicine: riding a tiger. J Ayurveda Integr Med 2010;1(1):13–5.

Ayurveda and Yoga—A Primer

INTRODUCTION TO AYURVEDA

India has a unique constellation of traditional and complimentary medical practices, which include Ayurveda, Yoga, Unani, Siddha, Sowa Rigpa, and Homeopathy (AYUSH). The Ministry of AYUSH, Government of India has a formal structure to regulate quality, education, and practice involving over 785,000 registered doctors, 8500 manufacturing units, 3500 hospitals, 500 colleges admitting nearly 30,000 students every year.

Ayurveda commonly referred as "knowledge of life" is one of the ancient yet living health systems with wide acceptance especially in South Asia. Classical Ayurveda has eight main branches: internal medicine; gynecology, pediatrics; general surgery; head and neck; toxicology; mental health; rejuvenation and longevity; and reproductive health.

Charaka Samhita, *Sushruta Samhita*, and *Ashtanga Hridaya* are three main Ayurveda classics, which give descriptions of over 700 herbs and 8000 formulations. Historically, Ayurveda has been an experiential, inclusive, progressive, and continuously evolving knowledge system with universal attributes. However, its progress was stalled for thousands of years due to a series of invasions. Ayurveda still receives acceptance from the general public and its prolonged use has led to folk medicine and home remedies for common ailments.

BASIC CONCEPTS

The epistemology of Ayurveda is based on harmonious relation between microcosm and macrocosm and theory of *Mahabhuta*. All objects in the universe are composed of five basic elements known as *Mahabhuta*; appearing in living systems as dynamic principles known as three *Dosha—Kapha, Pitta, Vata*. Dominance of *Dosha* determines individual constitution known as prakriti type. Every living body is formed with seven types of tissue known as *Dhatu*. The waste products of metabolism such as stools, urine, and sweat are known as mala. Specific dominance of *Mahabhuta* also appear as six *Rasa* which are types of taste. Both *Dosha* and *Rasa* can influence each other (Figure A1.1).

333

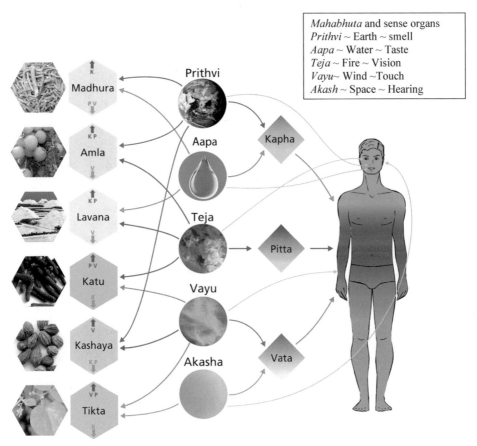

Mahabhuta and sense organs
Prithvi ~ Earth ~ smell
Aapa ~ Water ~ Taste
Teja ~ Fire ~ Vision
Vayu~ Wind ~Touch
Akash ~ Space ~ Hearing

FIGURE A1.1

Every substance in the Universe is formed from *Mahabhuta*, which manifest as *Dosha* in living systems and can be perceived by respective sense organs. *Mahabhuta* combinations can also be perceived as *Rasa* (taste). Each *Rasa* has specific effects - aggravation or alleviation of *Dosha*. The figure presents *Mahabhuta*, *Rasa* and their effects on *Dosha*.

The body has a network of channels known as *Srotasa*, which are responsible for free flow of nutrients and metabolites. *Agni* is biological fire, which is responsible for metabolic processes, digestion, absorption, and assimilation. Any undigested food can lead to toxic substances known as *Ama*, which is considered an important etiological factor for many diseases. Ayurvedic diagnosis known as nidana provides knowledge about pathogenesis. The six progressive stages of pathogenesis are elaborated for early diagnosis and prevention of diseases. Ayurveda classic *Madhava Nidana* contains over 3500 signs and symptoms and over 5000 disease types.

The principles of Ayurveda pharmacology and therapeutics are based on science of *Dravya* and *Guna*. There are 10 pairs of *Guna* with opposite attributes, which are used in Ayurveda therapeutics to balance *Dosha* (Figure A1.2).

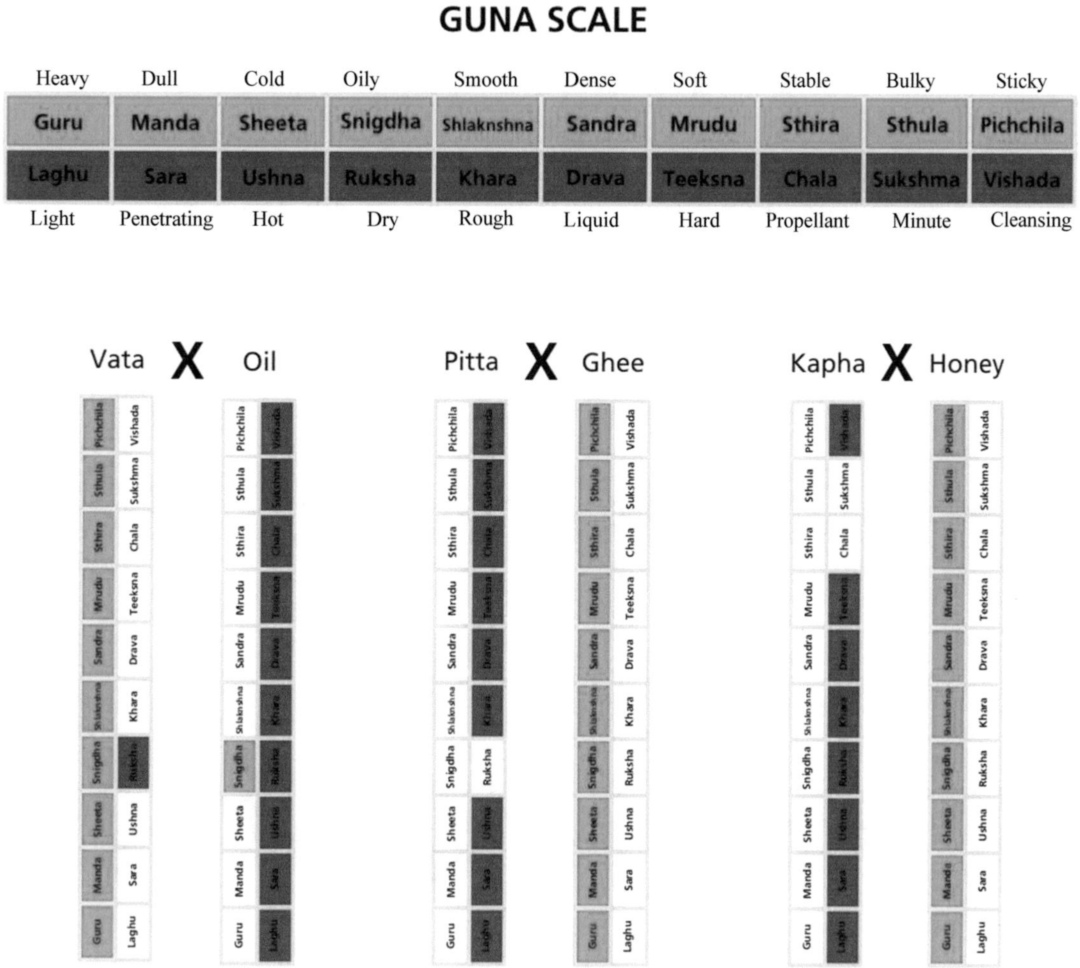

GUNA SCALE

Heavy	Dull	Cold	Oily	Smooth	Dense	Soft	Stable	Bulky	Sticky
Guru	**Manda**	**Sheeta**	**Snigdha**	Shlaknshna	**Sandra**	**Mrudu**	**Sthira**	**Sthula**	**Pichchila**
Laghu	**Sara**	**Ushna**	**Ruksha**	**Khara**	**Drava**	**Teeksna**	**Chala**	**Sukshma**	**Vishada**
Light	Penetrating	Hot	Dry	Rough	Liquid	Hard	Propellant	Minute	Cleansing

Vata **X** Oil   Pitta **X** Ghee   Kapha **X** Honey

FIGURE A1.2

Examples of how the concept of *Guna* can be used in therapeutics to manage *Vata*, *Pitta*, and *Kapha* vitiation by using substances which have opposite quality. Sesame oil can alleviate vitiated *Vata*, Ghee can alleviate vitiated *Pitta*, and honey can alleviate vitiated *Kapha*.

Ayurveda knowledge base contains over 3500 herbal, animal, and mineral materials such as *Dravya*, which are used to prepare medicinal formulations. Figure A1.3 illustrates sophistication and multifarious effects explained by Ayurveda with the example of the herb *Tinospora cordifolia* (Guduchi). Over 1100 formulations of Guduchi are available for treating more than 50 diseases.

Ayurveda focus is on health protection. Detailed advice on how to remain healthy is provided in a branch known as *Swasthavritta*. It includes a healthy daily routine known as *Dinacharya* and a seasonal routine known as

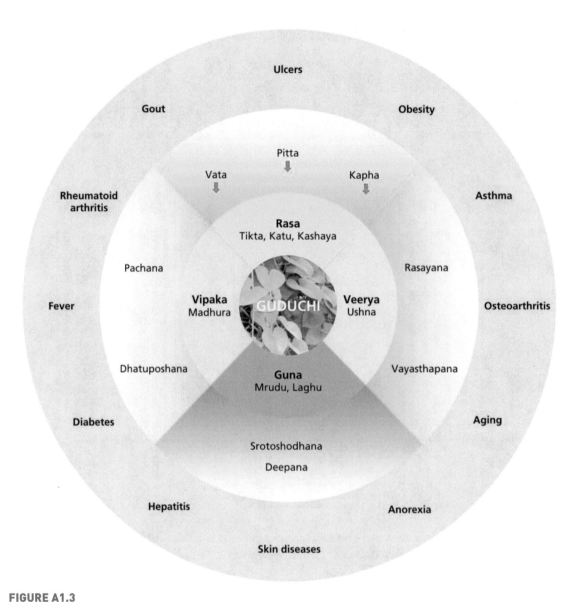

FIGURE A1.3

An example of a typical Ayurvedic drug *Tinospora cordifolia* (Guduchi) illustrating the relationship between the concepts of *Dravya*, *Guna*, *Dosha*, and *Rasa*, which can be used for therapeutic purposes.

Ritucharya, which describe how to align routine lifestyle in tune with nature. Many specialized drugs known as *Rasayana* are used for immunity and longevity. Body purification and detox procedures known as *Panchakarma* are used for rejuvenation and therapeutic purposes.

An optimal healthy state is known as *Swastha* when a person is *stable within self*; when *Dosha* are balanced, *Dhatu* are well nourished, waste products are effectively removed, srotasa are clean, and *Agni* is active with a feeling of bliss at the levels of body, mind, and spirit. Ayurvedic therapeutics is a logical and systematic process aimed at fine tuning internal homeodynamic balance in tune with the external environment with help of natural interventions. Ayurveda offers personalized management based on individual *Prakriti* types.

INTRODUCTION TO YOGA

Yoga literally means to integrate, to join, or to unite. It considers a person as a whole with body–mind–spirit and not in isolation. Yoga concentrates on strong immunity, which can resist several physical and psychological diseases. It suggests developing balanced behavior and stable personality with positive attitude. Yoga teaches behavioral methods for a healthy body and mind. Reducing Yoga merely as a *therapy* is not correct. In fact, healthy mind and body is a prerequisite for practice of Yoga.

Generally, the term Yoga is commonly used for exercises or *Asana*. Sanskrit lexicon describes various meanings of the term Yoga—union, addition, junction, total, way, abstract contemplation, performance, opportunity, contact, constellation, mixture, medicinal formulation, and a science of concentration and consciousness. Wordweb describes Yoga as "discipline aimed at training the consciousness for a state of perfect spiritual insight and tranquility that is achieved through the three paths of actions and knowledge and devotion." Great sage Patanjali who authored the most authoritative classic known as *Yoga Sutra*, defines Yoga as capability to achieve control over a distractible mind; *Bhagwad Geeta* defines Yoga as dexterity in action; *Yoga Vasistha* another ancient classic text describes it as techniques to soften and slow down the mind. Most of the scholars agree that Yoga is a way of life and an approach toward the ultimate goal. The goal of Yoga is to be in tune with reality, consciousness finally leading to liberation of the soul through perfection. Yoga is for conditioning the mind to remain alert in a cognitive state and to be aware of the surrounding situation as well as the inner self. Buddha calls this as *Nirvana*, Patanjali describes it as *Kaivalya*, and Veda names it as *Moksha*. Yoga describes consciousness-based paradigm and suggests ways to get relief from tensions, stresses, diseases, and miseries for complete health.

Great philosopher, Swami Vivekananda said "Every soul is potentially divine. The goal is to manifest this divinity within by controlling external and internal environment". Yoga suggests union of the soul with the divinity. One can attempt this goal through various ways based on individual nature, inclination, and preference. It could be through total devotion to work, or selfless worship, or meditation or a combination of any of these.

MAIN TYPES OF YOGA

Yoga is Indian heritage involving physical, mental, and spiritual practice, which aims to transform body and mind. The ultimate goal of Yoga leads to *Nirvana*. Four major types of Yoga are mentioned in Indian classics based on the path individuals wish to choose. Raja Yoga and Hatha Yoga are more relevant for this book as they deal with meditation, posture, and breathing practices, which are useful in health promotion. Raja Yoga deals with meditation, in which the mind is trained in eight steps as described by Patanjali. Hatha Yoga is about physical techniques supplementary to a broad construct of Yoga. It concentrates on body postures (*asana*), breathing techniques (*Pranayama*), and purification procedures (shuddhi kriya). Bhakti Yoga is loving devotion to a deity. Jnana Yoga deals with the highest and continued quest for pure knowledge and understanding of nature.

Present practice of Yoga is a mix of Hatha Yoga and Raja Yoga.

ANATOMICAL CONCEPTS OF YOGA

The anatomical concepts of Yoga are different than Ayurveda. The *Atman*, or soul, is considered to be covered by five layers of consciousness, or sheaths, known as *Kosha* (Figure A1.4). The outermost layer is known as *Annamaya Kosha*. *Anna* in Sanskrit means food. *Kosha* refers to physical body structure, which is dependent on nourishment from food. The next layer is known as *Pranamaya Kosha*. This layer is related to the vital energy force known as *Prana*, which is required for body functions. Next layer is *Manomaya Kosha*. This is a mental sheath that deals with knowledge and sense of self-existence. Next is *Vidnyanamaya Kosha*. This is about emotional and intellectual aspects and covers intellect and the five sense organs. Next layer is *Anandmaya Kosha* or sheath of bliss. This is the innermost layer on *Atman*. Removing this layer liberates *Atman* giving experience of samadhi and turya, which are the highest states of meditation. The Yoga practice helps to uncover these layers from the *Atman* and facilitates its liberation to experience the blissful nature of *Brahman*.

Another important concept of Yoga is *Chakra*, which is a part of the subtle (*Sookshma*) body (Figure A1.5). There are seven *Chakra*—Mooladhara (at the base of spine), Swadhisthana (in sacral bone), Manipura (at navel), Anahata (in chest), Vishuddha (in throat), Adnya (forehead), and Sahasrara (top of head). *Nadi* are the channels through which the life force or vital energy (*Prana*) moves. Several *Nadi* connect *Chakra*. The main three nadis are Sushmna (middle nadi that connects mooladhara to sahasrara chakra). Ida (right) and Pingala (left) run parallel to Sushumna and they have solar and lunar dominance, respectively. Kundalini, the supreme energy, travels from Muladhara through Sushumna. When it reaches Sahasrara, then the Yogi gets detached from the body and mind (Figure A1.6).

FIGURE A1.4

Concept of *Pancha Kosha*. Five subtle layers starting from physical to metaphysical dimensions.

FIGURE A1.5

Concept of *Chakra* in Yoga. Seven subtle anatomical regions, which are responsible for physiological and spiritual functions.

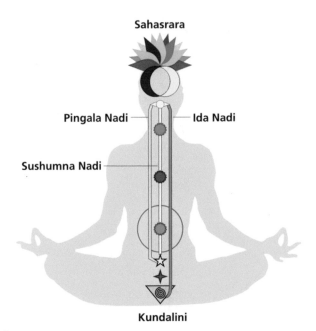

FIGURE A1.6

Concept of *Nadi* suggests energy meridians in the body. In the process of *Samadhi*, *Kundalini* from *Mooladhara Chakra* travels toward *Sahasrara* bringing the ultimate bliss.

THERAPEUTIC ASPECTS OF YOGA

Yoga considers disease as a disturbance that perturbs the equilibrium. Yoga describes various preventive, promotive, therapeutic, and rehabilitative measures for several diseases. The terms kriya or karma indicate cleansing, purificatory, and reconditioning therapies. Hatha Yoga describes six processes for purification of mind and body. They have manifold effects like panchakarma of Ayurveda and are aimed at removal of toxins and noxious causative agents. These procedures are also useful for a healthy person to prepare body and mind ready for Yoga practices. These procedures are as follows:

- Neti—nasal cleansing using water or a soft cloth. In this technique, sterile lukewarm isotonic water is poured in one nostril, and is released from the other.
- Dhauti is cleaning gastrointestinal tract, respiratory tract, external ears, and eyes. It is divided in to four parts which are internal cleaning, teeth cleaning, cardiac or chest region cleaning, and rectal cleansing.
- Nauli is movements of abdominal muscles that clean internal organs. The procedure requires proper training and practice. It is indicated for various digestive and metabolic diseases like constipation and obesity.

- Basti is a procedure for cleansing the colon by sucking in either air or water.
- Kapalbhati is breathing with rhythmic, short, and forceful exhalations with passive inhalation. The technique is believed to clean sinus, skull, and brain.
- Mudra and bandha are special postures described by *Hatha Yoga*. They consist of certain neuromuscular locks leading to changes in internal pressures.

A classic text by sage Patanjali known as *Yoga Sutra* advises eight stages for liberating the soul toward samadhi for union with divinity:

1. Yama deals with morality and code of conduct. It has five components—nonviolence (at the level of body, mind, and speech), truth, honesty, and not having desire for other's belongings and abstention.
2. Niyama is about personal conduct. It includes hygiene, contentment, self-study, asceticism, and surrender to Ishwara.
3. Asana is set of postures to retain health and preparing body and mind for breathing and meditation.
4. Pranayama is controlled breathing technique to gain control on life processes and energy or prana.
5. Pratyahara is withdrawal of senses from external objects.
6. Dharana is concentration of mind.
7. Dhyana is meditation.
8. Samadhi is the stage when the mind and the object of meditation merge together.

The philosophy and practice of Yoga requires physical and mental involvement. The learning of Yoga requires the right Guru (expert guide and mentor) who can demonstrate and correct the postures. Due to rising popularity, some components of Yoga are being repackaged, branded, and commercially propagated. Some asanas may have contraindications. For example, patients of hypertension should not undergo shirshasana and halasana. Procedures of Yoga therapies are considered safe when they are practiced with proper guidance.

Yoga deals with union of mind and body and maintaining harmony between an individual and surrounding environment. Yoga offers a path for preventive and promotive health. More than just health, it advocates a comprehensive approach toward life. Yoga way of life changes outlook, behavior and results in relaxation, balanced mind, and clarity in thought process. Yoga repackaged as different variants like mindfulness, salutogenesis, transcendental meditation, samadhi Yoga, power Yoga, kundalini Yoga, Iyengar Yoga, hot Yoga, and many such are being practiced as mind–body medicine, lifestyle medicine, and behavioral medicine. Yoga remains an attraction for practitioners and researchers due to its inclusive approach and simple, accessible, affordable, convenient, and safe procedures. Yoga intervention can be easily integrated with any medical system. Therefore, Yoga can play a key role in future integrative medicine.

During the United Nations General Assembly in September 2014 Indian Prime Minister stated that "Yoga embodies unity of mind and body; thought and action; restraint and fulfillment; harmony between man and nature; a holistic approach to health and wellbeing." He proposed an idea to observe World Yoga Day. This proposal received overwhelming support from 170 nations including all the countries of the European Union. As a result, starting from 2015, every 21st June will be celebrated as World Yoga Day.

AYURVEDA AND YOGA

Philosophies of Ayurveda and Yoga share a common perspective of the relationship between microcosm and macrocosm. Both concentrate on tuning between internal and external environments. The concepts, practices, and therapeutic approaches of Yoga and Ayurveda are complementary. Ayurveda focuses more on physical aspects; while Yoga is more about psychological, social, and spiritual dimensions of health. Ayurveda operates more at physical level with the aim to "maintain health" but Yoga drives the healthy person forward to the ultimate goal of life with a metaphysical view. Ayurveda aims at union of body, mind, senses, and soul; Yoga attempts to free the mind and experience the state of harmony and perfection.

Ayurveda has adopted the concept of mind control from Yoga.

Generally, Ayurveda operates controlling "external nature" while Yoga deals with internal nature. Controlling hunger, thirst, and evil desires is part of Ayurveda management. Yoga and Ayurveda assess effects of diet on mind in the same way in terms of satva, rajas, and tamas. However, physical health aspects of Ayurveda are elaborate as hinted by many of Yoga texts. Ayurveda suggests the modulation of pathophysiology by maintaining *Dosha* balance through daily and seasonal regime, lifestyle, exercise, diet, therapeutic procedures, and drugs. The special instructions about respecting natural urges are given importance. Elimination of wastes and toxic materials is another common thought between Ayurveda and Yoga. Yoga also suggests physical therapies and cleansing procedures.

Every individual has mind, which connects with senses. Ayurveda and Yoga have different concepts of mental health similar to modern psychiatry and psychology. Yoga suggests means of controlling mind through detached watchfulness toward our thoughts. Yoga suggests refraining from greed, grief, fear, anger, jealousy, attachment, and malice. Ayurveda reiterates similar concepts as sadvritta, which is code of conduct for good behavior.

FURTHER READING

The available knowledge is very vast. This primer is just an overview. We recommend following
books and papers for those who wish to know more about Ayurveda and Yoga.

[1] Valiathan MS. The legacy of Caraka. Orient Longman; 2003.

[2] Valiathan MS. Legacy of Sushruta. Orient Longman; 2007.

[3] Valiathan MS. Legacy of Vagbhata. Universities Press (India) Private Limited; 2009.

[4] Dwarkanath C. Introduction to Kayachikitsa. Popular Book Depot; 1959.

[5] Kutumbiah P. Ancient indian medicine. Orient Longman; 1999.

[6] Athawale VB. Pathogenesis in Ayurveda. Delhi: Chowkhambha Sanskrit Pratishthan; 2001.

[7] Vasant L. Ayurveda—the science of self-healing. Lotus Press; 1990.

Index

Note: Page numbers "f" indicate figures.

Printed in the United States
By Bookmasters